Mammalian development

a practical approach

TITLES PUBLISHED IN
THE
PRACTICAL APPROACH
SERIES

Affinity chromatography

Animal cell culture

Biochemical toxicology

Biological membranes

Carbohydrate analysis

Centrifugation (2nd Edition)

DNA cloning

Drosophila

Electron microscopy
in molecular biology

Gel electrophoresis of nucleic acids

Gel electrophoresis of proteins

H.p.l.c. of small molecules

Human cytogenetics

Human genetic diseases

Immobilised cells and enzymes

Iodinated density-gradient media

Lymphocytes

Lymphokines and interferons

Microcomputers in biology

Mitochondria

Mutagenicity testing

Neurochemistry

Nucleic acid and
protein sequence analysis

Nucleic acid hybridisation

Oligonucleotide synthesis

Photosynthesis:
energy transduction

Plant cell culture

Prostaglandins
and related substances

Spectrophotometry
and spectrofluorimetry

Steroid hormones

Teratocarcinomas
and embryonic stem cells

Transcription and translation

Virology

Mammalian development

a practical approach

Edited by
M Monk
MRC Mammalian Development Unit,
4 Stephenson Way, London NW1 2HE, UK

OXFORD · WASHINGTON DC

IRL Press Limited
PO Box 1,
Eynsham,
Oxford OX8 1JJ,
England

British Library Cataloguing in Publication Data

Mammalian development : a practical approach.—(Practical
   approach series)
   1.   Mammals—Development
   I.   Monk, M
   599.03   QL959

ISBN 1-85221-030-3 (hardbound)
ISBN 1-85221-029-X (softbound)

Cover illustration. Top right, human 8-cell embryo (Chapter 14, Figure 3);
bottom left, mouse 8-cell embryo (Chapter 2, Figure 1).

Printed by Information Printing Ltd, Oxford, England.

Preface

Mammalian molecular embryology has evolved from the application of a wide range of microtechniques to the study of mammalian development, mainly in the mouse. The advent of *in vitro* fertilization and embryo transfer in the human increasingly requires the application of these techniques to the human embryo to enable pre-implantation embryo culture, embryo biopsy, assays for assessing developmental potential and early diagnosis and possible treatment of genetic disease. In post-implantation fetuses, our current knowledge of the morphological changes accompanying normal growth and development now provides a sound basis for new techniques designed to manipulate specific cell lineages involved in organogenesis. The stage is now set for the study of the regulation of expression of specific genes throughout development.

The contents of this book cover the isolation and culture of embryos and a wide range of cellular and subcellular procedures applicable to the study of mammalian development — *in vitro* fertilization, marking and moving individual cells, gene transfer, nuclear transfer and analysis of membranes, intracellular organization, chromosomes, specific DNA sequences, proteins and enzymes. Most of the procedures have been developed in the mouse but will be applicable to mammals and mammalian cells in general.

I am most grateful to all the contributors for their willing and cheerful collaboration and support, and to IRL Press for their patience, expert help and advice generously given at all times during the preparation of this book.

Marilyn Monk

Contributors

N.D.Allen
AFRC Institute of Animal Physiology and Genetics Research, Department of Molecular Embryology, Babraham, Cambridge CB2 4AT, UK

S.C.Barton
AFRC Institute of Animal Physiology and Genetics Research, Department of Molecular Embryology, Babraham, Cambridge CB2 4AT, UK

R.Beddington
ICRF Developmental Biology Unit, Department of Zoology, South Parks Road, Oxford OX1 3RE, UK

P.R.Braude
Embryo and Gamete Research Group, Department of Obstetrics and Gynaecology, University of Cambridge Clinical School, Rosie Maternity Hospital, Cambridge, UK

A.C.Chandley
MRC Clinical and Population Genetics Unit, Western General Hospital, Crewe Road, Edinburgh EH4 2XU, UK

E.P.Evans
William Dunn School of Pathology, South Parks Road, Oxford OX1 3RE, UK

R.Haffner
Institute of Cancer Research, Chester Beatty Laboratories, Fulham Road, London SW3 6JB, UK

C.M.Hetherington
Biological Services, National Institute for Medical Research, The Ridgeway, Mill Hill, London NW7 1AA, UK

S.K.Howlett
AFRC Institute of Animal Physiology and Genetics Research, Department of Molecular Embryology, Babraham, Cambridge CB2 4AT, UK

J.McConnell
Department of Anatomy, University of Cambridge, Downing Street, Cambridge CB2 3DY, UK

A.McLaren
MRC Mammalian Development Unit, 4 Stephenson Way, London NW1 2HE, UK

M.Monk
MRC Mammalian Development Unit, 4 Stephenson Way, London NW1 2HE, UK

M.L.Norris
AFRC Institute of Animal Physiology and Genetics Research, Department of Molecular Embryology, Babraham, Cambridge CB2 4AT, UK

B.A.J.Ponder
Institute of Cancer Research, Haddon Laboratories, Clifton Avenue, Sutton, Surrey SM2 5PX, UK

H.P.M.Pratt
Embryo and Gamete Research Group, Department of Anatomy, University of Cambridge, Downing Street, Cambridge CB2 3DY, UK

W.F.Rall
Rio Vista International, Route 9, Box 242, San Antonio, TX 78227, USA

W.Reik
AFRC Institute of Animal Physiology and Genetics Research, Department of Molecular Embryology, Babraham, Cambridge CB2 4AT, UK

M.A.H.Surani
AFRC Institute of Animal Physiology and Genetics Research, Department of Molecular Embryology, Babraham, Cambridge CB2 4AT, UK

C.J.Watson
Department of Biochemistry and Microbiology, University of St Andrews, St Andrews KY16 9AL, UK

D.G.Whittingham
MRC Experimental Embryology and Teratology Unit, Woodmansterne Road, Carshalton, Surrey SM5 4EF, UK

K.Willison
Institute of Cancer Research, Chester Beatty Laboratories, Fulham Road, London SW3 6JB, UK

M.J.Wood
MRC Experimental Embryology and Teratology Unit, Woodmansterne Road, Carshalton, Surrey SM5 4EF, UK

Contents

**3. ISOLATION, CULTURE AND MANIPULATION OF
POST-IMPLANTATION MOUSE EMBRYOS** **43**
R.Beddington

Abbreviations

ABC	avidin–biotin peroxidase complex
ACh	acetylcholine
AD	adrenaline
APRT	adenine phosphoribosyltransferase
BrdU	bromodeoxyuridine
BSA	bovine serum albumin
CCD	cytochalasin D
ConA	concanavalin A
CNF	cuneiform nucleus
CPMV	cowpea mosaic virus
DA	dopamine
DAB	diaminobenzene
DAPI	4,6-diamidino-2-phenylindole
DBA	*Dolichos biflorus* agglutinin
DBH	dopamine-β-hydroxylase
DEPC	diethyl pyrocarbonate
DMEM	Dulbecco's modified Eagle's medium
DMSO	dimethyl sulphoxide
DNP	dinitrophenol
EC	embryonal carcinoma
EDC	1-ethyl-3(3-dimethylaminopropyl) carbodiimide
EDTA	ethylenediamine tetraacetic acid
EM	electron microscopy
FCS	fetal calf serum
FITC	fluorescein isothiocyanate
GABA	γ-aminobutyric acid
GAPDH	glyceraldehyde phosphate dehydrogenase
G6PD	glucose-6-phosphate dehydrogenase
HBSS	Hank's balanced salt solution
hCG	human chorionic gonadotropin
Hepes	N-2-hydroxyethylpiperazine-N'-2-ethanesulphonic acid
HPRT	hypoxanthine phosphoriboxyltransferase
HRP	horseradish peroxidase
5-HT	5-hydroxytryptamine
ICM	inner cell mass
IEF	isoelectric focusing
LH	luteinizing hormone
LM	light microscopy
MEM	minimal essential medium
NA	noradrenaline
nAChR	nicotinic acetylcholine receptor
NOR	nucleolar organizing region
ODAB	oxidized 3′,3-diaminobenzidine
PAP	peroxidase–anti-peroxidase
PBS	phosphate-buffered saline
PGK	phosphoglycerate kinase
PHA	phytohaemagglutinin

PHD	phenylhydrazine hydrochloride
PLL	poly-L-lysine
PMSG	pregnant mare's serum gonadotropin
PRPP	phosphoribosyl pyrophosphate
PVP	polyvinyl pyrrolidone
RIA	radioimmunoassay
SC	synaptonemal complex
SEM	scanning electron microscopy
SPF	specific pathogen free
SSC	standard saline citrate
TBS	Tris-buffered saline
TCA	tricarboxylic acid
TE	trophectoderm
TEM	transmission electron microscopy
TMB	tetramethylbenzidine
TRITC	tetramethylrhodamine isothiocyanate
VAB	veronal acetate buffer
WGA	wheat germ agglutinin

Introduction

From Aristotle onwards, the development of mammals has been a target for speculation and enquiry. But the first experimental, as opposed to purely descriptive, study was perhaps that of Walter Heape, who in 1890 transferred two cleaving rabbit eggs from an Angora donor to the oviduct of a Belgian Hare, to see whether the mother or the young would determine the length of gestation, and whether the foreign uterine environment would affect the breed characteristics. Both donor young were born alive. It took Heape seven more years of transfers before he achieved another success, an example of perseverance that should inspire those of us who have problems with our transfers today.

In spite of this lengthy history, we know far more about the development of sea-urchins, nematodes and *Drosophila*, even of fish, frogs and birds, than we do about the development of mammals. Because of the intra-uterine inaccessibility of the mammalian embryo, the subject of mammalian development has been derived more from reproductive biology and from genetics than from classical embryology. Perhaps that is why mammalian developmental biologists, unlike classical embryologists, use the term 'embryo' so ambiguously. Properly speaking, a two-cell stage or a blastocyst is not an embryo; the embryo is formed only at gastrulation, and develops into the fetus. Much of our understanding of the mechanisms of mammalian development relates to the pre-embryonic stages between fertilization and implantation, which are relatively easy to grow in culture, while many of the systems that have been most fruitfully studied involve extra-embryonic membranes rather than the embryo itself.

A recent and immensely valuable recruit to the study of mammalian development is of course molecular biology. The impact of recombinant DNA technology in particular has stimulated an explosive growth of the field, one manifestation of which is the present volume. Earlier manuals, for example 'Methods in Experimental Embryology of the Mouse' (1970) by K.A.Rafferty, and 'Methods in Mammalian Reproduction' (1978) and 'Methods in Mammalian Embryology' (1971), both edited by J.C.Daniel, give detailed and useful accounts of many of the experimental manipulations in use B.C. (before cloning) that are still relevant. The only previous laboratory manual to have appeared in the post-cloning era is the excellent 'Manipulating the Mouse Embryo' (1986), by B.Hogan, F.Costantini and E.Lacy. Though there is some inevitable overlap with the present volume, there is much that is complementary. Any self-respecting laboratory of mammalian development needs to have both books on its shelves.

The structure of the present book is straightforward. It tells you how to obtain your experimental material, how to manipulate it, and how then to analyse it. The basics include everything you always wanted to know about keeping mice (Chapter 1), methods for the recovery and handling of pre-implantation (Chapter 2) and post-implantation (Chapter 3) stages of mouse development, and methods for the recovery, fertilization and culture of human eggs (Chapter 14). Manipulations involve genes (Chapter 11), nuclei (Chapter 12) and cells (Chapter 6), as well as entire pre- or post-implantation stages (Chapters 2, 3, 13, 14), and include various methods of dissection and transfer as well as low temperature preservation. Analytical techniques include chromosome preparation, both meiotic (Chapter 4) and mitotic (Chapter 5), methods for the recognition of cell markers (Chapter 6), appropriately miniaturized biochemical assays (Chapter 7), protein electrophoresis systems (Chapter 8), cDNA library construction (Chapter 9) and *in situ* hybridization of nucleic acid probes to tissue sections (Chapter 10), a technique of enormous potential for the analysis of the molecular basis of mammalian development at the cellular level.

As we have seen, the study of mammalian development is rooted in reproductive biology and in genetics. We know a lot about reproduction in mice, rats, rabbits, guinea pigs, sheep and humans, but for genetics the mouse has been the mammal par excellence. One source of the strength of mammalian development in Britain, reflected in the fact that all the contributors to the present comprehensive manual are drawn from this country, lay in the development of mouse genetics, including developmental genetics, in Britain during the 1950's and early 1960's. R.A.Fisher in the Cambridge Genetics Department will be remembered for his monumental contributions to statistics, but his love was for his mice. Douglas Falconer, trained under Fisher, joined Waddington's Institute of Animal Genetics in Edinburgh, supported by the Agricultural Research Council. Mary Lyon moved from the Edinburgh Institute to Harwell, supported by the Medical Research Council. Grüneberg, author of 'The Genetics of the Mouse', was invited to University College London by J.B.S.Haldane in 1938, and was later also supported by the Medical Research Council. With four such nests of mouse geneticists, all well funded and highly productive, it is not surprising that the subject flourished in this country.

The mouse was not only predominant in mammalian genetics: it was also the species on which *in vitro* culture systems for pre-implantation stages were so successfully devised in the 1960's. Small wonder that Hogan *et al.*'s manual is explicitly devoted to mice alone, while the present book has in addition a chapter devoted entirely to the human. The problems, both practical and ethical, in working with human material are well set out in Chapter 14.

Yet we are an anthropocentric species. Funding bodies, and often research workers too, are more interested in the health and happiness of people than in the health and happiness of mice. Typically, a field expands when it is seen as socially relevant at the same time that technical advances are opening up exciting new research opportunities. The new possibilities of studying early human development set out in Chapter 14 should be set beside the urgent need to increase the success rate of IVF, to devise other methods of treating infertility, and to decrease the toll of genetic and chromosomal disorders. New perspectives unfold as the analysis of development expands to include the molecular level, while the advent of transgenic mice means that instead of studying the development of an embryo in a foreign mother (as Heape did), we can now study the expression of a gene in a foreign genome. These advances should make possible the application of genetic engineering to farm animals, and an increased understanding of the etiology of congenital malformations in our own species.

Whatever their underlying motivation, students of mammalian development cannot fail to be struck by the excitement and beauty of what they are studying. The mass of sperm, in a frenzy of activity; the exquisite symmetry of a pre-compaction 8-cell stage; the gleaming expanded mouse blastocyst; the pregnant uterus, that masterpiece of packing; the tiny fetus, with its beating heart and massive liver visible through the delicate transparent amnion. How does it all happen? I am confident that the generation of research workers who use this book will uncover many of the answers. They will also uncover more questions than they answer: but that is the way of science.

Anne McLaren

CHAPTER 1

Mouse husbandry

COLIN M.HETHERINGTON

1. INTRODUCTION

The mouse is the most commonly used laboratory mammal. Due to its small size and high fecundity it is relatively cheap to breed, buy and maintain. Many strains of mice have been bred in laboratories for over 50 years and there are now several hundred inbred, congenic, recombinant inbred, outbred and mutant strains. The genetics and characteristics of many of these strains are extensively documented (1,2).

2. HUSBANDRY

2.1 Source of mice and general health measures

A regular and reliable supply of healthy, genetically-defined animals is essential to a successful research project. Depending on circumstances it may be more effective to either purchase animals from a commercial breeder or to breed ones own animals. If a suitable source of commercially bred animals is available and occasional large batches of age-matched animals are required, the purchase of mice may be the most efficient and cost-effective approach. If animals are used on a more regular basis there may be advantages in breeding the animals. It is important that the animals are healthy at the outset and that they remain healthy. Maintenance of health in the long term depends on the management protocols which control the animals welfare and staff access.

Mice are kept under a variety of conditions. Animals from a basic facility with a minimum of procedures to prevent infections are known as 'conventional' animals, while those bred or maintained under conditions where precautions are taken to avoid infections are know as 'barrier' bred or maintained. The term specified pathogen free (SPF) is also used to describe barrier bred animals. In this instance health reports specify the pathogens which have been shown to be absent from the animals. These include ecto- and endoparasites, such as mites and intestinal worms and protozoa, certain bacteria, mycoplasmas and viruses. Gnotobiotic animals are bred in isolators and their microbiological flora is known. Axenic mice are also isolator bred and are free of all organisms (i.e. germ free).

The form of animal husbandry adopted depends on the type of facilities available and the nature of the experiments. If an experiment produces animals that are to be kept for a long period, or that are to form the basis of a breeding colony, some form of barrier system is essential. If no precautions are taken to prevent infections it is inevitable that the health of the animals will deteriorate with a resulting decrease in breeding performance or death. A minimal barrier system requires people entering the animal area to remove their laboratory coats worn in other areas, and to put on protective

1

clothing. It is essential that they wash their hands thoroughly before handling the animals and the wearing of surgical gloves and a change of footwear or overshoes should be considered. A more stringent barrier includes the requirement for staff to shower on entering the unit and a complete change of clothes.

Whatever the precautions adopted it is sensible that the breeding animals receive the highest protection and staff not directly involved with breeding should not enter the breeding area. Whenever possible breeding colony staff should not be involved with other animals of a potentially lower health status. Precautions for experimental animals will inevitably be less stringent to allow access to scientists and their technicians. In general however the greater the number of people entering an area the greater the risk of the introduction of an infection.

2.2 **Caging**

Mouse cages are available in various shapes, sizes and designs. The two most useful sizes are probably those with a floor area of 200 cm² or 1200 cm² and 12 cm high (internal measurements). The former provides accommodation for breeding pairs or trios or up to five adult mice while the latter is used for up to 25 stock mice depending on their size. Boxes are available made of clear polycarbonate, opaque polypropylene or metal. They may have solid or grid bases. There is little to suggest which the mice prefer, and cages are chosen to suit the management system and the scientists' preference. Polypropylene cages with stainless steel grid lids and solid bases are the most common type but if, for example, cages are to be autoclaved regularly, metal cages may be more appropriate. Cages with grid bases can be used for stock but are not recommended for breeding cages. Special cages have been developed for wild mice to reduce the need for handling during cleaning but it is possible to keep them in standard mouse cages. Cages may be held on wall-mounted racks, allowing easy floor cleaning and the recapture of escaped animals, or on trolleys designed for the purpose.

2.3 **Bedding**

Bedding is usually of woodchips or shavings. The wood must be free of insecticides and herbicides. Allergies to wood dust can develop in staff and consideration should be given to using dust-free sawdust and to the staff wearing masks when handling the dry bedding. Other absorbent materials are available for bedding, but problems can arise when they are used in breeding cages as they can desiccate the pups. Shredded paper or tissues can be provided for nesting material.

2.4 **Diets**

Nutrition is a very important factor in the breeding and maintenance of mice in good health and fertility. Various commercial mouse diets are available. Some are manufactured for breeding animals and others for stock. They are supplied as pellets, of various shapes and sizes, or powders. Expanded pelleted diets result in less wastage by the animals, but food hoppers hold a smaller weight of food. Diets for autoclaving are supplemented with nutrients to compensate for the inevitable degradation that occurs. Some diets are coated by the manufacturer to prevent the pellets sticking together during autoclaving. The physical changes to diet pellets brought about by autoclaving depends both on the constitution of the diet and the sterilizing and drying cycle of the autoclave. A diet should always be tested in the autoclave that will normally be used to ensure that it is acceptable both nutritionally and physically after sterilization. It may be necessary to spread the diet on trays in the autoclave to prevent clumping of the pellets. Some diets may emerge so hard that it is impossible for them to be eaten by weanling mice. Diets sterilized by irradiation are commercially available. It is important to ensure that the packaging of irradiated diets is intact and the contents are therefore sterile before use.

2.5 Water

Water can be provided by water bottles or by automatic watering systems. Whichever system is used it is vital that leaking water is avoided. There is a real risk of mice, especially young animals, drowning or dying from the effects of a soaking if a bottle or nozzle of an automatic system leaks.

The domestic water supply should be checked to determine whether it requires additional treatment before use. This is particularly true for a barrier maintained breeding facility. In Britain the domestic supply is normally adequate. Acidification, chlorination, ultraviolet light or filtration can be used to improve the microbiological status of the supply (3). Water for mice in gnotobiotic isolators must be sterile.

2.6 Animal room environment

Forced air ventilation is required to remove odours and allergens. The room temperature should be maintained at $21 \pm 2°C$ and the relative humidity at $55 \pm 10\%$. The current trend is towards greater environmental control and full air conditioning is recommended. It is important that air is distributed evenly throughout the room and that the animals are not exposed to drafts. The number of air changes per hour that are required is dependent on the stocking density and the effectiveness of the air distribution within the room. Sixteeen air changes per hour in a room containing $150-175$ mice/m² of floor space is adequate provided the cages are cleaned regularly. It should be noted that the microclimate in the cage may differ significantly from that of the room with significantly higher temperature, humidity and ammonia concentration in the cage.

To ensure a regular supply of embryos at a constant development stage at the same time of day throughout the year, a constant lighting regime is required. Mouse rooms should therefore have no windows. The light cycle most commonly used is 12 h dark/12 h light. See also Section 5.2.

2.7 Fumigation of rooms

Inevitably with time the health status of an animal colony will deteriorate. In an experimental facility without breeding animals the health status can be restored if healthy animals are issued into a clean fumigated or disinfected room, and the infected room is gradually emptied as experiments are terminated. It is essential during the period that the clean and infected room are both in use that a strict barrier is maintained between them. Once the room has been emptied it can be thoroughly cleaned and disinfected or fumigated. The method adopted will depend on the ability to isolate the room and its ventilation system from the rest of the building. If the room can be effectively sealed, formaldehyde fumigation, which is probably the most effective method, is feasible. Formaldehyde may be generated in various ways. Among these is heating a solution of formaldehyde in a suitable container with an electric heating element, or mixing two parts formaldehyde with one part of potassium permanganate and allowing the oxidative reaction to volatilize the formaldehyde. A relative humidity of 75% or more will increase the effectiveness of the procedure. Once the formaldehyde has been liberated the room should be sealed and left for 24 h. Great care is required during fumigation and it is recommended that a suitable respirator is worn. 10 ml of formaldehyde are required for every cubic metre of room. If the formaldehyde is to be liberated by heat a 10% solution of the commercially available formaldehyde should be used in order to raise the humidity simultaneously.

Alternatively, or if the room cannot be sealed, it can be disinfected using an agent such as Tegodor, a combination of formaldeyde, glutaraldehyde and quaternary compounds (Th. Goldschmidt Ltd., York House, Station Rd, Harrow, Middlesex.) with a pressure spray or fogger. A respirator is again advised.

3. HANDLING MICE

3.1 **Approach**

(i)   Lift the mouse from the cage with the base of the tail held between thumb and forefinger, and transfer to a grid cage top (*Figure 1a*) or a surface that it can

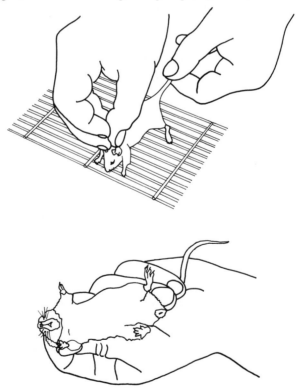

Figure 1. Correct way to grasp and hold a mouse.

grasp (e.g. the sleeve of a laboratory coat). The mouse's natural response is to move away from the handler and while pulling gently backwards on the tail the mouse can be pinned to the grid between the thumb and bent forefinger of the free hand.

(ii)   Allow the skin of the neck to slip in a controlled fashion between thumb and finger until the scruff is held tightly behind the ears.

(iii) Grip the tail between the 4th and 5th fingers and the palm of the hand (*Figure 1b*). If the handler tries simply to grasp the scruff or does not hold it tight behind the ears he is liable to be bitten.

Procedures such as ear clipping, s.c. and i.p. injections and palpation can be carried out with the mouse held as in *Figure 1*.

When working with mice in an isolator they may be picked up by the base of their tail using a pair of large forceps with rubber sleeves on their tips.

Wild mice are more difficult to handle because they jump. It is advisable to place the cage in a deep box before removing the lid. When transferring animals between cages it may be found easier to coax them into a bottle than to handle them.

3.2 Identification

There are two common systems used to identify mice, ear clipping with or without toe clipping. Combining ear and toe clipping it is possible to achieve identifying numbers up to 9999 (*Figure 2*). There is always the risk of an ear being torn or chewed, and records should be kept of the animals in each cage. Newborn mice may be identified by toe clipping.

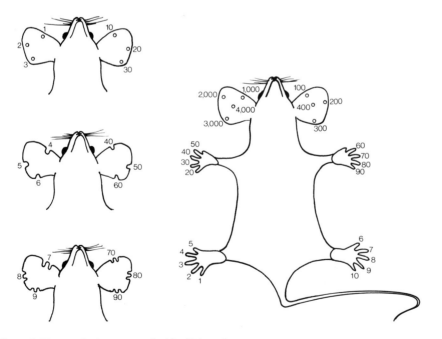

Figure 2. Two numbering systems for identifying mice.

3.3 Injections

A 25 or 27 gauge (0.5 or 0.4 mm) needle should be used. Care must be taken to ensure that there are no air bubbles trapped in the syringe or needle.

Intraperitoneal injections are given while holding the animal as in *Figure 1b* on its back with the head slightly down. The site of injection should be $2-3$ mm lateral to the umbilicus.

Subcutaneous injections can be in the scruff of the neck close to the handler's thumb while holding the mouse firmly on a surface that it can grasp.

3.4 Euthanasia

Mice may be easily killed by cervical dislocation.
(i)   Pick up the animal to be killed by the base of its tail as described in Section 3.1.
(ii)   Instead of grasping the scruff of the neck, firmly pinch the neck while pulling steadily on the base of the tail. Alternatively hold a rod or ruler across the neck while the tail is pulled.

If large numbers of animals are to be killed a euthanasia chamber can be used consisting of a container linked to a cylinder of carbon dioxide. Place the animals to be killed in the container and displace the air by carbon dioxide. Care must be taken to ensure that the animals are dead prior to disposal and it should be noted that mice $1-8$ days old take longer to die than adults. See also Section 8.

4. BREEDING

Mice will breed from 6 to 8 weeks of age and continue for about 200 days depending on the strain. The gestation period of the mouse is 20 days and parturition is often associated with a post-partum oestrus. Concurrent lactation and pregnancy can result in delayed implantation (i.e. embryos remain viable but blocked at the blastocyst stage during lactation).

Mice are commonly bred in pairs, trios (one male with two females) or harems (one male with more than two females). It is advisable not to put more than one male in a breeding cage as the males of a number of strains will fight. It should be noted that for the maximum production per female, pair mating is best. Other mating ratios produce a higher number of animals weaned per cage, but the pre-weaning mortality will be higher. Mice are weaned at approximately 3 weeks of age and should be removed from the breeding cage before the birth of the subsequent litter. To prevent fighting, male mice are caged together from weaning. Adult males from different cages should not be mixed.

Breeding records are kept so that unproductive animals can be identified and culled. Breeding females with litter intervals of over 40 days should be culled.

The breeding system adopted depends on the number of animals required and whether they are of an inbred or outbred strain. An inbred mouse strain must be maintained by full sib mating (brother × sister) while an outbred strain must be maintained in such a way that inbreeding is minimized. For details of various systems of breeding consult references 1 and 4.

Wild mice may be kept and bred in standard mouse cages, but one must remember their jumping ability.

5. EXPERIMENTAL PROCEDURES

5.1 **Stud males**

Stud males are test mated to ensure fertility and a record kept of their mating performance to ensure that they are working effectively. The record should include not only the fact that they mate regularly but also that matings are fertile.

5.2 **Timed matings**

The oestrus cycle of the mouse is normally 4 days, and females are receptive to mating only at oestrus.

If a regular daily supply of mated animals is required it is sufficient to set up a series of cages with one male and three females in each, bearing in mind that in theory one in four females should be in oestrus on a particular day. A ratio of one male to three females per cage will usually produce the best results in the majority of strains.

If a group of mated animals is required for a particular day it is possible to increase the probability of mating using the Whitten effect (5). When a group of female mice is placed with a male their oestrus cycles become synchronized and the majority of animals will mate on the third night. This phenomenon can be used to increase the number of animals mating on a particular day. A male is caged on one side of a barrier separating him from the females in the same cage for 2 days. The females are allowed to mate on the third day (6). Alternatively it is possible to check females and mate up only those animals that are in oestrus. The oestrus cycle may be followed by observing changes in the appearance of the vagina. At oestrus the vagina is gaping, the tissue is pink, slightly moist and with pronounced striations.

When mice mate the ejaculate of the male coagulates in the vagina of the female producing a copulation plug. By checking female mice for the presence of a plug it is possible to determine whether the animal has mated. Sometimes the plug is easily visible but normally a probe is used to detect its presence in the vagina. The plug remains in the vagina for varying lengths of time (up to 18 h). The sooner the animals are checked after mating the less the likelihood of the plug falling out prior to detection. Mice normally mate during the middle of the dark period of the diurnal cycle. Various systems of defining the time of mating and the age of the embryos are in use. The day of finding the vaginal plug may be designated day 0, day ½, day 1, or the first day of pregnancy in different laboratories.

The time of mating can be changed relative to the working day by changing the diurnal light/dark cycle. Thus by completely reversing the cycle, mating will occur around midday. Changes in the diurnal cycle allow different developmental stages of embryos to be available during normal working hours.

If animals are brought in from another establishment whose diurnal light/dark cycle is different or unknown a few days should be allowed for the mice to acclimatize.

5.3 **Pseudopregnancy and vasectomy**

Pseudopregnant animals are produced in a similar system as in Section 5.2 but the mating is with vasectomized males.

Male mice are easiest to vasectomize at about 6 weeks of age before they become fat. The males to be vasectomized should be of a strain or genotype with a good breeding performance. Vasectomy is carried out under anaesthesia through a single transverse ventral incision. It is possible to locate the vas deferens by manipulating the fat pad and without exteriorizing the testes. A short length of each vas deferens should be removed to ensure that sperm cannot pass between the severed ends. Ligation of the vas is unnecessary. The incision in the skin can be closed using suture thread or suture clips (Michell 12mm). The latter should be removed after 12 days. There is no need to sew the body wall.

The vasectomized animals should be test mated to check that they are sterile.

5.4 **Superovulation**

If large numbers of pre-implantation embryos are required at one time the gonadotropins PMSG (pregnant mare's serum gonadotropin as Gestyl from Organon or Folligon from Intervet Laboratories Limited) and hCG (human chorionic gonadotropin from Sigma or Intervet Laboratories Limited) can be administered to females prior to mating in order to increase the ovulation rate. Successful induction of superovulation depends on several variables including age, weight and strain of mouse, and the time of injection of the gonadotropins.

The best age for superovulation is usually in the range of 3−5 weeks; the optimum for a given genotype being within 4−6 days. Body weight is often a better indication of development in certain strains and if commercially bred animals are used weight may be the only available measure of developmental age.

The time of administration of the gonadotropins relative to each other and the light cycle of the mouse room affects both the number of eggs ovulated and their developmental uniformity. A 40−46 h interval between the PMSG and hCG injections will usually be found optimal. Ovulation normally occurs between 10 and 13 h after hCG injection. To obtain the best synchronization and yield it is important that the hCG is administered before the release of endogenous luteinizing hormone (LH) which is regulated by the diurnal light/dark cycle. It is generally assumed that endogenous LH release occurs 15−20 h after the midpoint of the second dark period following PMSG administration. Thus a feasible injection schedule for animals held in a room with a light period from 0800 h to 2000 h would be PMSG between 1500 and 1600 h and hCG 46 h later between 1300 and 1400 h. This would be about 3 − 4 h before endogenous LH release.

The dose of both PMSG and hCG should generally be 2−5 IU i.p. Different strains and hybrids respond quite differently to a given dose.

After hCG injection the females should be caged individually with proven stud males and checked for vaginal plugs on the following morning.

5.5 **Anaesthetics**

Although mice are probably the most frequently anaesthetized animals in laboratory work, it is still difficult to achieve true surgical anaesthesia without some mortality. Apart from anaesthetic overdose the most common cause of death during anaesthesia is hypothermia. There are marked variations in response to anaesthetics associated with strain, age, sex, weight and nutritional status.

Among the anaesthetics that are commonly used are pentobarbitone, tribromoethanol

(Avertin) and fentanyl-fluanisone (Hypnorm, Crown Chemical Co. Limited) with midazolam (Hypnovel, Roche) (7,8).

Pentobarbitone can be prepared from powder at 5 mg/ml in warm normal saline or as a 1/10 dilution of the commercial stock solution (60 mg/ml). The effective dose varies with so many factors that a control group of the same strain, age, sex and weight should be anaesthetized prior to the experimental animals to determine the correct dose. The dose required is normally in the range $0.7-1.0$ mg for a 25 g mouse.

Tribromoethanol may be made as follows.

(i)   Prepare a 100% solution by mixing 10 g, of tribromoethyl alcohol with 10 ml of tertiary amyl alcohol.

(ii)   For use dilute the stock solution to 2.5% with sterile water or normal saline.

(iii)   Keep both the stock and diluted solutions at 4°C in the dark.

There may be considerable variation between preparations and, as with pentobarbitone, variation between animals, so a test group should again be used. The dose required should be about 3.0 mg for a 25 g mouse. Tribromoethanol has been reported to be irritant to the viscera of mice and this can cause intestinal problems and death. This side effect seems to be strain dependent.

Fentanyl-fluanisone with midazolam probably provides the best available surgical anaesthesia. The commercial preparations should each be doubly diluted with sterile water (for injection) and then equal volumes mixed together. The resulting mixture contains 0.5 mg/ml midazolam, 2.5 mg/ml fluanisone and 0.079 mg/ml fentanyl citrate. This mixture is stable at room temperature for at least 8 weeks. A 25 g mouse requires an injection i.p. of $0.1-0.2$ ml of this mixture.

Ether is irritant to the mouse respiratory tract and can cause excessive mucous secretion. It is difficult to maintain a consistent level of surgical anaesthesia with ether and it is in addition highly explosive. It is not recommended.

Chloroform should not be used under any circumstances. Not only is it hepatotoxic but trace concentrations have been shown to interfere seriously with the breeding performance of male mice.

It is essential that anaesthetized mice are kept warm on a heated pad under a light or preferably in an incubator until they recover consciousness.

5.6 Re-derivation by hysterectomy

Caesarian re-derivation of a mouse breeding colony is the most effective method to restore its health status. A breeding colony of the required or higher health status is needed to provide foster mothers.

If a successful operation is to be carried out it is essential that newborn litters in the foster colony are synchronized with full term pregnant females of the strain to be re-derived. The animal to undergo the Caesarian is selected by palpation. Timed matings can be used to aid synchronization, but palpation is always advisable owing to the normal spread in the time of parturition.

A full term pregnant female may be palpated while held as shown in *Figure 1b*. the uterus is gently felt between the thumb and forefinger of the free hand. At around the 18th and 19th day of pregnancy the individual conceptuses can be felt as small spheres. Immediately prior to parturition the fetuses lie in an elongated position and can be moved slightly within the uterine lumen. This is the time at which the Caesarian operation should be performed. An animal that is overtly unhealthy or has already started to give birth should not be used.

An isolation area is used to house the foster mothers and re-derived litters until it can be confirmed that their health status has not been compromised as a result of the procedure. The actual operation can be performed in various ways depending on circumstances and the final health status to be achieved. Most commonly it is carried out on the open bench or in a laminar flow sterile cabinet.

Prior to the operation a sterile screw-capped container is prepared containing a disinfectant solution of 10% formaldehyde and 10% liquid soap, as a wetting agent, in sterile water at 36°C. A 500 ml beaker containing about 250 ml of the same solution is also required. The abdomen of the pregnant mouse should be shaved.

(i)   Kill the mouse by cervical dislocation, immediately dip it in the disinfectant solution in the beaker, blot to remove excess liquid and transfer to a sterile operating surface.

(ii)   Expose the body wall by grasping the skin of the abdomen immediately above and below the midline with fingers and thumb and pulling simultaneously towards the head and feet (this procedure can be made easier if a small midline knick is first made in the skin with a pair of scissors).

(iii)   Make a longitudinal incision in the body wall and expose the uterus.

(iv)   Hold the vagina with forceps and cut caudally. While gradually raising the uterus, trim the mesentry away. Take care not to puncture the uterine wall. In this way the uterus is removed without touching any surfaces. Provided the skin is peeled well out of the way any areas which the uterus might touch should in any case, be sterile.

(v)   Transfer the uterus to the disinfectant in the screw-capped container and take to the vicinity of the foster mother. This might be via a dunk tank or entry port to an isolator or simply to a clean isolation room.

(vi)   Tip the uterus into a suitable dish, remove from the disinfectant and blot lightly.

(vii)   Place it on a sterile surface, slit it longitudinally along the anti-mesometrial surface, and remove the fetuses from their membranes with the aid of a second pair of scissors and forceps. The umbilical cord should be crushed and torn, rather than cut, to prevent bleeding.

(viii) Dry the pups with tissues and gently stimulate with a tissue or surgical swab until they are pink and breathing regularly. Breathing can be stimulated by gently squeezing the tail, and if the pup squeaks it is most likely to survive.

The complete procedure from cervical dislocation of the mother to delivery of the last pup should take less than 4−4.5 min. It is possible to foster pups with a mother whose natural litter would be 2 − 3 days old but the success rate is lower.

The foster mothers and re-derived litter should be held in isolation until their health status has been shown to be satisfactory. If some of the mother's natural young have been left in the litter these can be used for health screening.

5.7 Fostering

Fostering may be used during hysterectomy re-derivation and in situations where the natural mother is unable to rear her own litter due to her own health, genetic defects, or because the litter is too small to maintain lactation. Mice may be fostered at birth or at any time prior to weaning. The natural litter of the foster mother should be about the same age as the animals to be fostered. The strain of the foster colony is chosen for good mothering ability and preferably the coat colour should be different from the animals to be fostered, or the mating arranged so that the natural young of the foster mother are not of the same coat colour as the pups to be fostered.

If the foster mother's young are a different coat colour from the pups to be fostered it is possible to leave a few natural young with the mother and add those to be fostered. If more than one foster mother is available the risk of cannibalism or neglect can be reduced by introducing a few pups to be fostered into each litter. If there is no alternative to using a foster mother whose natural litter is the same colour as the mice to be fostered great care must be taken to mark the fostered animals by toe clipping or all the natural young must be removed. When fostering a new born animal toe clipping is not recommended as it may encourage cannibalism. In these circumstances it is essential to ensure that the foster mother has finished giving birth and that all her natural young are removed. The genetic authenticity of the re-derived animals should always be confirmed after weaning.

6. TRANSPORTATION

Mice can be easily transported by road, rail or air. Suitable transport boxes usually made of polypropylene or cardboard with and without filters on the air vents are commercially available (Williton Box Co. Ltd., Williton, Somerset). Transport isolators may also be used to maintain the health status of the animals. It is important to ensure that the boxes or other containers conform to the requirements of the shipping agent (e.g. IATA regulations for air and British Rail regulations).

The animals are packed with sufficient food and a source of moisture for the complete journey allowing for possible delays. Water is provided as moistened food contained in a thin plastic sachet (9) or 1% reconstituted gelatine in a suitable container. Alternatively, for short journeys, a piece of potato or apple is provided, but remember when sending animals abroad that there may be regulations governing the importation of vegetable matter.

The labels on the boxes should include the address and telephone number of the sender and the consignee, a description of the contents (species, strain, age, number) and the number of boxes in the consignment. In order to avoid the boxes being opened in transit the label should state that sufficient food and water has been provided for the journey. This is especially important if animals of a particular health status are being shipped in filtered boxes or transport isolators as inadvertent opening of the container may render the animals valueless. Also a label is included stating clearly: Urgent − Live Animals.

The person due to receive the animals should be consulted to ascertain the documentation required. This may include a statement confirming that the animals are healthy and fit to travel and an import licence. Some countries require a health certificate signed by a Veterinary Inspector of the Ministry of Agriculture while others will accept a statement signed, for example, by a veterinary surgeon or the head of the animal facility. If animals are to be sent abroad all the necessary documentation must be arranged prior to shipping and the necessary papers must travel with the animals.

7. IMPORTATION

Whenever animals are to be brought into an animal house from an outside source it is advisable to keep them in strict isolation for 28 days while health checks are carried out. This is especially important if the mice have come from a source that is not usually used and if the health status of the colony of origin is unknown.

If animals are to be imported into the UK an importation licence issued by the Ministry of Agriculture under the Rabies (Importation of Dogs, Cats and other Mammals) Order (1974) is required. It is also necessary for the animals to be kept in premises registered under this Act and for the animals to be moved from the port of entry to the registered premises in a registered vehicle. The mice must then be kept in quarantine for 6 months. When mice are imported from abroad it must either be possible to carry out experiments in the quarantine facility or a breeding colony should be imported. It is permissible to remove mice born in quarantine 3 weeks after weaning with the permission of the Ministry of Agriculture.

8. LEGISLATION GOVERNING EXPERIMENTAL ANIMALS

It is essential to ensure prior to starting an experiment that you will not be contravening any national or international regulations governing the use of animals for scientific purposes. In the UK the relevant legislation, administered by the Home Office is the Animals (Scientific Procedures) Act 1986. Mice must be obtained from breeding or supplying establishments designated under the Act and the experimental animal facilities must also be designated. If breeding is part of the experimental protocol then the premises must be designated for both breeding and scientific procedures. The experimentor must also possess the necessary licences (e.g. Project licence and Personal licence). In addition it should be noted that this Act also sets out in Schedule 2 approved methods of euthanasia.

9. REFERENCES

1. Festing,M.F.W. (1979) *Inbred Strains in Biomedical Research*. Macmillan Press.
2. Festing,M.F.W. (1987) *International Index of Laboratory Animals*. 5th edition, Laboratory Animals Ltd, PO Box 101, Newbury, Berkshire, UK
3. Small,J.D. (1983) In *The Mouse in Biomedical Research*. Foster,H.L., Small,J.D. and Fox,J.G. (eds), Academic Press, Vol. 3, p. 90.
4. *The UFAW Handbook on the Care and Management of Laboratory Animals* (1967) Third edition, Livingstone, Edinburgh.
5. Whitten,W.K. (1956) *J. Endocrinol.*, **14**, 160.
6. Ross,M. (1962) *J. Anim. Technicians Assoc.*, **13**, 1.
7. Green,C.J. (1979) *Animal Anaesthesia*. Laboratory Animals Ltd, London.
8. Flecknell,P.A. and Mitchell,M. (1984) *Lab. Anim.*, **18**, 143.
9. Peters,A.G. and Bywater,P.M. (1983) *Anim. Technol.*, **34**, 71.

CHAPTER 2

Isolation, culture and manipulation of pre-implantation mouse embryos

HESTER P.M.PRATT

1. INTRODUCTION

There are two broad areas of developmental research that require *in vitro* culture of pre-implantation embryos, namely, the study of the embryo as an autonomously differentiating system (1,2), and the use of the early embryo as a vehicle for gene insertion in the production of transgenic mice (see Chapter 11). This chapter describes in some detail the procedures for isolation of eggs and pre-implantation embryos, and for the manipulation and observation of embryos at various stages of development.

Three points should be made at the outset.

(i)   Mouse development has tended to become the paradigm of pre-implantation mammalian embryogenesis. This has arisen because, with the exception of the rabbit and possibly the human (see Chapter 14), the mouse is the only species that can be cultured *in vitro* over the entire pre-implantation period. However, it should be noted that conclusions derived from mouse embryos are not necessarily applicable to mammals in general.

(ii)   Embryos tend to develop more slowly *in vitro* in the chemically defined media currently in use than they do *in vivo*, suggesting that optimal growth and development may require specific factors from the reproductive tract.

(iii) A mention must be made of the benefits of accurate and regular observation. Familiarity with the material generates the skill necessary to distinguish between normal and abnormal development and to assess the developmental stage that the embryos have reached.

2. RECOVERY OF EMBRYOS

Conditions for the removal and culture of embryos need not necessarily be aseptic but should be dust free and as clean as possible. Since cultures are unlikely to be maintained for more than 4 or 5 days they can generally be kept free of bacterial contamination by conducting all manipulations in sterile media with flame-pulled glass mouth pipettes under dissecting microscopes in a dust-free area, for example surrounded by Perspex dust covers. Details of reagents, media and equipment are given in Section 5.

Eggs and embryos (particularly during the 1-cell and 2-cell stages) are extremely sensitive to environmental stress. During conventional culture there are likely to be fluctuations in temperature necessitated by frequent removal from the incubator for observation. The most obvious consequence of such shifts in temperature is an artefactual lengthening of the G2 and M phases of the cell cycle (3). One way to overcome this problem is to equip all dissecting microscopes with heated stages set to $36 \pm 1 °C$, and to pre-equilibrate all media at $37 °C$.

There is no generally accepted optimal culture medium and different laboratories have adopted slightly different formulations to give developmental rates comparable with *in vivo* growth. The essential common features are bicarbonate buffer, presence of lactate and pyruvate and a macromolecule [generally bovine serum albumin (BSA)]. The differences lie in the concentrations of these and other components as well as variations in the final osmolarity of the media (discussed in refs 4, 5).

The exact procedure employed to recover embryos varies according to the developmental stage and specific details will be dealt with in sequence. In broad outline the process involves collecting the embryos into warm ($37 °C$) Hepes-buffered recovery medium containing BSA (M2 + BSA) using a flame-pulled glass Pasteur pipette attached to a mouth piece. Debris is removed by repeated passage of the embryos through warm ($37 °C$) wash drops of M2 + BSA followed by a rinse in the bicarbonate-buffered culture medium (M16 + BSA). Embryos are finally transferred into drops of M16 + BSA ($\sim 50 \, \mu l$) under paraffin oil in plastic Petri dishes, and incubated at $37 °C$ in an atmosphere of 5% CO_2 in air. This method of culture is the most convenient if repeated access is required for observation or manipulation. In a busy laboratory large numbers of Petri dishes can accumulate quite rapidly and the most convenient and safest way of storing and retrieving them from the incubator is to stack the dishes on narrow Perspex trays which extend the depth of the incubator. If frequent access to embryos is necessary it is advisable to have these in a second incubator designated for short-term cultures thereby ensuring that the long-term cultures remain undisturbed and enjoy greater stability of CO_2 levels and temperature.

2.1 Fertilized and unfertilized eggs

2.1.1 *In vivo mating*

The procedures for setting up matings in response to natural or experimentally induced ovulation (superovulation) are dealt with in Chapter 1, Section 5. Our experience shows that stud males only mate efficiently if they are used regularly but not more frequently than alternate nights. Superovulation is a useful device for increasing the numbers of eggs and embryos obtained from young ($3-4$ weeks) mice (for discussion of the factors involved see ref. 6). To obtain good synchronization of fertilization and hence of subsequent development, the hormone injection schedule should ensure that mating occurs prior to ovulation (i.e. fresh eggs are fertilized immediately) and that ovulation is timed to occur in response to the injected hormone and not the surge of endogenous luteinizing hormone (LH). An appropriate schedule is presented in Chapter 1 (Section 5.4). If embryos of developmental ages different from these are required, the necessary deviation from this injection regime will produce a more asynchronous population due to the spread in time during which ovulation and fertilization will occur. Methods for synchronizing such embryos and their component cells are dealt with in Sections 3.1 and 3.2. It is always wise to question the normality of embryos developing from superovulated eggs and to use the appropriate criteria under study to compare them with naturally ovulated eggs.

To recover fertilized or unfertilized eggs carry out the following procedure

(i)   Kill the female mice by cervical dislocation, dissect out the oviducts and place them in warm (37°C) saline or M2 + BSA (Section 5). Manipulate the oviducts in a plastic Petri dish or glass cavity block on the heated stage of a binocular dissecting microscope. If ovulation has occurred within the past 10 h (approximately) the masses of cumulus cells surrounding the eggs will be clearly visible through the thin distended walls of the ampulla.

(ii)   Release the cumulus masses in M2 + BSA culture medium by tearing the walls of the ampulla with fine watchmaker's forceps.

(iii)   Remove as much medium as possible (without flattening the eggs by drawing the meniscus too low) and replace it with approximately 1 ml of warm (37°C) hyaluronidase solution (Table 4). The cumulus cells will disperse gradually to release the enclosed eggs over the next 2−4 min. This process can be hastened by gentle pipetting of the egg masses through a wide bore glass mouth pipette. Both cold and hyaluronidase treatment can activate unfertilized eggs so eggs must be kept warm and exposed to hyaluronidase for the minimum length of time.

(iv)   After the eggs have settled to the bottom of the dish, replace as much of the hyaluronidase solution as possible with warm M2 + BSA, collect the eggs by mouth pipette and wash them free of cumulus cells by repeated passage through further washes of M2 + BSA.

(v)   In cases where thousands of eggs need to be collected (e.g. for RNA extraction, see Chapter 9), it is more efficient to remove the cumulus cells by using a nylon mesh (mesh size ∼20−30 μm) inserted into the base of the barrel of a 5 or 10 ml plastic syringe. Wet the mesh with medium, pass the medium containing the eggs through the mesh and collect the flow-through containing cumulus cells in a Petri dish. Recover the eggs by inverting the mesh over a second Petri dish and passing fresh medium through it. The eggs that are released should be largely free of cumulus cells but any contaminating cells can be removed using a mouth pipette.

(vi)   If eggs are not recovered from oviducts until 10 h after injection of human chorionic gonadotropin (hCG) or later, endogenous enzymes will have liberated cumulus masses. In this case the best method for recovering eggs is to use the flushing method described in Section 2.2 for cleavage stages followed by hyaluronidase treatment if necessary.

Fertilized eggs can be distinguished from those that are unfertilized within hours of fertilization by their prominent second polar body (*Figure 1*, panel 5). (Frequently the first polar body degenerates in many strains of mice.) These fertilized eggs can then be staged developmentally with reference to the presence and positions of the male and female pronuclei (*Figure 1*, panels 6−10). The sequence of events during the 1-cell cycle (detailed timings are given in *Table 1*) is initiated by fertilization and is independent of the time of ovulation (8). It is advisable to avoid fertilization of eggs ovulated

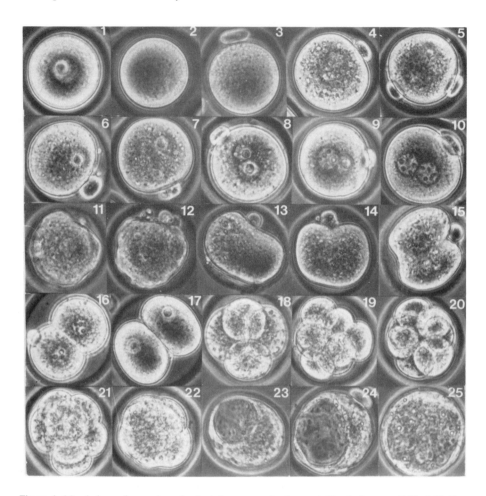

Figure 1. Morphology of normal pre-implantation mouse development. For timings see *Table 1*.(1) Pre-ovulatory oocyte with germinal vesicle intact. (2) Pre-ovulatory oocyte showing breakdown of germinal vesicle. (3 and 4) Extrusion of first polar body followed by ovulation and fertilization. (5) Resumption of meiosis by female set of chromosomes and extrusion of second polar body. (5 and 6) Decondensation of sperm nucleus and formation of male pronucleus. (7) Formation of nuclear membrane around haploid set of female chromosomes to form female pronucleus, which is subcortical, near the second polar body and smaller than the male pronucleus. (8−10) Migration of pronuclei to centre of egg. DNA replication. (11 and 12) Breakdown of pronuclear membranes. Ruffling of embryo surface indicating reorganization of the cytoskeleton preparatory to cleavage. (13) Elongation of embryo. (14 and 15) Formation of 'waist'. (16) Newly formed 2-cell embryo with visible nuclei. (17) Later stage 2-cell embryo with visible nuclei. (18−25) Later stages of pre-implantation development. 18: 4-cell embryo. 19−20: 6- to 8-cell embryos. 21: compacting 8-cell embryo. 22: compacted 8- to 16-cell embryo. 23 and 24: early blastocyst. 25: fully expanded blastocyst. (For more details see *Figure 2*.) (Reproduced from ref. 7.)

more than 12 h previously because of the increased incidence of spontaneous parthenogenetic activation of aged eggs (9) due in part to the gradual disorganization of the cytoskeleton (10). Controlled experimental parthenogenetic activation of eggs is dealt with in Chapter 12.

2.1.2 *In vitro fertilization*

This method of obtaining eggs and embryos is convenient for two reasons. It provides a more synchronous population than *in vivo* mating because fertilization, which initiates development (8), can be timed precisely. Secondly, the procedure can be divorced from the normal day/night cycle, and thereby allow insemination of eggs at a time which permits study of a subsequent developmental stage during normal working hours. The detailed procedure is given in Chapter 13, Section 6.2. Our own procedure is similar and will not be reproduced in this chapter.

2.2 Cleavage stages (2-cell to 8-cell)

Development from the 2-cell stage to the late 8-cell/early 16-cell (morula) stage (*Figure 1*, panels 16−22) occurs in the oviduct over a period of approximately 47 h. The second cell cycle, like the first, is relatively long (~20 h) whereas the third and subsequent cycles adopt the more conventional somatic cell cycle length of 12 h. Thus populations of a particular stage should be recovered from oviducts at predictable times after hCG injection or the inferred time of fertilization *in vivo*. However, this assumption does not take account of the substantial inter- and intra-embryo variation in developmental timing (assessed with respect to morphological and/or cell cycle stage) which is always encountered.

The heterogeneity between embryos is influenced by the asynchrony of fertilization (8) while the variation in cell cycle stage within an embryo appears to be due to intrinsic and heritable differences in the blastomeres (16). Since a complete understanding of developmental mechanisms can only be obtained if the sequence and timing of events is known, methods for obtaining synchronized embryos and their constituent blastomeres have to be developed. These methods are discussed in Sections 3.1 and 3.2.

The cleanest way of recovering embryos from the oviduct is to use a fine needle with attached syringe and to flush them out. This requires some practise but once the skill is acquired it is far preferable to the alternative method which is to shred the oviduct with fine watchmaker's forceps and collect the embryos from among the debris.

To remove embryos from oviducts carry out the following steps.

(i)   Remove the oviduct by first cutting through the top of the uterus, leaving a small portion of it attached to the oviduct, and then dissecting the oviduct away from the ovary. This must be done very carefully to ensure that the opening of the oviduct remains intact since it is extremely difficult to insert a needle once it has been damaged. The most convenient needles for flushing oviducts are 30 gauge metal ones which have been cut and polished to give a straight (not bevelled) and smooth end.

(ii) Place the cut oviducts in a drop (just enough medium to cover the oviducts) of M2 + BSA (*Table 5*) in a plastic Petri dish on the heated stage of a binocular dissecting microscope. If too much medium is used the oviducts may float around making the flushing manipulations much more difficult.

(iii) Use watchmaker's forceps to probe the oviduct and locate the end (this has the appearance of a ribbed sleeve in many strains of mice). Then insert the flushing needle attached to a 1 ml syringe filled with M2 + BSA.

Table 1. Timing of developmental events[a].

	Period of event		
	Hours post-insemination		*Hours post-hCG injection*
First cell cycle (8)			
(1-cell)			
Completion of meiosis II			
GVBD and ovulation		–	~12
G_1 phase	2–11		
Second polar body extruded		1–2	
Male pronucleus forms		4–7	
Female pronucleus (smaller) forms		5–8	
Pronuclei migrate centrally		8–10	
S phase	11–17		
G_2/M phase	17–20		
Dissolution of pronuclear membranes		16.5–19.5	
Cytokinesis	18.5–21.5		~32

	Hours post-cleavage		*Hours post-hCG injection*
Second cell cycle (11)			
(2-cells)			
G_1 phase	0–1		
1st transcription of embryonic genes		1–1.5	
S phase	1–6		
2nd transcription of embryonic genes		7–10	
G_2 phase	6–18		
M phase	18–20		
Cytokinesis	18–20		~48–50
Third cell cycle (3)			
(4-cells)			
G_1 phase	0–1		
S phase	1–8		
G_2 + M phase	8–13		~58–60
Fourth cell cycle (3, 12, 13)			
(8-cells)			
G_1 phase	0–2		
S phase	2–9		
cytoplasmic polarization		2–7	
cell flattening (compaction)		3.5–7.5	
surface polarization		5–9.5	
G_2/M phase	9–12		
Cytokinesis (decompaction during mitosis in intact embryos)	11–14		~72

Table 1 continued.

	Period of event	
	Hours post-inseminiation	*Hours post-hCG injection*
Fifth cell cycle (13, 15)		
(16-cells)		
Recompaction after division		
(intact embryos)	0−5	
Cytokinesis (outer cells)	11−12	∼84
(inner cells)	13−14	
Sixth cell cycle (13, 14)		
(32-cells)		
cavitation begins		
(intact embryos)	0−2	90−98
Subsequent cell cycles		
Fully expanded blastocyst		∼110
Hatched blastocyst		∼120

[a]This table lists the sequence of readily observable morphological events during pre-implantation mouse development *in vitro* in relation to phases of individual cell cycles (where known) and provides timings with respect to the previous cleavage and time of hCG injection.
Data on cell cycles 1−5 are derived from single cells or pairs of cells synchronized to the previous division whereas later cycles are from intact embryos and hence show greater variation.
Cycles 1 and 2 are for F_1 hybrid [(C57 BL ♀ × CBA ♂) F_1 ♀ × CFLP ♂] embryos whereas later cycles are for MFI outbred mice. All timings are derived from groups of embryos and the period of any event incorporates both the range of times over which the event is initiated within the population as well as its duration.

(iv)   Flush $0.1−0.2$ ml of medium through the needle. If successful the oviduct should swell and medium containing embryos should be discharged through the cut end of the uterus.

(v)   Separate the embryos from other cellular debris by mouth pipette, wash them through a warm pre-equilibrated drop of M16 + BSA (*Table 5*) and culture them in the same medium under paraffin oil at 37°C in 5% CO_2 in air.

2.2.1 *2-Cell block*

At this point it is necessary to interpose an important note of caution. For the majority of mouse strains, eggs or embryos placed in culture at any stage before the mid 2-cell stage will arrest development in the G_2 phase of the second cell cycle—a phenomenon referred to as the 2-cell block (discussed in ref 17). If recovery from the oviduct is delayed until embryos have reached the late 2-cell stage (∼48 h post-hCG) then development *in vitro* in simple defined media is unimpaired. There are however, a few strains of mice (mostly inbred strains or hybrids between inbred strains) whose eggs and embryos do not block at the two-cell stage (*Table 2*). In our laboratory we use (C57BL ♀ × CBA/Ca ♂) F_1 hybrid females.

In a comparison between blocking and non-blocking strains of mice it was shown that the potential of embryos to arrest at the 2-cell stage is determined solely by the

Table 2. Strains of mice that produce eggs which do not exhibit the '2-cell block' phenomenon[a].

Strain	References
(C57BL/10ScSn/Ola ♀ × CBA/Ca/Ola ♂) F_1	18
(C57BL × CBA T6 T6) F_1	19
(C57BL × CBA − LAC) F_1	20
(C57BL/10J × SJL/J) F_1	21
B6A F_1	17
(C57 × SJL) F_1	17

[a]50% or more of fertilized eggs develop to the blastocyst stage.

genotype of the egg and occurs irrespective of the paternal or embryonic contribution (18). In practical terms the 2-cell block phenomenon means that studies of the 1-cell to 2-cell transition including *in vitro* fertilization are best undertaken using eggs from a non-blocking strain [although the main features of development at this stage do not differ qualitatively between eggs from blocking and non-blocking strains (18)]. Furthermore if a blocking strain of mouse is used (if the mice are outbred this is most likely to be the case) then care must be taken to ensure that 2-cell stage embryos are recovered from the oviducts late enough in the second cell cycle to ensure their normal development to blastocysts *in vitro*. This is likely to be approximately 36−40 h post-hCG but it is best determined by experiment on the particular strain of mouse available.

2.3 Morulae

The term 'morula' is imprecise but generally refers to a compact aggregate of blastomeres (*Figure 1*, panels 21, 22). Since this is a purely morphological description it can embrace all stages in development from compacted 8-cells to the 16- to 32-cell stage just prior to blastocoel formation. By observing the development of embryos *in vitro* during this period it is possible to make a rough assignment of their developmental stage since the component blastomeres flatten upon one another (compact) at the 8-cell stage but then decompact during mitosis to the 16-cell stage and then subsequently recompact again (*Figure 2*, panels 4−6; for timings see *Table 1*). Late morulae are generally regarded as 16- to 32-cell embryos which have not cavitated and it is at this stage (~ 84 h post-hCG) that they pass through the uterotubal junction from the oviduct into the uterus. When recovering embryos at this stage it is therefore advisable to flush not only the oviducts but both uterine horns as well (see Section 2.4).

2.4 Blastocysts

Embryos enter the uterus and start to cavitate (form a blastocoel) at about 94 h post-hCG when they contain approximately 32 cells (i.e. when most of the cells have reached the end of their fifth cell cycle, *Table 1*). The blastocysts continue to expand as their component cells progress through the next two cell cycles (*Figure 1*, panels 23−25, *Figure 2*, panels 9−10). When they contain approximately 128 cells (in the region of 120 h post-hCG) the zona pellucida starts to thin and blastocysts 'hatch'. (*Figure 2*, panel 11), attach to the uterine epithelium and begin the process of implantation. Once this process is underway they cannot be dislodged by flushing the uterine horns. It is therefore advisable to collect blastocysts well in advance of hatching and attachment

and allow them to expand and hatch *in vitro* in a simple defined medium (e.g M16 + BSA, *Table 5*) or hatch and outgrow in a more complex medium [e.g. Dulbecco's modified Eagle's medium (DMEM) + 10% fetal calf serum (FCS), see Section 3.5.2].
To recover blastocysts carry out the following steps.

(i)   Dissect individual uterine horns away from underlying mesentery and cut at the uterotubal junction and cervix. Lay each horn on to filter paper dampened with M2 + BSA (*Table 5*) and remove any excess blood.

(ii)   Insert a 25-gauge needle attached to a syringe into the lumen of the uterine horn at the ovarian end and flush $0.2-0.5$ ml of M2 + BSA through into a plastic Petri dish or glass cavity block. This procedure does not need to be done under a dissecting microscope.

(iii) Collect blastocysts into warm (37°C) M2 + BSA wash them through warmed and equilibrated drops of M16 + BSA (*Table 5*) and incubate in culture drops of M16 + BSA under oil at 37°C in 5% CO_2 in air.

(iv)   Blastocysts will expand, hatch and stick to an appropriately coated plastic Petri dish (e.g. Falcon type) but will not undergo proper attachment and trophectodermal outgrowth in M16 + BSA. A more complex medium containing serum (e.g. DMEM + 10% FCS) is required for this subsequent development (see Section 3.5.2).

3. MANIPULATION OF EMBRYOS AND CELLS

3.1 Staging of embryos

The cellular and molecular events of pre-implantation development occur in a precise sequence. Very little is known about the control of the timing of these events but it is clear that there are at least two underlying clocks. One of these is the cell cycle itself, other events are regulated in sequence and time independently of the cell cycle (2). Analysis of timing mechanisms requires a synchronized population of cells or embryos.

As already discussed in Section 2.1.1 the population of embryos recovered from an *in vivo* mating will be heterogeneous, and hours post-hCG injection, or hours after the inferred time of fertilization, are inadequate as means of staging embryos. The solution to this problem is to select one of a number of morphological criteria, observe the embryos in culture at defined intervals (e.g. every 30 min or every hour), and select those embryos that have attained a particular stage within the preceding chosen time interval. In this way cohorts of embryos can be gathered which are synchronized to a particular developmental transition. The following events are readily observable in intact embryos and provide a means of synchronizing them — (for timings see *Table 1*).

(i) *Cleavage.* (*Figure 1*, panels $11-21$). Cell numbers are easily counted up until compaction occurs at the 8-cell stage when cell outlines become obscure. Fertilization *in vitro* will improve the synchrony of cleavage to the 2-cell stage, however intrinsic heterogeneity in cleavage times will probably necessitate resynchronization to each subsequent cell division.

(ii) *Compaction.* (*Figure 1*, panels 21, 22; *Figure 2*, panels $1-4$). This process of cell flattening generally occurs at $3.5-7.5$ h after the onset of the 8-cell stage (4th cell cycle) and provides a useful transition point since, with experience, it is possible to assess complete compaction objectively.

(iii) *Division to the 16-cell stage.* (end of 4th cell cycle beginning of 5th cell cycle) (*Figure 2*, panels 5, 6). This stage is marked by decompaction prior to mitosis and recompaction after division is completed. Serial observations of late 8-cell embryos

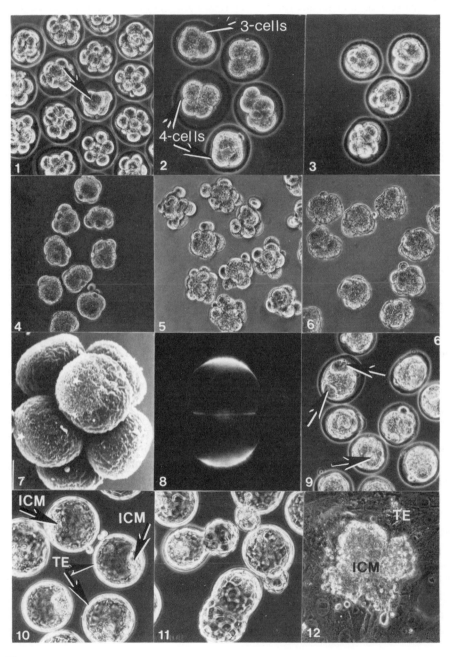

can therefore be used to generate sub-populations of newly arisen 16-cell embryos.

(iv) *Blastocoel formation.* (6th cell cycle). (*Figure 2*, panels 9, 10). The earliest manifestation of fluid transport occurs when the embryo has approximately 32 cells (*Table 1*) and involves the formation of a fluid-filled vacuole (or occasionally two vacuoles which enlarge and subsequently coalesce). Selection for this feature can therefore provide populations of embryos which are just starting to form blastocysts and where the majority of cells will be in their sixth cell cycle.

3.2 Staging of isolated cells

Cleavage is asynchronous and the degree of asynchrony increases progressively as pre-implantation development proceeds. Although the morphological criteria described in Section 3.1 may be adequate for staging and synchronizing embryos for some purposes, a precise analysis of developmental mechanisms can only be undertaken on cells of known age. The one unambiguous method of obtaining such cells it to disaggregate embryos to single cells and observe the cultures at a defined time interval as discussed in Section 3.1 and group all those cells that have divided during that time. This procedure requires a method for removing the zona pellucida and a method for disaggregating cells.

3.2.1 *Removal of the zona pellucida*

It is possible to remove the zona mechanically but chemical methods are preferable when large numbers of embryos are involved. Two reagents are in general use, one is an acidified Tyrode's solution (22) and the other a solution of pronase (23). In both cases the composition and osmolarity are crucial to success and it is advisable to test any new batch on a few embryos before committing the entire culture to the treatment.

(i)   Transfer embryos in small groups into warm (37°C) acid Tyrode's (*Table 3*) or pronase solution (*Table 4*). The temperature (37°C) is particularly important for successful zona removal with acid Tyrode's.

Figure 2. Key developmental transitions (and some anomalies) during formation of a blastocyst. (For timings see *Table 1*). (**1**) 8-cell embryos prior to compaction. One embryo starting to compact (arrow). (**2**) Compacted embryos with less than 8 cells. Compaction of embryos with less than 8 cells can occur amongst the slowly dividing group in a population where the majority are compacting at the 8-cell stage. This observation demonstrates that compaction is not linked to a particular cell cycle. Though some are potentially viable such embryos are probably best discarded. (**3**) Non-integrated blastomeres in compacted embryos. Compaction takes 4−5 h and does not occur synchronously in all blastomeres hence some compacting embryos may contain blastomeres that are not integrated into the main mass. Any embryo not fully compacted approximately 8 h after the onset of the 8-cell stage should be discarded as abnormal. (**4−6**) Sequence showing decompaction of compacted 8-cell embryos during division to the 16-cell stage and recompaction afterwards (see Section 3.1). The zonae pellucidae have been removed so that surface changes can be visualized more clearly. (Kindly provided by Julia Chisholm.) (**4**) Compacted 8-cell embryos. (**5**) Decompacting 8-cell embryos. (**6**) Recompacted 16-cell embryos. (**7**) Polarization of surface microvilli on individual blastomeres of an 8-cell embryo viewed by SEM (see Section 4.1.2). (**8**) Polarization of surface microvilli on a pair of 8-cell blastomeres revealed by FITC−Con A staining (see Section 4.1.1). (**9**) Embryos within a few hours of the onset of fluid accumulation. Normal embryos may exhibit more than one cavity initially (e.g. arrowed embryo) but these ultimately coalesce into one. Forms in which this does not occur should be discarded. (**10**) Early blastocysts. ICM and TE indicated. Discard any that are not expanding and do not exhibit a clear ICM. (**11**) Hatching blastocysts showing the various morphologies encountered. (**12**) Embryonic outgrowth showing polyploid TE and knot of ICM.

Table 3. Composition of acid Tyrode's (22).

	(g/100 ml)
NaCl	0.80
KCl	0.02
$CaCl_2$ or ($CaCl_2 2H_2O$)	0.02 (0.0265)
$MgCl_2\ 6H_2O$	0.01
D-glucose	0.10
PVP	0.40

Preparation
Adjust the pH to 2.5 with 1 M HCl (N.B. Analar only). Millipore the solution and store in 1 ml aliquots at 4°C. Use within 2 days.

Table 4. Reagents

Reagent	Supplier	Preparation	Storage
Water (Analar)[a]	BDH Ltd Poole, Dorset, UK		
Hormones PMSG (Folligon)	Intervet PO Box 5830AA, Boxmeer, The Netherlands	1 vial (1000 IU) make up to 20 ml with 0.9% sterile saline (50 IU/ml)	Store at −20°C in 5 ml aliquots. Thaw at 4°C. Discard after 1 week at 4°C.
hCG (Chorulon)	As PMSG	1 vial (1500 IU) make up to 30 ml with 0.9% sterile saline (50 IU/ml)	As PMSG
Bovine serum albumin[b] (BSA) (crystallized and lyophilized)	Sigma Chemical Co. Fancy Road, Poole, Dorset BH17 7NH UK Cat No. A4378		Store desiccated at 4°C.
Fetal calf serum[b] (FCS)	Flowlabs Ltd, PO Box 17, Second Avenue, Industrial Estate Irvine, Ayrshire KA12 8NB, UK	Store at −20°C on receipt. Thaw, heat-inactivate at 56°C for 30 min.	Store at −20°C in 10 ml aliquots.
Oil (light, colourless liquid paraffin oil)	BDH Ltd Product No. 29436	Autoclave for 20−30 min at 15 p.s.i. allow to cool, filter through Whatman filter paper.	Store at room temperature or at 37°C for use in culture dishes.
Pyruvate (sodium salt) crystalline type II	Sigma Chemical Co., Cat No. P2256		Store desiccated at 4°C
Lactate (sodium salt) 60% syrup	Sigma Chemical Co. Cat No. L1375		Store at 4°C
Streptomycin sulphate	Glaxo Labs. Ltd Greenford, UK		
Benzyl penicillin (Crystapen)			
Polyvinylpyrrolidone (PVP)	BDH Ltd		
Phenol red	BDH Ltd	0.13 g in 10 ml of isotonic $NaHCO_3$	4°C

Table 4 continued.

Reagent	Supplier	Preparation	Storage
Phosphate buffered saline (PBS) tablets (Dulbecco A)	Oxoid Ltd, UK Cat No. BR14a	One tablet in 100 ml of double-distilled water. Autoclave.	4°C
Pronase (for removing zona pellucida)	Calbiochem-Behring Corp., PO Box 22, Bishops Stortford Herts CM22 7RQ, UK Cat No. 537088	Make up 0.5% pronase in PBS with PVP added to 10 mg/ml. Mix well and centrifuge	Filter sterilize, aliquot the supernatant and store at −20°C;
Hyaluronidase type II from sheep testes (for removing cumulus cells)	Sigma Chemical Co. Cat No. H2126	Dissolve 20 mg of hyaluronidase and 0.2 g of PVP in 20 ml of PBS	Millipore filter and store in 1 ml aliquots at −20°C
Hepes	Sigma Chemical Co. Cat No. H3375 Calbiochem-Behring Hepes (Ultrol) Cat No. 391338		
DMEM (for trophectoderm outgrowth)	Flow Labs Ltd.	Add L-glutamine 20 mg/ml, penicillin 60 μg/ml, streptomycin 50 μg/ml and heat-inactivated FCS 10%.	Store at 4°C

[a]A purer grade may be necessary for culturing human embryos (see Chapter 14, Section 4.3).
[b]Buy a small sample of each new batch, test for toxicity and if acceptable buy a large quantity.

(ii)   Pipette embryos gently up and down in the solution and observe them under the microscope. In the case of acid Tyrode's the process only takes a matter of seconds, the zona swells and then thins and it is advisable to remove the embryos to warm M2 + BSA (*Table 5*) before the zona is dissolved completely. Over-exposure of the embryos to the low pH will result in lysis of some or all of the cells within minutes. Early stages (up to and including early 4-cell) are more sensitive to acid Tyrode's treatment (especially early in the cell cycle) than later stages. One-cell embryos are most viable if zonae are removed immediately after fertilization. Zona removal with pronase takes longer (3−4 min) and the enzyme is diluted and effectively inhibited by transferring the embryos to warm M2 + BSA.

(iii) Wash the embryos in M16 + BSA (*Table 5*) culture medium and incubate at 37°C in 5% CO_2 in air under oil.

There are two important points to note. First removing the zona may affect surface, and possibly cytoskeletal, components (24), so the necessity of a subsequent 'recovery' period *in vitro* should always be considered. Second, once divested of their zonae, embryos of all stages are very sticky and tend to adhere to one another as well as to most culture dishes. Zona-free embryos are therefore cultured individually in small drops of M16 + BSA in non-adherent Petri dishes (e.g. Sterilin).

3.2.2 *Disaggregation of cells*

In order to disaggregate embryos efficiently with minimal damage to the cells it is

Table 5. Composition of culture media (M16, M2).

1. Stock solutions for making M16 and M2

	Component	*g/100 ml*
Stock A:		
(10 × conc)	NaCl	5.534
	KCl	0.360
	KH_2PO_4	0.162
	$MgSO_47H_2O$	0.294
	Na lactate 60% syrup	2.608 (3.2 ml)
	Glucose	1.000
	Penicillin	0.060
	(10^5 IU)	
	Streptomycin	0.050
Stock B:		
(10 × conc)	$NaHCO_3$	2.106
Stock C:		
(10 × conc)	Na pyruvate	0.36
Stock D:		
(10 × conc)	$CaCl_22H_2O$	2.52
Stock E:		
(10 × conc)	Hepes (Ultrol)	5.957
	adjust to pH 7.4 with	
	2 M NaOH before	
BSA	making up to 100 ml.	0.400

Preparation

Millipore filter (0.22 μm pore size) all stocks and store at 4°C in Falcon plastic tubes. Stocks A, D, E will keep for 3 months, B and C for 2 weeks.

2. To make up 10 ml of M16 + BSA or M2 + BSA from stock solutions

Stock (*ml*)	*A*	*B*	*C*	*D*	*E*	*Phenol red* (*mg*)	*Water* (*Analar*) (*ml*)	*BSA* (*mg*)	*pH*
M16	1.0	1.0	0.1	0.1	–	0.1	7.8	40	7.6
M2	1.0	0.16	0.1	0.1	0.84	–	7.8	40	7.4

Preparation

Adjust pH as necessary. Millipore filter (0.22 μm pore size). Store in convenient sized aliquots in Falcon tubes at 4°C. For Ca^{2+}-free M16 and M2 omit stock solution D, add 0.1 ml of 15 mg/ml NaCl. Add 6 mg/ml BSA.

necessary to use a micropipette with a flame-polished end and an internal diameter just in excess of the diameter of the cells to be disaggregated. The micropipettes are easy to pull by hand using 1.2 mm tubing (Section 5.3.6) although an electrode puller can be used and the squaring and polishing is done using a microforge. Pipetting aids for capillary pipettes provide suitable holders for the micropipettes which can be attached to a conventional mouth piece. Early cleavage stages (2-cell to late 8-cell) are disaggregated simply by exposure to media deficient in Ca^{2+} which combats the Ca^{2+}-dependent cell adhesion system (2).

(i)   Incubate embryos (without their zonae) for approximately 10 min in Ca^{2+}-free M2 + 6 mg/ml BSA (*Table 5*) under oil at 37°C in air.

(ii)   Pipette small groups of embryos $(3-4)$ up and down through a micropipette with a bore slightly less than the diameter of the embryos. Carry out all the manipulations with the culture dish on the heated stage $(37°C)$ of a binocular dissecting microscope. Set aside any single cells resulting from this manipulation and subject the remaining aggregates and other embryos to further pipetting until disaggregation is complete. In some cases pairs of cells will be seen to be attached by a cytoplasmic bridge (the 'midbody'), which represents the remains of the previous cleavage furrow and hence indicates the original point of contact between the cells as well as the orientation of the cleavage plane (12).

(iii)   Wash isolated cells through warm drops of M16 + BSA (*Table 5*) in non-adherent dishes (e.g. Sterilin) and culture them in small drops either singly or in groups. In the latter case care must be taken to ensure that the cells do not touch one another as they will tend to stick together.

(iv)   If cells of known age are required, culture the isolated cells and examine them at hourly or half hourly intervals for division. Pool pairs of cells that have divided within a chosen time interval and culture until they attain the appropriate age. If single cells are required subject these pairs of blastomeres to a second cycle of disaggregation as previously described. As before, the midbody provides a topographical marker for the previous cleavage furrow.

The 16-cell stage is the first stage at which two cellular phenotypes (polar and apolar) can be distinguished (discussed in Section 4.1). By collecting the progeny of single 8-cell blastomeres *in vitro* it is possible to separate these two subpopulations since the pairs of 16-cell blastomeres are generally of different sizes and in most cases the larger cell is polar (presumptive trophectoderm, TE) and the smaller cell is apolar (presumptive inner cell mass, ICM) (26, and Section 4.1).

From the mid to late 16-cell stage onwards embryos become progressively more difficult to disaggregate due to the development of focal tight junctions and adherens junctions between outside cells which are not susceptible to disruption by Ca^{2+} depletion alone. Methods for dissociating cells are therefore more drastic and not necessarily compatible with long-term viability. Incubation for 15 min at $37°C$ in 25 mg/ml trypsin and 10 mg/ml EDTA in Ca^{2+}-free M2 + 6 mg/ml BSA followed by extensive pipetting using a flame-polished micropipette has been found satisfactory for late 16-cell embryos and early blastocysts (13, 25). For later blastocysts a preliminary incubation in the microfilament-disrupting drug cytochalasin D (CCD, 0.5 μg/ml at $37°C$ for 10 min) which causes rounding up of cells and loosening of intercellular adhesions can increase the accessibility for the trypsin/EDTA solution. Successful disaggregation of late blastocysts is dependent upon complete removal of the CCD prior to trypsin treatment.

Disaggregation of isolated ICMs is discussed in Section 3.4.

3.3 Aggregation of embryos and isolated cells

One method of constructing chimaeric (or 'allophenic') mice is to aggregate two or more genotypically distinct morulae together and re-implant the composite embryo into the uterus of a pseudopregnant foster mother (23, and see Chapter 6, Section 2.2).

This technique involves nudging two or more zona-free 8- to 16-cell embryos together in culture until they stick together. Adhesion can be assisted by incubation in a lectin, phytohaemagglutinin, (PHA; 1:20 of Gibco stock solution) during the contact period.

Some experimental procedures require reaggregation of individual, pairs or groups of synchronous or asynchronous cells. This can be achieved by taking advantage of their intrinsic 'stickiness' and simply pipetting the cells close together in warm (37°C) M2 + BSA with the culture dish on a heated microscope stage. Again the lectin PHA (Gibco stock diluted 1:20 in M2 + BSA) can be used to aid aggregation. This has no detected effect on subsequent development and may be used to facilitate the adhesion of cells in defined orientations (12). A 1−2 min exposure to PHA at 37°C in M2 + BSA is generally sufficient to ensure adhesion between the cells which can then be washed free of lectin, rinsed and transferred with a clean pipette to small drops of M16 + BSA under oil at 37°C in 5% CO_2 in air for aggregation. Analysis of the morphology of the resulting cell clusters [assisted by 4, 6-diamidino-2-phenylindole (DAPI) staining of nuclei, Section 3.6.2] has been used to assess the influence of cell interactions on cell fate during development (26).

3.4 Isolation of inside cells from morulae or inner cell masses from blastocysts

Many procedures used for isolating clusters of viable inside cells from either morulae (> 16 cells) or blastocysts (> 32 cells) are based on the application of cytotoxic reagents to the exterior of the embryo which will lyse the outer cells but not reach the inner cells due to the barrier provided by the tight junction permeability seal. The most commonly used procedure involves cytotoxic antibody followed by complement (immunosurgery) but ionophores have also been employed.

The immunosurgery procedure is based on the method described originally by Solter and Knowles (27, discussed in ref. 28).

(i)   Incubate embryos with or without their zonae, in warm (37°C) rabbit anti-mouse species antiserum (heat-inactivated at 56°C for 30 min and diluted 1:10 with M2, *Table 5*) for 2−4 min (morulae and early blastocysts) or 10 min (expanded blastocysts).

(ii)   Wash embryos three times through warm M2 + BSA *(Table 5)* then incubate them for 5−7 min in complement diluted between 1 in 5 and 1 in 10 in phosphate-buffered saline (PBS). Serum is the source of complement and the type used is important. Rat serum is the least toxic (29) but guinea pig serum can be used. Rabbit serum is invariably toxic. Absorb the complement with agarose (30), store it at −70°C and only thaw it immediately prior to use.

(iii) After incubation in complement, transfer the embryos to fresh M2 + BSA and leave them for a further 20−30 min at 37°C during which time the outer cells should be seen to bleb and lyse.

(iv)   Remove these outer cells by pipetting the embryos through a flame-polished micropipette with an internal diameter slightly larger than the clump of inside cells or ICM.

In order to disaggregate clusters of inside cells of ICMs it is necessary to decompact the aggregates first and this can be done by a 20-min incubation in warm Ca^{2+}-depleted M2 + BSA or M2 + BSA containing 0.5 μg/ml CCD. The cells can then be disaggregated in this medium using a flame-polished micropipette with a diameter smaller than the cluster of cells (14).

An alternative method for isolating advanced ICMs is to allow blastocysts to hatch and outgrow (see Section 3.5.2) and then remove the knot of ICM cells that develops on top of the layer of trophectoderm outgrowth, using a mouth pipette.

3.5 Isolation of trophectoderm

Trophectoderm (TE) is difficult to separate from ICM without recourse to elaborate microdissection methods as described by Gardner (31). However two other methods can be employed.

3.5.1 *Isolation as vesicles or from blastocysts*

Mouse ICM is much more susceptible to lethal damage by X-irradiation or anti-metabolites than TE and treatment of embryos (starting from the 2- to 8- cell stage) with a variety of these reagents has been used to generate trophectodermal vesicles devoid of any apparent ICM (discussed in ref. 28). In addition, early cleavage stage blastomeres disaggregated and cultured individually or in small groups frequently give rise to 'false blastocysts' which appear to be composed of only TE (26 and references cited therein). The purity of TE generated by these approaches is highly questionable.

An alternative procedure yields rafts of trophectodermal epithelium from the mural region of the blastocyst, and involves forming 'giant' blastocysts by aggregating together up to 10 morulae as described in Section 3.3. When these 'giant' blastocysts have expanded they are sufficiently large that a razor blade, scalpel or sharp glass needle can be used under a dissecting microscope to sever the mural TE from the other pole of the blastocyst containing ICM and polar TE (32). If transferred to culture these rafts of epithelium may re-seal and expand as vesicles of mural TE.

3.5.2 *Isolation from outgrowths*

Embryos are cultured to the hatched blastocyst stage in M16 + BSA (*Table 5*) and then transferred to plastic Petri dishes (Falcon) or plastic coverslips in a more complex medium (e.g. DMEM) supplemented with heat-inactivated 10% FCS, penicillin (60 μg/ml) and streptomycin (50 μg/ml).

Embryos will attach to the culture dish within 24 h and outgrowths of polyploid TE will start to spread on to the substratum within the next day (*Figure 2*, panel 12). Such cultures retain viability and continue to proliferate for approximately 1 week but changing the medium may extend this period. The ICM develops on top of the outgrowth as a knot of cells which can be removed using a mouth pipette.

It is possible to obtain further post-implantation development by careful manipulation of the substrata and media used (33, see also Chapter 3, Section 6).

3.6 Counting of cells

Cell numbers can be assessed directly if the embryos or cell aggregates can be disaggregated easily. Otherwise cell nuclei of embryos or aggregates can be stained and embryos disaggregated as discussed below.

3.6.1 *Giemsa staining of nuclei (34)*

(i)   Expose whole embryos (with or without their zonae), ICMs or other clusters of cells, to 0.9% sodium citrate for periods ranging from 1 to 10 min during which time the cells should swell and loosen their contacts. (The later the stage of intact embryos the longer they require in the sodium citrate solution.)

(ii)   Transfer embryos or cells to acetone-cleaned slides, fix by dropping a single drop of ethanol:acetic acid (3:1 v/v) onto them and air dry.

(iii)   Millipore the Giemsa stain (5% in PBS, *Table 4*) immediately before use and stain the samples for 10 min before counting the nuclei.

This method is not suitable for late blastocysts — but see Section 3.6.2.

3.6.2 *DAPI staining*

(i)   Stain nuclei with the vital fluorescent dye DAPI at 100 μg/ml in culture medium for 30 min at 37°C (26).

(ii)   Disaggregate embryos (up to early blastocysts) or cell clumps in calcium-free media or 0.9% sodium citrate, air dry, fix with 70% ethanol and count the nuclei using a fluorescent microscope equipped with appropriate light source and filters.

To count the nuclei of TE and ICM in late blastocysts carry out the following procedure.

(i)   Carry out immunosurgery (as described in Section 3.4) until the TE cells start to swell in the complement.

(ii)   Wash the blastocysts carefully in M2 + BSA (*Table 5*), incubate in DAPI (100 μg/ml in M2 + BSA for 30 min, 37°C) wash again, and place the blastocysts in a small drop of M2 + BSA on an acetic acid-cleaned microscope slide.

(iii)   Shell the ICM out of the TE using a flame-polished micropipette and transfer it to a drop of 0.9% sodium citrate.

(iv)   Allow the TE to dry down and after 15 min partially disaggregate the ICM and allow it to dry down.

(v)   Fix the cells by pipetting 70% alcohol on to them and allow them to air dry.

(vi)   Count fluorescent ICM and TE nuclei as described above.

Differential counts of ICM and TE cells in intact blastocysts have also been undertaken using polynucleotide-specific fluorochromes (35).

3.6.3 *Microdensitometry of Feulgen-stained nuclei*

Cell numbers can also be obtained from counts of nuclei stained by the Feulgen reaction. In addition, DNA content of stained nuclei can be assessed on an integrating microdensitometer (e.g. Vickers M85) by comparison with cell nuclei of different ploidy derived from air-dried smears of mouse liver (3,11).

(i)   Incubate embryos or cell aggregates for 15 min in Ca^{2+}- and Mg^{2+}-free Hank's medium (Gibco) at 37°C, dry them in air on clean microscope slides, fix in ethanol:acetic acid (3:1 v/v) for 5 min, followed by ethanol:acetic acid:formaldehyde (40%) fixation (85:5:10 by vol) for 1 h, dry in air again and store them at -20°C;

(ii)   Prepare Schiff's basic stain for the Feulgen reaction as follows. Add 4.0 g of Basic fuchsin to 800 ml of boiling double-distilled water, and then cool for 30 min. Add 12 g of potassium metabisulphite in 120 ml of 1M HCl and leave the solution overnight. Decolourize the solution by stirring with activated charcoal for 1 h, filter and then store it in a dark bottle at 4°C.

(iii)   Stain samples with Schiff's reagent. First soak in 5 M HCl for 55 min at 26°C, then wash in double-distilled water for 5 min and then stain in freshly prepared and filtered Schiff's reagent for 2 h in the dark.

(iv)   Wash three times for 10 min in freshly prepared sulphurous acid (consisting of equal volumes of 1% potassium metabisulphite and 0.1 M HCl). Rinse in tap water.

(v)   Dehydrate in graded alcohols and finally in xylene.

(vi)   Mount samples in Depex, store at 4°C, and analyse them within 2 weeks on an integrating microdensitometer (e.g. Vickers M85) by measuring light absorption at a wavelength of 560 nm. A fixed and stained sample of mouse liver analysed simultaneously in all experiments will provide reference values for 2C and 4C amounts of DNA.

4. THE USE OF LABELLED LIGANDS TO STUDY THE ORGANIZATION OF CELLS AND EMBRYOS

A variety of fluorescent, peroxidase-labelled, or radiolabelled, antibodies, lectins and other ligands have been applied to the mouse embryo in studies of the organization, composition and function of cells at different stages (for comprehensive reviews see 36−39).

4.1 Cell labelling and surface morphology

Lectins and antibodies have been used to obtain quantitative data on the detailed surface topography and molecular organization of embryo membranes (reviewed 40) but are more commonly applied as general membrane markers which reveal asymmetries in the distribution of surface microvilli on individual blastomeres (12). In normal embryos this surface polarity is indicative of a profound reorganization of the embryo which develops at the 8-cell stage and is thought to underly the process of cell divergence (2).

Different types of information can be obtained depending upon the age of the embryos and whether they are disaggregated before or after staining. General surface markers (e.g. lectins or anti-mouse species antibodies) applied to embryos up to the 8-cell stage will stain all blastomeres homogeneously prior to compaction and in a polarized fashion thereafter. This is true whether or not the embryos are disaggregated prior to staining. At the 16-cell stage, compaction of the embryos restricts access of stain to the outside cells and so, after staining of intact embryos, disaggregation and identification of labelled (outer) and unlabelled (inner) cells is a means of determining cell position and numbers in the intact embryo. If staining is carried out after disaggregation all cells stain but two phenotypes, one polar and one non-polar, can be distinguished. If the two approaches are combined and intact embryos are stained with a reagent conjugated to one fluorochrome [e.g. fluorescein isothiocyanate (FITC)], followed by disaggregation and staining with a second reagent conjugated to a different fluorochrome [e.g. rhodamine (TMRTC)], the polar population (prospective TE) can be assigned to the external position, and the non-polar population (prospective ICM) can be assigned to the internal position (2,25).

The staining procedures outlined above are described in Section 4.1.1. Stained cells or embryos may be examined directly, or following scanning electron microscopy (Section 4.1.2), to provide information on the inside or outside position of the cells within the embryos, their phenotypes and hence presumptive fates (2,25), as well as their numbers. Staining also enables the recovery of inside and outside cells for subsequent developmental analysis (Sections 3.2, 3.3, 3.4).

4.1.1 *Staining of embryos with antibodies or lectins*

Staining can be either a one-step procedure ('direct' — using labelled antibody or lectin) or a two-step procedure ['indirect' — using two interacting reagents (generally antibodies) where the first is unlabelled and the second labelled]. Internalization and other forms of redistribution of label can be prevented by including 0.02% sodium azide in the dilutions of antisera, as well as in the washing and mounting media, or by fixing the cells in 3.7% formaldehyde in PBS (glutaraldehyde fixation makes the cells autofluorescent), followed by extensive washing. However, cells will no longer be viable after fixation or azide treatment.

(i) *Direct labelling.*

(1)   Incubate unfixed embryos or cells at room temperature for 15−30 min in small drops of FITC- or TMRTC-conjugated lectin or antibody. Appropriate antibody dilutions (in M2, see *Table 5*) are determined by experiment. Whole sera, IgG fractions and Fab fractions of IgG have all been used (25). Concanavalin A (Con A) is the most frequently used lectin at concentrations of 0.5−1.0 mg/ml in M2 + BSA (*Figure 2*, panel 8).

(2)   Wash embryos by repeated passage through drops of M2 + BSA (phenol red must be omitted because it fluoresces). Cells and embryos can be viewed alive or fixed immediately after staining.

(3)   The most convenient slides for mounting embryos for viewing are tissue typing slides (Baird & Tatlock, details in Section 5). Place each slide in a container of adequate depth and immerse the slide in liquid paraffin oil. With the slide and its container under a binocular microscope use a mouth pipette to dispense a small volume of M2 + BSA under the oil and into the centre of each well. Add small groups of the labelled embryos to these drops. Reduce the volume of the drops, remove the slide from the oil, drain it slightly and apply a glass coverslip. Wipe the slide free of oil and view under a fluorescence microscope fitted with an incident source and filters appropriate for the fluorochromes used.

(ii) *Indirect immunofluorescence.* For this method use an unlabelled antibody as the first layer and proceed as in (1) and (2) above. Then incubate the embryos at room temperature for 15 − 30 min in the second layer (an FITC- or TMRTC- conjugated antibody or IgG directed against the first antibody) diluted appropriately in M2. Then follow steps (2) and (3).

4.1.2 *Scanning electron microscopy*

Scanning electron microscopy (SEM) has been used to examine surface morphology (*Figure 2*, panel 7, and ref. 41) as well as the topographical distribution of membrane determinants detected immunocytochemically (42). Surface binding ligands may be combined with a visible marker [e.g. *Staphylococcus aureus* which binds to the Fc portion of some immunoglobulins via its component protein A (42,43) or other forms of insolubilized protein A, lectins or antibodies].

The SEM procedure is essentially similar to that used for larger specimens and only those adaptations necessary for embryos or their component cells will be dealt with here. Samples are processed stuck down to glass coverslips and great care is taken to eliminate dust and other particles by Millipore filtration of all solutions (filter pore size = 0.22 μm) immediately before use (41).

(i)   Notch glass coverslips (9 mm × 9 mm are a convenient size) to specify the side which carries the embryos. Clean by soaking in 100% ethanol, wipe with lint-free paper and air dry for at least 1 h. Soak the clean coverslips in a freshly prepared and Millipored solution of poly-L-lysine (PLL, Sigma Type 1b, 1 mg/ml in double-distilled water) for approximately 1 h. Immediately prior to use rinse the coverslips two or three times in 0.1 M cacodylate buffer pH 7.4 and place them in individual wells of a multi-well plastic tissue culture dish (Nunc) containing 0.1 M cacodylate buffer pH 7.4.

(ii)   Fix zona-free embryos or blastomeres in 6% glutaraldehyde (Sigma) in 0.1 M cacodylate buffer pH 7.4 for 1 h at room temperature.

(iii) Transfer embryos to 1% glutaraldehyde in 0.1 M cacodylate buffer pH 7.4 and mouth pipette them on to the centre of a prepared coverslip. The multi-well dishes will fit on to the stage of a binocular dissecting microscope and if angled mouth pipettes are constructed the entire procedure can be viewed down the microscope. The use of 1% glutaraldehyde fixative during transfer assists firm attachment of the cells to the PLL-treated glass substrate. Ensure that the adhesion is successful by using the pipette to pass streams of buffer over the attached cells. Adhesion generally occurs readily and if not the quality of the PLL coating should be questioned. Take care not to touch the cells with the end of the pipette or allow them to touch one another as they will be impossible to dislodge without damage.

(iv)   Dehydrate the samples immediately through graded alcohols which are freshly Millipored before use (30 min in 20%, 40%, 60% ethanol, overnight at 4°C in 70% ethanol, then 30 min each in 80%, 90%, 95% and two exposures to dry 100% Analar ethanol).

(v)   Critical point drying, gold coating and mounting for viewing then follow conventional procedures (42).

4.1.3 *Freeze-fracture*

This technique, used in combination with agents which bind to specific membrane components and induce identifiable lesions in membrane replicas [e.g. the sterol-binding antibiotic filipin, (32)], enables the topography of the membrane to be examined with an approximately 10-fold increase in resolution compared with SEM (32, 44). The processing of material is as conventionally described (45) apart from the fact that the small size of the embryos necessitates their being mounted in specially constructed holders (44) or embedded in a matrix [e.g. gelatin (32)]. It is not necessary to remove the zona.

4.2 Labelling of intracellular components

4.2.1 *Immunocytochemistry at the light microscope level*

Early mouse embryos and their component cells are most easily manipulated during fixation and immunocytological staining if stuck down to coverslips. A method first used for lampbrush chromosomes has been subsequently adapted for visualization of cellular organelles and cytoskeletons (46) using, for example, antibodies to actin (46), tubulin (41), microtubule organizing centres (47), nuclear matrix lamins (48), clathrin and lysosomes (48).

Where it is necessary to relate intracellular organization to surface polarity, zona-free embryos or cells can be labelled with FITC− or TMRTC−Con A followed by washing [see Section 4.1.1 (i)] prior to immunocytological staining.

Embryos or cells are fixed, extracted and stained in specially constructed glass chambers consisting of a glass washer (1.5 mm thick and 25 mm external diameter) to which a glass coverslip has been attached using paraffin wax (46). The procedure for immunocytochemistry is as follows.

(i)   Coat the coverslip for 5 min at room temperature with 0.1 mg/ml Con A, or, if the cells are already stained with Con A to identify surface polarity, with a 1:20 dilution of PHA (Gibco).

(ii)   Rinse the coverslip with M2 (see *Table 5*) + 4 mg/ml polyvinylpyrrolidone (PVP) and fill the chamber with M2 + PVP.

(iii)   Pipette the cells onto the coverslips where they should attach firmly and immediately.

(iv)   Close the chambers with a second coverslip and stack them in a 50 ml polycarbonate centrifuge tube with a hole drilled through its base and a Perspex plunger inserted in the bottom third of the tube. Up to four chambers can be inserted per tube each interleaved with a 25 mm Millipore pre-filter.

(v)   Adhere the cells on to their glass coverslips by centrifuging them at 500 *g* for 10 min at 20°C in a HB-4 rotor of a Sorvall RC-5 centrifuge.

(vi)   After centrifugation insert a rod through the hole in the base of the tube and push the plunger up gently until the chambers emerge and can be removed.

(vii)   Take off the top coverslip which served as a lid. Fix the cells with 3.7% formaldehyde in PBS for 30−45 min at 20°C, wash four times in PBS, extract with 0.25% Triton X-100 in PBS for 10 min at 20°C and finally wash twice in PBS. This procedure provides the best preservation of cells and penetration of antibodies and induces a minimum of reorganization of intracellular proteins during extraction of the fixed cells (46). However the sequence of fixation and permeabilization must be tested for each determinant. For example, when staining microtubule organizing centres, extraction prior to fixation is necessary to reduce the cytoplasmic background (47).

(viii)   Remove the coverslips from their chambers and transfer them into PBS containing 0.1% Tween-20 (PBS−Tween) before incubating them with the specific antibodies or other reagents.

(ix)   Incubate extracted cells at room temperature for approximately 20 min in an appropriate dilution of the specific antibody in PBS−Tween + 3 mg/ml BSA, wash twice in PBS−Tween and stain for 20 min in the second (labelled) antibody diluted in PBS−Tween + 3 mg/ml BSA.

(x)   Wash the cells again in PBS−Tween, remove the coverslips from the chambers, mount in Citifluor, (City University London) which reduces fading of fluorescein label, and view on a fluorescence microscope. Where peroxidase-labelled antibody is used as a second layer the peroxidase activity is revealed using the method of Graham and Karnovsky (49).

4.2.2 *Intracellular ultrastructure examined by transmission electron microscopy*

The following procedures are used to prepare samples for transmission electron microscopy (TEM). All manipulations are done under a dissecting microscope using glass cavity blocks and glass mouth pipettes (angled if necessary). BEEM capsules (Polaron Equipment Ltd, Watford, Herts) are used for embedding with a low viscosity resin (e.g. Spurr).

(i)   Fix the embryos or blastomeres for 15−30 min at room temperature in freshly prepared 3% glutaraldehyde in 0.1 M sodium cacodylate (pH 7.3). Wash twice in the cacodylate buffer.

(ii)   Post-fix in 1% OsO_4 (osmium tetroxide) in 0.1 M sodium cacodylate (pH 7.3) for 30 min at room temperature. OsO_4 vapour is highly toxic and this manipulation is best undertaken by incubating embryos in small drops of OsO_4 under oil in plastic Petri dishes (set up in a fume cupboard).

(iii)   Wash the embryos or blastomeres in double-distilled water, stain them *en bloc* with saturated aqueous uranyl acetate for 30 min at room temperature.

(iv)   Dehydrate them through a graded alcohol series. Embryos tend to stick to the surfaces of micropipettes and cavity blocks when 70% alcohol is reached and it may be more practical to remove and replace the alcohol rather than transfer embryos between cavity blocks.

(v)   Because of its low viscosity Spurr resin has been found satisfactory as an embedding medium. Run a test polymerization for each batch since some of the resin components are labile. Make up Spurr resin according to the manufacturer's instructions. Store at −20°C for a maximum of 2 weeks and allow to reach room temperature before use.

(vi)   Infiltrate embryos in 1:1, Spurr:100% alcohol (v/v), 1−2 h at room temperature, followed by 2:1, Spurr:100% alcohol (v/v), 1−2 h at room temperature. Use an angled mouth pipette to localize a few embryos in a small drop of infiltration mix in the centre of the flat base of a BEEM capsule, fill up the capsule with Spurr and polymerize at 60°C for 16 h.

(vii)   Cut ultra-thin sections and deposit them on copper grids (grids with thin bars, e.g. Gildergrids G200 HS are preferable).

(viii) Make up staining solutions.
   (a) Alcoholic uranyl acetate: a saturated solution in 50% ethanol. Store in a dark bottle.
   (b) Lead citrate: add 1.33 g of lead nitrate and 1.76 g of sodium citrate to 30 ml of CO_2-free double-distilled water in a 50 ml volumetric flask. Shake continuously for 1 min and intermittently for 30 min. Add 8 ml of 1 M NaOH (Analar solution not pellets). Make up to 50 ml with CO_2-free double-distilled water.

(ix)   Place the required number of drops (one per grid) of Millipore-filtered alcoholic uranyl acetate on to a square of pink dental wax in a plastic Petri dish. Place one grid on to each drop, section side down. Cover with a lid and leave for 10−15 min.

(x)   Remove each grid in turn and wash off the stain with running drops of CO_2-free double-distilled water. Dry the grids on clean Velin tissue or hardened filter paper.

(xi)   Centrifuge the lead citrate solution for 10 min at 5000 r.p.m. Place drops of the clear supernatant on to dental wax in a Petri dish as before. Put some NaOH pellets around the edge and place one grid on each drop, section side down. Do not breath over the drops as CO_2 will cause lead precipitation. Stain for 5 min. Wash off the stain with CO_2-free double-distilled water and dry.

(xii)   View by transmission electron microscopy.

4.3 Cell lineage markers

The ideal cell lineage marker should be cell localized, cell autonomous, stable, ubiquitous, easy to detect and developmentally neutral. Few markers available currently fulfil all these criteria (reviewed in 50) but some have been successfully applied to the mouse (see Chapter 6). Two cell lineage markers have been applied recently to the pre-implantation mouse embryo. Short-term labelling has been achieved by incubating isolated blastomeres or intact embryos with FITC (0.5 mg/ml in ungassed M16, pH 7.7, containing 4 mg/ml PVP for 10 min at room temperature, followed by washing in M2 + BSA and subsequent culture in M16 + BSA (51). Cells must be labelled in suspension otherwise they will stick to the dish and lyse. This procedure, however, has the disadvantage common to many cell marking techniques that the cells selected for labelling need to be isolated from their neighbours and hence disaggregation and reaggregation steps are necessary. A more ideal marker would be one which labelled an entire population within intact viable embryos without disrupting their integrity.

This criterion appears to be fulfilled by a monodisperse suspension of fluorescein-coated carboxylated latex microparticles (yellow green Fluoresbrite; 0.2 μm particle diameter) (13). The label is readily incorporated into cleavage-stage cells by the endocytic route, is clearly identifiable in living, fixed and sectioned preparations and has no apparent effect on development. Furthermore, these microparticles are available in various sizes and can be obtained coupled to other fluorochromes. Isolated blastomeres can be labelled and their fate followed after reaggregation with unlabelled cells or the reagent may be used to label the entire outside cell population of an intact embryo from the 16-cell stage onwards.

The procedures for labelling embryos with fluorescent microparticles are as follows.

(i) Incubate zona-free embryos or cells in a 1:50 dilution of a stock suspension of microparticles in M2 + BSA (*Table 5*) for $5-15$ min at 37°C, wash them through $2-3$ drops of M2 + BSA and either analyse them immediately, or reaggregate them with other unlabelled blastomeres (see Section 3.3) and return them to culture for subsequent analysis. Where necessary carry out disaggregation as described in Section 3.2.2.

(ii) Intracellular labelling with fluorescein-coated latex microparticles can be preceded or followed by surface labelling using rhodamine-conjugated Con A (TMRTC−Con A, Polysciences). Incubate embryos or cells in TMRTC−Con A at 1 mg/ml in M2 + BSA to identify outside cells in zona-free intact embryos (1 min for 16-cell embryos; 5 min for blastocysts; at 37°C) or to reveal surface polarity in isolated cells (5 min, 37°C). After incubation wash embryos or cells in $2-3$ changes of M2 + BSA.

(iii) To analyse the fluorescence, fix the cells or embryos in 4% formaldehyde in PBS for 10 min, wash them in M2 + BSA and mount them in this medium in the wells of a tissue typing slide (Baird and Tatlock, see Section 4.1.1). Use a fluorescence photomicroscope for examining specimens and take photographs on Kodak Tri-X 35 mm film.

(iv) Determine the position and developmental fate of fluorescein−latex-labelled cells within blastocysts following a cell lineage experiment by three alternative methods.

 (a) Analyse whole embryos.

 (b) Surface label intact embryos with TMRTC−Con A (as a marker for trophectoderm) disaggregate the blastocyst (see Section 3.2.2) fix the cells (4% formaldehyde in PBS for 10 min) and score the cells for their FITC and/or TMRTC labelling patterns.

 (c) Process blastocysts for serial section analysis by TEM. JB-4 (Polysciences), a water-soluble resin, has been found to be the optimal embedding medium. (See ref. 13 for details of processing technique.)

5. REAGENTS, MEDIA AND EQUIPMENT

5.1 Reagents

Use the best grade of chemicals available. Designate a set of spatulas for dispensing tissue culture chemicals only and ensure that they are cleaned after every use.

The reagents in *Table 4* have been found satisfactory in our laboratory.

5.2 Media

The media in *Table 5* are used successfully in our laboratory. M2 is a Hepes-buffered medium (pH 7.4 at 37°C) for recovering and handling embryos out of the incubator (based on the recipe in ref. 52). M16 is the CO_2-buffered medium used for culture (based on the recipe in ref. 53). Both media have been modified as described in ref. 54 and

Table 6. Trouble shooting.

Possible causes	Tests and remedy
Problem: no eggs superovulated or eggs degenerate in the oviduct.	
1. Females too old	Use younger mice (3.5−4 weeks old).
2. Hormones have deteriorated	Make up fresh stocks.
	Buy new batches.
Problem: eggs remain unfertilized after mating.	
3. Males infertile	Change stud males
4. Ovulation and mating not synchronized	Check timing of light/dark cycle in mouse house. (Mating occurs at midpoint of dark cycle which is generally 19.00−07.00). Use pre-pubertal mice (3.5−4 weeks) preferably (no endogenous LH surge). Do not inject hCG earlier than 14.00.
Problem: eggs or embryos lyse soon after release into medium.	
5. Osmolarity incorrect (embryos swell or shrink).	Check osmolarity and make up fresh media if necessary.
6. Dirty glassware (e.g. cavity blocks)	Use plasticware as alternative.
7. Sterilizing filters, syringes and other plasticware used in preparation of media are toxic.	Incubate media in contact with relevant items then check for embryo viability.
8. Media and/or oil is toxic.	Check BSA and oil (see 14).
9. Volatile solvents	Check if paint or solvents have been used recently in the room. Check viability of embryos when manipulations are carried out elsewhere.
Problem: eggs or embryos lyse soon after removal of the zona.	
10. Embryos exposed to the zona-removing solutions for too long (especially acid Tyrode's).	Hasten the procedure. Use smaller batches of embryos. Ensure acid Tyrode's is at 37°C. Wash embryos in M2 + BSA rapidly and adequately. Remove zonae using pronase instead of acid Tyrode's (or vice versa).
11. Solutions are toxic.	Check osmolarity. Ensure pH adjustment is made with Analar HCl. Solutions stored for too long (do not freeze acid Tyrode's). Discard and renew. Change chemical stocks.
Problem: embryos arrest in culture at 2-cell stage.	
12. 'Blocking' strain used and embryos removed from oviduct too early.	Change mouse strain. Do not remove embryos until approximately 32 h after hCG injection or later.
13. Fluctuations in temperature and/or pH. Check 5−9.	Pre-equilibrate and pre-warm all media Use heated blocks and heated stages on microscopes.

Problem: embryos arrest in culture at later stages (frequently the 16-cell stage) after a period of normal, if slowed, development. (Embryos without zonae or disaggregated cells are particularly sensitive to toxic components) (55). Check 5−9 and 13.

Table 6 continued.

Possible causes	Tests and remedy
14. Media are toxic.	Check BSA. Compare different batches and sources for toxicity. Check oil for toxic contaminant. (a) Extract oil with medium overnight in 37°C CO_2 incubator. Compare embryo viability in media overlain with extracted and unextracted oil. (b) Compare embryo viability in $5-10$ μl drops versus 50 μl drops. If the oil is con-taminated embryos often degenerate more quickly in small drops. Screen other batches of oil using zona-free 8-cell embryos in small drops of medium and buy a successful batch in bulk. Change chemical stocks for media, acid Tyrode's and pronase solutions.
Zona removal at fault Check 10 and 11.	
15. Plastic culture dishes contaminated See also 6 and 7.	Screen other batches and buy in bulk.
16. Unidentified flaw in procedure remains.	Compare embryos and reagents with those from another laboratory.

are made up from stock solutions given in *Table 5*. Osmolarity of M2 and M16 is in the range $285-290$ mOsmol. DMEM + 10% FCS is used as a blastocyst outgrowth medium.

5.3 Equipment

5.3.1 *Glassware*

A special set of glassware should be designated for media preparation, rinsed and soaked in double-distilled water immediately after use, washed in double-distilled water only and dried in a hot oven.

5.3.2 *Microscopes*

It is important to set a room aside for embryo manipulation and culture to minimize contamination. An air conditioner to maintain normal room temperature is advisable. Binocular dissecting microscopes (e.g. Model M5 Wild Heerbrugg) with magnifica-tion ranging from approximately ×6 to ×50 are essential. They should be housed in dust-free Perspex hoods and ideally fitted with heated stages and warming blocks set to 36 ± 1°C.

5.3.3 *Incubators*

A 37°C incubator maintaining 5% CO_2 in air is necessary. The CO_2 monitor can either be integral to the incubator or an additional piece of equipment (e.g. CO_2 Monitor Con-troller. Gow-mac Instrument Company). Alternatively the embryo culture dishes can be placed in sealed flasks pre-equilibrated with a mixture of 5% CO_2 in air. In cases where frequent access to cultures is required (e.g. for serial observations) separate long-term and short-term incubators are advisable.

5.3.4 *Laminar flow hood*

It is advisable to prepare all media in a laminar flow hood (e.g. Slee Model HLF/H. Supplier: South London Electrical Equipment Ltd) which should not be housed in the same room as the microscopes and incubators.

5.3.5 *Dishes for manipulation and culture*

Glass cavity blocks (1.25 inches square with concave depression, Hoslabs) are convenient for collecting embryos from large volumes of fluid because the embryos or cells will tend to roll to the centre of the dish.

Pre-packed, sterile Falcon (Becton Dickinson & Co) plastic Petri dishes, tubes and pipettes are used for conventional culture. When a non-adherent surface is required Sterilin Petri dishes are employed (Plasticware supplier: Hawfell).

5.3.6 *Glass pipettes*

Variation in composition and thickness of the glass and pipette diameter of some types available can introduce unnecessary difficulties when flame-pulling disposable Pasteur pipettes for routine handling of intact embryos. We find 'Volac' pipettes (Fisons Scientific) to be suitable.

Precision glass capillary tubing is necessary for making micropipettes for disaggregating embryos and handling single cells (e.g. GC 100−15: Clark Electromedical Instruments).

The micropipettes are pulled and flame polished using a microforge (e.g. Microforge de Fonbrune Etablissements Beaudoin). Pipetting aids (AR Horwell, Cat Nos. 7091 00 and 7091 10) make suitable adaptors for mouth pipettes.

5.3.7 *Needles for flushing oviducts*

30-Gauge Luer lock metal needles (Holborn Surgical Instruments Co Ltd) are used with the bevelled ends cut flat and the ends polished.

5.3.8 *Tissue typing slides*

(Supplier: Baird and Tatlock). Cat No. 403/0522 Type J_1. Cover glasses 70 × 46 mm, Cat No. 403/0523.

6. ACKNOWLEDGEMENTS

Much of the methodology described in this chapter is derived from work in Dr Martin H.Johnson's laboratory, Department of Anatomy, Cambridge, during the past 10 years and I should like to express my gratitude to all members of the laboratory past and present for their contribution to it. Special thanks are due to Ms Gin Flach for her efficient technical and organizational skills, to Dr Martin Johnson and Dr Tom Fleming for helpful criticisms and to Mrs Shelagh Eggo for typing the manuscript. The work described has been supported by grants from MRC, CRC and EMBO.

7. REFERENCES

1. Johnson,M.H., McConnell,J. and Van Blerkom,J. (1984) *J. Embryol. Exp. Morphol.*, **83**, Suppl 1−6, 197.
2. Johnson,M.H., Chisholm,J.C., Fleming,T.P. and Houliston,E. (1986) *J. Embryol. Exp. Morphol.*, **97**, Suppl. 97.
3. Smith,R.K.W. and Johnson,M.H. (1986) *J. Reprod. Fertil.*, **76**, 393.
4. Brinster,R.L. (1971) In *Biology of the Blastocyst.*, Blandau,R.J. (ed.), University of Chicago Press, p. 303.
5. Biggers,J.D., Whitten,W.K. and Whittingham,D.G. (1971) In *Methods in Mammalian Embryology*. Daniel,J.C. (ed.), W.H.Freeman and Co, San Francisco, p. 86.
6. Gates,A.H. (1971) In *Methods in Mammalian Embryology*. Daniel,J.C. (ed.), W.H.Freeman and Co., San Francisco, p. 64.
7. Pratt,H.P.M., Bolton,V.N. and Gudgeon,K.A. (1983) *Molecular Biology of Egg Maturation. CIBA Foundation Symposium No. 98.* Pitman Books, London, p. 97.
8. Howlett, S.K. and Bolton,V.N. (1985) *J. Embryol. Exp. Morphol.*, **87**, 175.
9. Kaufman,M.H. (1983) *Early Mammalian Development: Parthenogenetic Studies.* Cambridge University Press.
10. Webb,M., Howlett,S.K. and Maro,B. (1986) *J. Embryol. Exp. Morphol.*, **95**, 131.
11. Bolton,V.N., Oades,P. and Johnson,M.H. (1984) *J. Embryol. Exp. Morphol.*, **79**, 139.
12. Johnson,M.H. and Ziomek,C.A. (1981) *J. Cell Biol.*, **91**, 303.
13. Fleming,T.P. and George,M.A. (1987) *Wilhelm Roux's Arch.*, **196**, 1−11.
14. Chisholm,J.C., Johnson,M.H., Warren,P.D., Fleming,T.P. and Pickering,S.J. (1985) *J. Embryol. Exp. Morphol.*, **86**, 311.
15. MacQueen,H.A. and Johnson,M.H. (1983) *J. Embryol. Exp. Morphol.*, **77**, 297.
16. Graham,C.F. and Lehtonen,E. (1979) *J. Embryol. Exp. Morphol.*, **49**, 277.
17. Biggers,J.D. (1971) In *Biology of the Blastocyst.* Blandau,R.J. (ed.), University of Chicago Press, p. 319.
18. Goddard,M.J. and Pratt,H.P.M. (1983) *J. Embryol. Exp. Morphol.*, **73**, 111.
19. Kaufman,M.H. and Sachs,L. (1976) *J. Embryol. Exp. Morphol.*, **35**, 179.
20. Kaufman,M.H. and Sachs,L. (1975) *J. Embryol. Exp. Morphol.*, **34**, 645.
21. Whitten,W.K. and Biggers,J.D. (1968) *J. Reprod. Fertil.*, **17**, 399.
22. Nicolson,G.L., Yanagimachi,R. and Yanagimachi,H. (1975) *J. Cell Biol.*, **66**, 263.
23. Mintz,B. (1971) In *Methods in Mammalian Embryology*, Daniel,J.C. (ed.), W.H.Freeman and Co., San Francisco, p. 186.
24. Sobel,J.S. (1984) *J. Cell Biol.*, **99**, 1145.
25. Handyside,A.H. (1980) *J. Embryol. Exp. Morphol.*, **60**, 99.
26. Johnson,M.H. and Ziomek,C.A. (1983) *Dev. Biol.*, **95**, 211.
27. Solter,D. and Knowles,B.B. (1975) *Proc. Natl. Acad. Sci. USA*, **72**, 5099.
28. Snow,M.H.L. (1978) In *Methods in Mammalian Reproduction*. Daniel,J.C. (ed.), Academic Press, New York, p. 167.
29. Nichols,J. and Gardner,R.L. (1984) *J. Embryol. Exp. Morphol.*, **80**, 225.
30. Cohen,A. and Schlesinger,M. (1970) *Transplantation*, **10**, 130.
31. Gardner,R.L. (1978) In *Methods in Mammalian Reproduction*. Daniel,J.C. (ed.), Academic Press, New York, p. 137.
32. Pratt,H.P.M. (1985) *J. Embryol. Exp. Morphol.*, **90**, 101.
33. Hsu,Y.-C. (1978) In *Methods in Mammalian Reproduction*. Daniel,J.C. (ed.), Academic Press, New York, p. 229.
34. Tarkowski,A.K. (1966) *Cytogenetics*, **5**, 394.
35. Handyside,A.H. and Hunter,S. (1984) *J. Exp. Zool.*, **231**, 429.
36. Johnson,M.H. (1981) *Int. Rev. Cytol. Suppl.*, **12**, 1.
37. Jacob,F. (1979) *Curr. Top. Dev. Biol.*, **13**, 117.
38. Solter,D. and Knowles,B.B. (1979) *Curr. Top. Dev. Biol.*, **13**, 139.
39. Wiley,L.M. (1979) *Curr. Top. Dev. Biol.*, **13**, 167.

40. Wolf,D.E. (1983) In *Development in Mammals*. Johnson,M.H. (ed.), Elsevier Amsterdam, Vol. **5**, p. 187.
41. Maro,B. and Pickering,S. (1984) *J. Embryol. Exp. Morphol.*, **84**, 217.
42. Johnson,M.H. and Ziomek,C.A. (1982) *Dev. Biol.*, **91**, 431.
43. Surolia,A., Pain,D. and Khan,M.I. (1982) *Trends Biochem. Sci.*, **7**, 74.
44. Magnuson,T., Demsey,A. and Stackpole,C.W. (1977) *Dev. Biol.*, **61**, 252.
45. Navaratnam,V., Thurley,K.W. and Skepper,J.N. (1982) In *Progress in Anatomy*. Harrison,R.J. and Navaratnam,V. (eds), Cambridge University Press, Vol. 2, p. 201.
46. Maro,B., Johnson,M.H., Pickering,S.J. and Flach,G. (1984) *J. Embryol. Exp. Morphol.*, **81**, 211.
47. Maro,B., Johnson,M.H., Webb,M. and Flach,G. (1986) *J. Embryol. Exp. Morphol.*, **92**, 11.
48. Johnson,M.H. and Maro,B. (1985) *J. Embryol. Exp. Morphol.*, **90**, 311.
49. Graham,R.C. and Karnovsky,M.J. (1966) *J. Histochem. Cytochem.*, **14**, 291.
50. West,J.D. (1984) In *Chimeras in Developmental Biology*. Le Douarin,N. and McLaren,A. (eds), Academic Press, New York, p. 39.
51. Ziomek,C.A. (1982) *Wilhelm Roux's Arch.*, **191**, 37.
52. Fulton,B.P. and Whittingham,D.G. (1978) *Nature*, **273**, 149.
53. Whittingham,D.G. (1971) *J. Reprod. Fertil. Suppl.*, **14**, 7.
54. Goodall,H. and Maro,B. (1986) *J. Cell Biol.*, **102**, 568.
55. Fleming,T.P., Pratt,H.P.M. and Braude,P.R. (1987) *Fertil. Steril.*, **47**, 858−860.

CHAPTER 3

Isolation, culture and manipulation of post-implantation mouse embryos

ROSA BEDDINGTON

1. INTRODUCTION

It is only after the mammalian embryo has implanted in the uterus that differentiation and organization of the fetus occurs. The divergence of tissue lineages preceding implantation segregates three precursor populations: one that will give rise to the trophoblast, another that generates extra-embryonic endoderm and a third, the epiblast (embryonic ectoderm), from which is derived the entire fetus and the extra-embryonic mesoderm (1). In general, any experiment on the pre-implantation embryo which makes the epiblast chimaeric, even the addition of a single marked epiblast cell to the blastocyst, will result in chimaerism in each and every tissue of the fetus. Only during or after epiblast diversification can experimental interference preferentially modify or distinguish particular fetal tissues (2). Therefore, in order to study the development of a particular fetal constituent experiments must be conducted on the post-implantation embryo.

No method for the precise manipulation of the embryo *in utero* has been devised and so most experiments are carried out either *in vitro* or in an ectopic site *in vivo*, neither of which is likely to provide optimal conditions for normal differentiation and morphogenesis. In this chapter the most common isolation, grafting and culture procedures used to study the origin and development of the early fetal primordia will be outlined. However, particular emphasis will be placed on the more recent techniques designed to study the differentiation of cells in their normal environment in the intact embryo.

2. THE ANATOMY OF GASTRULATION AND EARLY ORGANOGENESIS

The mouse blastocyst starts to implant in the uterus on the 5th day of gestation. On the 6th day, due to the growth of the inner cell mass (ICM) downwards into the blastocoel and accumulation of extra-embryonic ectoderm above it, the embryo acquires a cylindrical shape with a distinct junction between the embryonic and extra-embryonic regions (*Figure 1a*). Subsequently, the pro-amniotic cavity forms in the epiblast and enlarges to extend into the extra-embryonic region. At this stage the embryonic region consists of two epithelial layers: the visceral endoderm on the outside and the epiblast inside, lining the proamniotic cavity (*Figure 1b*).

Gastrulation commences on the 7th day with the appearance of the primitive streak (*Figure 1c*). This is located in the midline of the posterior third of the embryonic region and defines the future anteroposterior axis of the fetus. The streak is the route through

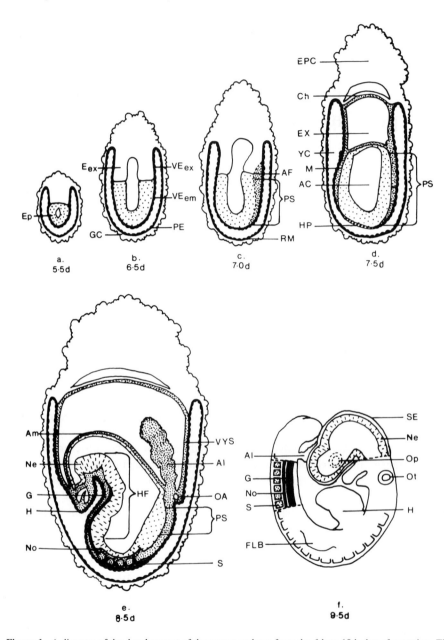

Figure 1. A diagram of the development of the mouse embryo from the 6th to 10th day of gestation. The age of embryos is given in days post-fertilization representing the following developmental stages (*Table 2*); **a**, stage 2; **b**, stage 3; **c**, stage 5; **d**, stage 7, **e**, stage 10; **f**, stage 13. Ep, epiblast; E_{ex}, extra-embryonic ectoderm; GC, trophoblast giant cells; VE_{ex}, extra-embryonic visceral endoderm; VE_{em}, embryonic visceral endoderm; PE, parietal endoderm; RM, Reichert's membrane; AF, posterior amniotic fold; PS, primitive streak; EPC, ectoplacental cone; Ch, chorion; EX,exocoelom; YC, yolk cavity; AC, amniotic cavity; M, mesoderm; HP, head process; Am, amnion; Ne, neurectoderm; G, gut; H, heart; No, notochord; HF, headfold; S, somite; OA, omphalomesenteric artery; Al, allantois; VYS, visceral yolk sac; FLB, forelimb bud; Ot, otic capsule; Op, optic evagination; SE, surface ectoderm.

which the epiblast invaginates to produce a third layer of loosely interconnected cells, the mesoderm. The mesoderm spreads laterally and anteriorly between the epiblast and visceral endoderm to form a complete intermediate layer in the embryonic region. Cells emerging from the anterior aspect of the streak have a more epithelial-like organization and are distributed along the midline of the anterior part of the cylinder. This quasi-epithelium, first recognizable on the 8th day, is known as the head process (*Figure 1d*) and is the probable precursor of the notochord, pre-chordal plate and, by virtue of intercalating into the outer visceral endoderm layer, of part, although probably not all, of the gut endoderm (3).

Mesoderm from the more posterior part of the streak moves into the extra-embryonic region and, as it does so, the junctional region between the epiblast and the extra-embryonic ectoderm is pushed into folds, the amniotic folds. These folds, the most prominent of which is the posterior amniotic fold (*Figure 1c*), bulge into the proamniotic cavity and eventually fuse. Meanwhile, lacunae form within the extra-embryonic mesoderm which proceed to coalesce creating a further cavity, the exocoelom, which is lined by mesoderm and separates the extra-embryonic ectoderm from the epiblast (*Figure 1d*). Thus the egg cylinder comes to be divided into three distinct chambers divided off from each other by the future amnion and chorion (*Figure 1d*).

Gastrulation, in the sense of continued ingression through the streak, persists up to and during the 10th day of gestation, but towards the end of the 8th day the first signs of organogenesis are apparent. A comprehensive account of organogenesis is not appropriate here but a brief account of the formation of the major organ primordia will be given. More detailed descriptions of the differentiation of the various organ systems can be found elsewhere (4, 5).

The ectoderm anterior to the primitive streak forms the neural plate which is slightly indented along the midline. This depression, known as the neural groove, becomes more accentuated as the neural folds form on the 9th day (*Figure 1e*). Subsequently, the folds fuse, transforming the neural plate into a closed tube running the length of the embryo. Effectively, this tube is the primordium of the central nervous system. The surface ectoderm on either side of the neural tube is flattened into cuboidal or squamous epithelium and constitutes the precursor of the epidermis. It is at the junction between the surface ectoderm and neurectoderm that the neural crest cells arise. These cells migrate ventrally and are dispersed throughout much of the body where they give rise to a variety of different tissue types, such as melanocytes and elements of the peripheral sensory and autonomic nervous system.

Beneath the neural groove runs the notochord (*Figure 1e*) and on either side of it the paraxial mesoderm. The paraxial mesoderm of the head region is not overtly segmented but in the trunk, and eventually the tail, there is overt metamerism in the form of somites (*Figure 1e* and *f*). These segmental blocks of paraxial mesoderm form in an orderly craniocaudal sequence beginning on the 9th day, new somites being added to the pre-existing columns from the unsegmented plate of mesoderm just anterior to the primitive streak. Lateral to the somites the mesoderm is divided into two layers, one associated with ectoderm and the other with endoderm, the two layers being separated by the embryonic coelom.

Anteriorly, the mesoderm beneath the headfold becomes organized into the heart

Table 1. Recovery of embryos.

Dissecting out embryos

1.   Remove the uterus from the mouse and place it in a 50 mm Petri dish containing PB1 + 10% FCS[a].
2.   Using watchmakers' forceps, tear the uterine muscle along the antimesometrial aspect of each implantation site (*Figure 2a*) to expose the decidua. Slide the forceps along the mesometrial uterine muscle to detach the decidua (*Figure 2b*). Transfer the decidua to a clean Petri dish containing the same medium.
3.   Hold the mesometrial (blunt) end of the decidua with forceps. Insert the point of closed forceps into the midline, approximately a third of the way down the decidua from the mesometrial end, pushing the forceps right through until they touch the bottom of the dish (*Figure 2c*). Open the forceps. The decidua should split in half (*Figure 2d*).
4.   Gently push the embryo away from the decidual tissue. When the embryo is almost entirely free, take hold of the ectoplacental cone and lift it out of the decidua.

Removal of Reichert's membrane

1.   Immobilize the embryo by inserting the closed point of one pair of forceps into the yolk cavity just beneath the ectoplacental cone. Bring the closed points of the second pair of forceps to the same position (*Figure 2e*) and, pressing firmly, draw them towards the embryonic pole (*Figure 2f*). This should tear the Reichert's membrane so that the egg cylinder emerges.
2.   The Reichert's membrane will retract towards the ectoplacental cone and most of the membrane can be removed by gently tearing it off.

---

[a]The composition of PB1 is given in Chapter 13, Section 3.1.1.

(*Figure 1e*) which begins to contract rhythmically towards the end of the 9th day. Blood islands appear in the extra-embryonic mesoderm lining the visceral yolk sac and by the 10th day a complex yolk sac circulation is established, blood entering the visceral yolk sac via the omphalomesenteric artery (*Figure 1e*) and being returned to the heart in the omphalomesenteric vein. At the posterior end of the primitive streak the mesoderm gives rise to the allantois (*Figure 1e*), which is first discernible on the 8th day as a discrete bud. The allantois grows through the exocoelom and fuses with the chorion, thus providing a direct link between the embryo and the ectoplacental cone. This lays the foundations for the differentiation of the chorioallantoic placenta, which eventually becomes the major nutritional and respiratory organ of the conceptus.

The exact origin and time of the delamination of gut endoderm remains unclear. It is derived from the epiblast and probably starts to emerge early in gastrulation, the head process and other invaginated epiblast cells replacing all or part of the pre-existing visceral embryonic endoderm. The duration of cell recruitment into this new external definitive endoderm layer is unknown. The sheet of gut endoderm assumes a characteristic tubular structure at the anterior and posterior ends of the embryo although, unlike the neural tube, this is not due to a fusion of folds but rather to its folding back on itself such that the hindgut and foregut are present initially as blind ending tubes. Consequently, the gut remains open to the yolk cavity longer in the trunk region than elsewhere (*Figure 1e*).

3. RECOVERY AND STAGING OF EMBRYOS

3.1 **Recovery**

All manipulations on the early post-implantation embryo require that the embryo be dissected free of the decidua and that Reichert's membrane, which does not readily

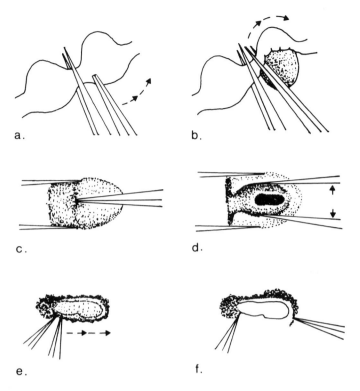

Figure 2. Diagram of the dissection for recovery of an 8th day embryo and the removal of Reichert's membrane. The various manipulations are described in *Table 1*.

expand in culture, be reflected or removed. The methods used are given in *Table 1* and *Figure 2*. Two pairs of sharpened watchmakers' forceps are necessary for this dissection. PB1 medium (6) containing 10% fetal calf serum (FCS) instead of bovine serum albumin is suitable for storing and manipulating embryos at room temperature.

3.2 Staging of embryos

Early post-implantation development is characterized by the rapid production of new tissues and dramatic morphological changes. Therefore, careful staging of embryos is critical to the interpretation of any experiment. Most mammalian embryologists have been guilty in the past of using age as a measure of developmental stage; this has proved to be an inadequate and often misleading criterion. Not only can marked variation in embryonic development be observed even within litters but, during such a dynamic phase of morphogenetic change, embryos only a few hours apart in age may be significantly different in their developmental state. Therefore, it is recommended that embryos be classified according to distinctive morphological features and certain physical dimensions to provide a more accurate and sensitive means of standardizing experiments. *Table 2* provides a possible scheme for staging living embryos, although the criteria used may have to be refined or modified according to the nature of the experiment.

Table 2. Staging scheme for living 6th to 11th day mouse embryos[a].

Day of gestation[b]	Stage	Morphological features	Dimensions (mm)[c]
6th day (M)	1	Egg cylinder formed. Distinct embryonic/extra-embryonic junction.	0.1 × 0.1
6th day (A)	2	Formation of pro-amniotic cavity.	0.2 × 0.2
7th day (M)	3	Pro-amniotic cavity extends into extra-embryonic region.	0.25 × 0.25
7th day (A)	4	Formation of primitive streak.	0.3 × 0.4
7th day (A)	5	Appearance of amniotic folds	0.3 × 0.4
8th day (M)	6	Closure of the amnion.	0.35 × 0.4
8th day (M)	7	Expansion of amniotic cavity. Head process.	0.4 × 0.4
8th day (A)	8	Allantoic bud. Neural groove.	0.5 × 0.4
8th day (A)	9	Overt headfold. Heart primordium.	0.55 × 0.5
9th day (M)	10	1−7 somites. Neural folds. Foregut portal. Axis unturned. Heart contracting.	1.0
9th day (A)	11	8−12 somites. Fore- and hind-gut portals. Axis turning. Otic placode. 1 branchial arch. Visceral yolk sac circulation.	1.5
10th day (M)	12	13−20 somites. Complete turning. Closing anterior and open posterior neuropores. Fusion of cephalic folds. Two branchial arches. Lens placode. Optic evagination. Olfactory placode.	2.0
10th day (A)	13	21−29 somites. Forelimb bud. Tail bud. Posterior neuropore open. Three branchial arches. Optic evagination. Thickened olfactory placode.	3.0
11th day (M)	14	30−34 somites. Hindlimb bud. Whole neural tube closed.	4.0
11th day (A)	15	35−39 somites. Limb bud outgrowth. Lengthened tail. Bulging cerebral vesicles. Olfactory pits. Umbilical herniation.	5.0

[a]See ref. 7.
[b](M) = morning; (A) = afternoon.
[c]Dimensions for 6th to 8th day embryos are approximate measures of the largest diameter (d) × height (h) of the *embryonic* region of the conceptus. Dimensions for 9th to 11th day embryos represents the approximate crown/rump length of the fetus.

Table 3. Isolation of the posterior part of the primitive streak.

1.	Orientate the embryo with glass needles so that its long axis is perpendicular to the shafts of the needles.
2.	Using the tip of the left hand needle to prevent the embryo from moving, place the second needle across the embryonic/extra-embryonic junction along the intended line of cut (*Figure 3A,a*). Press the needle down until its shaft touches the bottom of the dish. Move the needle to and fro until the two fractions of the embryo come apart. It may be necessary to repeat this process several times using either needle. Separate the two pieces, detaching them from the bottom of the dish if necessary.
3.	Orientate the embryonic fraction of the cylinder so that the primitive streak is parallel to the shaft of the needle.
4.	Make a longitudinal cut in the same manner as described above as close to the posterior axis as possible (*Figure 3A,b*).
5.	Subsequently, make a transverse cut at the desired position along the primitive streak (*Figure 3A,c*). The fragment will flatten out and the lateral wings can be trimmed by making two further cuts on either side of the streak (*Figure 3A,d*). The endoderm and most of the mesoderm can be removed by carefully picking the tissue away with the tip of the needle.
6.	Transfer the primitive streak fragment to a glass cavity dish using a fine, hand-drawn Pasteur pipette. Because embryonic fragments tend to stick to glass, it is always advisable to use siliconized (Repelcote; Hopkin and Williams) Pasteur pipettes for transferring embryos or embryonic fragments from one solution to another. Stickiness may also be reduced by using cooled medium.

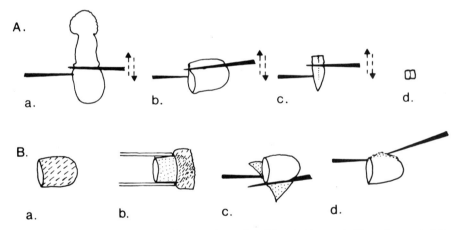

Figure 3. (A) Microsurgical isolation of the posterior portion of the primitive streak. The various manipulations are described in *Table 3*. **(B)** Microsurgical and enzymatic isolation of epiblast. Details of the method are given in *Table 4*.

4. ISOLATION OF DEFINED TISSUES OR REGIONS

4.1 Microsurgical dissection of embryos

Either siliconized glass needles (Section 9.3) held in Leitz instrument holders or sharpened tungsten needles may be used for dissection. These instruments are used to make straight cuts across the embryo in order to subdivide it into particular regions or fragments. Bacteriological plastic Petri dishes (e.g. from Sterilin) are preferable to glass because the embryos are less inclined to skid about. The isolation of the posterior part of the primitive streak of 8th day embryos is provided as an example in *Table 3* and *Figure 3A* but the basic technique can be applied to the isolation of any part of the

Table 4. Isolation of ectoderm, mesoderm and endoderm from the primitive streak stage mouse embryo.

1.	Microsurgically isolate the embryonic region of the egg cylinder (Section 4.1).
2.	Transfer the embryonic fraction (*Figure 3Ba*) to a filtered (Millipore Sartorius filter, pore size 0.45 μm) solution of 0.5% (w/v) trypsin (BDH), 0.25% (w/v) pancreatin (Difco) in calcium- and magnesium-free Tyrode Ringer's saline[a], pH 7.6−7.7, at 4°C. Incubate at 4°C for 10−20 min.
3.	Transfer the embryonic fractions in a minimal volume of digestion mixture to a large volume of PB1 + 10% FCS[b] (e.g. 5 ml in a 50 mm Petri dish).
4.	Gently pipette the embryos in and out of a hand-drawn, flame-polished, siliconized pipette, whose internal diameter is slightly smaller than that of the embryonic cylinder (*Figure 3B,b*). Each time draw the embryo into the pipette at the 'cut surface' end. The endoderm will peel off the cylinder and detach.
5.	Usually, once the endoderm is removed, the underlying mesoderm will flatten out as two distinct wings, attached to the epiblast only along the primitive streak. Isolate these wings by cutting them off as close to the primitive streak as possible (*Figure 3B,c*). If the mesoderm wings fail to 'unfold' after removal of the endoderm either further pipetting or careful teasing with needles should detach the mesoderm in the anterior and lateral regions.
6.	Remove remaining loosely adherent cells along the streak by picking them off with the point of a needle (*Figure 3B, d*). However, by the very nature of the primitive streak it is not possible, without removing it altogether, to ensure that there are no contaminating mesoderm cells left in this region (see *Figure 4*).

[a]Tyrode Ringers saline. Grams per litre: NaCl 8, KCl 0.3, NaH_2PO_4. $5H_2O$ 0.093, KH_2PO_4 0.025, $NaHCO_3$ 1, glucose 2.
[b]See Chapter 13, Section 3.1.1 for composition of PB1.

embryo at any stage. The general principle is to divide the embryo into progressively smaller pieces while retaining a means of identifying the region one intends to isolate.

4.2 Enzymatic separation of tissues

Proteolytic enzymes, in conjunction with microsurgical dissection, may be used to separate the germ layers. The most efficient procedure, devised by Levak-Svajger *et al.* (8), utilizes a mixture of trypsin and pancreatin and achieves not only clean separation but also prevents the tissues becoming unduly sticky and difficult to manipulate (9). An example of enzymatic separation is given in *Table 4* and *Figure 3B* and 8th day epiblast isolated by this procedure is shown in *Figure 4*. In practice, this method can be used for the isolation of a variety of tissues, even from quite advanced embryos, if the incubation period is varied according to the job in hand. For example the visceral yolk sac can be separated into its constituent endoderm and mesoderm components following a 3 h incubation at 4°C (10).

5. CULTURE AND ECTOPIC TRANSFER OF ISOLATED TISSUES AND FRACTIONS

5.1 Culture

The differentiation or further development of particular fractions of the embryo can be studied in culture. A variety of media and substrata have been used but these will not be reviewed here because most experiments involve straightforward tissue or organ culture techniques which are well described elsewhere (e.g. ref. 11). In general, the type of culture conditions used depends on what sort of information is required. Both

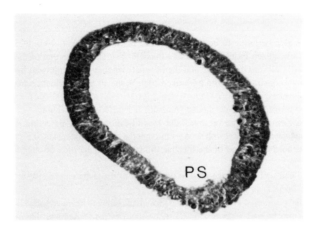

Figure 4. Section of 8th day epiblast isolated by combined enzyme treatment and microsurgery. PS = primitive streak.

standard tissue culture and organ culture techniques tend to disturb the normal relationships of the tissue layers and, therefore, are applicable only where disruption to morphogenesis is irrelevant. Such disruption may be useful in the isolation of particular cell types from an explant of heterogeneous or multipotent tissue. For example, neural crest cells can be isolated by explanting 9th day isolated neural tube into a standard tissue culture dish (Falcon) containing Eagle's minimal essential medium + 15% FCS + 2−4% chick embryo extract (12). After a few days a population of adherent migratory cells is found surrounding the initial explant, which by various criteria, including contribution to melanocytes *in vivo* (13), appear to be a relatively pure population of neural crest cells. Where it is important to retain some degree of normal morphogenesis, fractions should be grown in suspension culture. Often a bacteriological quality surface is not sufficient to prevent adhesion and spreading. Indeed, isolated tissues will sometimes spread out even on the meniscus of hanging drops of media. Therefore, it is advisable to coat the surface of the dish with a 1% solution of agar made up in the synthetic medium of choice for subsequent culture (14). For example, specific fractions of the 8th day embryo, left attached to the extra-embryonic region, have been grown in agar-coated dishes containing Dulbecco's modified Eagle's medium (DMEM) + 50% rat serum (Section 6.1). Under these conditions the fractions remain in suspension and undergo quite extensive morphogenesis.

5.2 Ectopic transfer

A variety of isolated fractions or tissues from post-implantation mouse embryos will continue to grow and differentiate if placed in a well vascularized site in an immunologically compatible adult host (2). Xenogeneic sites such as the chick chorioallantoic membrane have been used but in general development is somewhat less predictable than when syngeneic mice serve as hosts. Despite the abnormal morphogenesis observed in such grafts, the range of differentiated tissues formed in such an 'experimental teratoma' can be used to provide an estimate of the developmental potential of the transferred tissue. It should be noted that if trophectoderm or extra-embryonic ectoderm

Table 5. Ectopic transfer beneath the testis capsule.

1.	Isolate the region or tissue to be transferred and place it in PB1 + 10% FCS[a].
2.	Hand pull a siliconized Pasteur pipette to a diameter just larger than that of the tissue. Suck a column of paraffin oil into the pipette and then a small volume of air followed by some PB1 + 10% FCS. To obtain consistent paraffin oil-braked control, the air bubble should remain in the thin pulled part of the pipette.
3.	Anaesthetize a male syngeneic mouse (Avertin; 0.1 ml/5 g body weight).
4.	Under some form of low power binocular magnification (e.g. a dissecting microscope) make a transverse mid-ventral incision with scissors through the skin, approximately 0.5 cm long, immediately rostral to the normal position of the bladder. Make a similar incision through the body wall in the same position.
5.	Using a fine pair of forceps take hold of the fat pad of the testis on one side and pull the testis out through the opening in the body wall.
6.	Note the initial position of the air bubble and pick up the tissue to be transplanted in the Pasteur pipette.
7.	Holding the pipette and a syringe needle (21-gauge) in the same hand make a small hole in the testis capsule with the needle and insert the tip of the pipette into this hole. Slide the pipette under the testis capsule for a distance of approximately 4 mm.
8.	Gently expel the graft while withdrawing the pipette slightly. It may be possible to see the transferred tissue as it emerges onto the surface of the testis. If not, use the position of the air bubble as an indication of whether or not the tissue has left the pipette. To be on the safe side, so long as there is not too much medium in the pipette, the air bubble can be blown to the tip of pipette but do not blow it out into the testis. Withdraw the transfer pipette completely.
9.	Return the testis, together with its fat pad, to the abdominal cavity. A similar transfer can be made to the contralateral testis if required.
10.	Sew up the body wall and sew or wound clip the overlying skin.

[a]See Chapter 13, Section 3.1.1 for composition of PB1.

are included in the graft there is a likelihood that the haemorrhagic response evoked by these tissues will obscure other differentiation (15). Of particular interest in the mouse is the observation that up to the 8th day of gestation the isolated embryonic region will give rise to transplantable tumours when placed beneath the testis or kidney capsule (16,17). These tumours are characterized by the presence of embryonal carcinoma cells, which are poorly differentiated cells similar in morphological and other aspects to epiblast and primordial germ cells. The procedure for ectopic transfer beneath the testis capsule is given in *Table 5.*

6. CULTURE OF WHOLE EMBRYOS

In principle, a satisfactory means of culturing post-implantation embryos affords an opportunity for the experimental study of embryonic cells in their normal environment. Therefore, it is highly desirable that there should be reliable and efficient culture systems to support the normal development of embryos over relatively long periods. Unfortunately, the mouse embryo has proved less tolerant to explantation *in vitro*, during gastrulation and organogenesis, than has the rat (18,19). Regimes which support apparently normal growth and development of rat embryos for several days (20) are much less successful when mouse embryos are used. For example, unlike the rat, the development of pre-primitive streak stage mouse embryos is extremely unpredictable *in vitro* and although development is much improved if embryos are explanted after the primitive streak has formed it seems to be a general finding that normal growth and development cannot be sustained much beyond 24 or 36 h in culture. Nonetheless, during a phase of such rapid change, even 24 h is sufficient to observe extensive differentiation and morphogenesis and so despite its present inadequacies whole mouse embryo culture remains an invaluable tool for the study of post-implantation development.

The procedures described below for the culture of post-implantation embryos are derived entirely, albeit with some modifications, from those designed for rat embryos (18). Undoubtedly, there is still much room for improvement and, therefore, the protocols detailed below should be seen more as prototypes rather than definitive conditions.

6.1 Media

Embryos explanted on the 7th, 8th or 9th day of gestation can be grown in rat serum diluted 1:1 with DMEM (Flow Laboratories). For older embryos, 100% serum is advisable. The method of preparation of rat serum is given in *Table 6*. It is particularly important that serum should be obtained from blood in which the fibrin clot has formed in plasma rather than whole blood. If whole blood is allowed to clot before separation of the serum, early post-implantation rat embryos suffer a higher incidence of abnormalities and reduced growth (21). Neither the sex nor the strain of donor rats appears to be important (18).

The quantity of media required will depend upon the stage and number of embryos explanted and the duration of culture. No systematic analysis has been conducted with mouse embryos but, as a rough guide, 2.5 ml of medium will support the growth and development for 24 h of four 8th day embryos, three 9th day embryos or two 10th day embryos. It is convenient to pre-equilibrate the medium for several hours, or overnight, in a humidified, gassed incubator (5% CO_2 in air) at 37°C. This will remove most of the remaining ether from the serum (see *Table 6*) and reduce the amount of

Table 6. Preparation of rat serum.

1.	Etherize a rat.
2.	Make a ventral midline incision through the skin and body wall. Pull out the intestines and lay them to one side.
3.	The dorsal aorta lies in the midline adjacent to the vena cava. It is the smaller and paler of the two blood vessels. Using a 10 or 20 ml syringe attached to a syringe needle (21 gauge/1½ inches), insert the needle (bevel downwards and pointing towards the heart) into the dorsal aorta.
4.	Gently draw out the blood. If the blood is withdrawn too quickly the red blood cells will rupture and produce deleterious haemolysis in the serum. Depending on the size of the rat one should expect to obtain 8 − 15 ml of blood.
5.	Gently transfer the blood into a centrifuge tube. Spin *immediately* at 1200 *g* for 5 − 10 min. The blood cells are pelleted and a fibrin clot forms in the plasma layer. (At this stage the separated blood can be left to stand at 4°C for at least 18 h.)
6.	Squeeze the fibrin clot with fine forceps, without disturbing the red blood cells, to release trapped serum. Centrifuge at 1200 *g* for 5 min.
7.	Using a Pasteur pipette, decant and pool the serum from different bleeds into another centrifuge tube.
8.	Spin at 1200 *g* for 5 min. Decant the serum into a further tube. Aliquot into convenient volumes (i.e. 2.5 or 5.0 ml).
9.	Heat inactivate the serum in a water bath at 56°C for 30 min. Slightly unscrew the caps of the tubes to allow some of the ether to evaporate.
10.	Add antibiotics when the serum has cooled (benzyl penicillin, 100 IU/ml; streptomycin sulphate, 100 μg/ml). This is not essential. If the serum and media are handled with care contamination of cultures is extremely rare in the absence of any antibiotics. Store at −20°C for up to 3 months.

Table 7. Preparation of media.

1.	Thaw out an appropriate volume of rat serum.
2.	Add an equal volume of DMEM (Flow Laboratories) supplemented with glutamine if the synthetic medium does not contain it. Mix.
3.	Filter the medium (Millipore or Sartorius filter; pore size: 0.45 μm) and aliquot 2.5 or 3 ml into the tubes or bottles used for culture. The five basic requirements of culture vessels is that they should be sterile, of suitable size for rotation on a roller system (Section 6.2), of adequate volume to accommodate a gas phase of approximately 20 ml, gas tight, and embryos should not stick to them. Therefore, either siliconized glass bottles with ground glass stoppers (non-toxic and scrupulously clean), or Universal culture tubes (Sterilin) may be used. To ensure that they are gas tight, a seal of high vacuum silicone grease is advisable.
4.	Pre-equilibrate the medium in 5% CO_2 in air at 37°C for several hours.
5,	After transferring embryos, gas the medium with the appropriate gas mixture (Section 6.3) by gently blowing the gas onto the surface of the medium for about 1 min. Seal the cap or stopper and replace on rollers at 37°C.

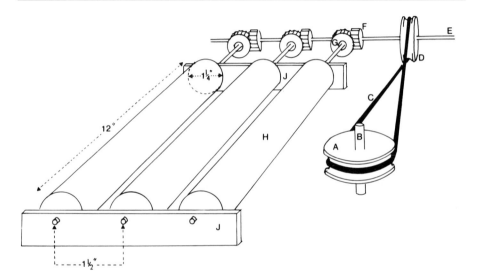

Figure 5. The basic components of a suitable roller culture system. **A.** Pulley fixed to drive shaft, **B.** Drive shaft attached to external continuous motor (30 r.p.m.), **C.** Drive-belt, **D.** Pulley fixed to second drive shaft, **E.** Second drive shaft, **F.** Bevel gear fixed to second drive shaft, **G.** Bevel gear fixed to drum axle, **H.** Rotating drum (diameter, 1.25 inches; length 12 inches), **J.** Fixed plates holding drum axles.

time spent gassing the medium by hand. *Table 7* provides the protocol for the preparation of media.

6.2 Roller system

With rat embryos, rotation of culture vessels, which ensures continuous exposure of all the medium to the gas phase, and, therefore, a constant oxygen tension, produces normal development over longer periods than does static culture (18). Early post-implantation mouse embryos will develop in static culture if explanted at the primitive streak stages but normal development is sustained for only 24 h or less (22). Older

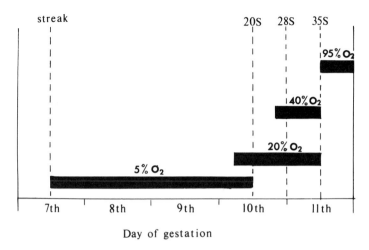

Figure 6. Percentage oxygen required in the gas phase for embryos at different developmental stages, streak = primitive streak, S = somites.

embryos and more extended cultures benefit from a rotating system and, judging from the experience with rat embryos, the same is probably true for primitive streak stage embryos, although no direct comparison between static and rotating culture systems has been undertaken at this stage in the mouse.

Rotation is achieved by either placing bottles or tubes on parallel rollers or by attaching the vessels to a rotating disc. To ensure a sufficient depth of culture medium to fully immerse the embryos, it may be necessary to mount either the tubes or the rollers at a slight angle so that the medium stays in the bottom of the culture vessel. The speed of rotation should be of the order of $30-60$ r.p.m. Roller systems mounted in a small $37°C$ incubator are available commercially from B.T.C. Engineering, Cambridge, UK. Alternatively, it is quite straightforward to construct one's own roller system which can be fitted into a standard incubator. The drive motor should be fitted outside the incubator to prevent any local heating effect. *Figure 5* illustrates the basic design of such a system.

6.3 Gas requirements

Progressively older embryos require a higher oxygen tension in the medium and, therefore, different gas mixtures are used for different stages of embryo. Experiments with rat embryos suggest that this increased demand for oxygen is due both to a change in metabolism, from a predominantly glycolytic source of energy to utilization of the Kreb's cycle and electron transport system, and to a necessary compensation for the absence of a functional chorioallantoic placenta which serves *in vivo* as an important organ of respiratory exchange (18). The appropriate percentage of oxygen in the gas phase at different developmental stages is shown in *Figure 6*. At all stages the mixture must contain 5% CO_2, the final volume being made up with N_2. At stages where there is some overlap in requirements for different mixtures it is better to use the higher oxygen mixture if cultures are to continue beyond 24 h.

Table 8. Some parameters for evaluating the development of cultured embryos.

Gross morphology
Degree of axial turning
Number of pairs of somites
Extent of neural fold fusion
State of the allantois (e.g. fused or unfused with the chorion)
Presence or absence of specific structures (e.g. limb buds, otic placode, etc.)
Relative size of different structures (e.g. head, somites, etc.)

Histology
Presence or absence, and morphology of specific structures
Cell packing density
Mitotic index of different tissues
Dead cell index of different tissues

Growth and metabolism
Crown/rump length (or other convenient measure of physical size)
Protein content (determined by the colorimetric method of Lowry *et al.*; ref. 23.
Incorporation of radioisotopes (e.g. [^{3}H]thymidine, [^{35}S]methionine)

Physiology
Presence or absence, and rate of heart beat
Presence or absence, and extent of visceral yolk sac circulation

---

6.4 Assessment of development

Obviously, the validity of any experiment involving cultured embryos depends upon the normality of the differentiation, morphogenesis and growth of embryos developed *in vitro*. Therefore, it is important to select suitably stringent criteria for evaluating their development. *Table 8* lists certain useful parameters which can be used. It is by no means an exhaustive list. Studies on particular aspects of gastrulation or organogenesis will require their own set of parameters according to the specific system or process being studied. Wherever possible such parameters should be compared with those of embryos of the same age recovered *in vivo*. Furthermore, unoperated controls should be included in every experiment as a further standard with which to compare experimental embryos and as a routine screen of the culture system.

7. INTRODUCTION OF CELLS OR LABEL INTO EMBRYOS IN VITRO

The injection of a suitable label into embryonic cavities can be used to mark superficial cells lining the cavity and to follow their ingression. This is a valuable technique for the study of the fate of epiblast cells invaginating through the primitive streak or for the migration of neural crest cells (24). Injection of other substances, such as specific enzymes or enzyme inhibitors, may also be useful in assessing the importance of particular molecules in various morphogenetic or differentiation events. Injection of labelled cells into the embryo, either into isotopic or heterotopic sites, allows one to follow the development of specific tissues or isolated regions of the embryos (25,26). The general strategy for producing '*in vitro* chimaeras' is shown in *Figure 7*.

It is beyond the scope of this chapter to describe all the manipulations possible on post-implantation embryos. The simplest two-instrument cell injection technique for both primitive streak stage and for early somite stage embryos will be described. More

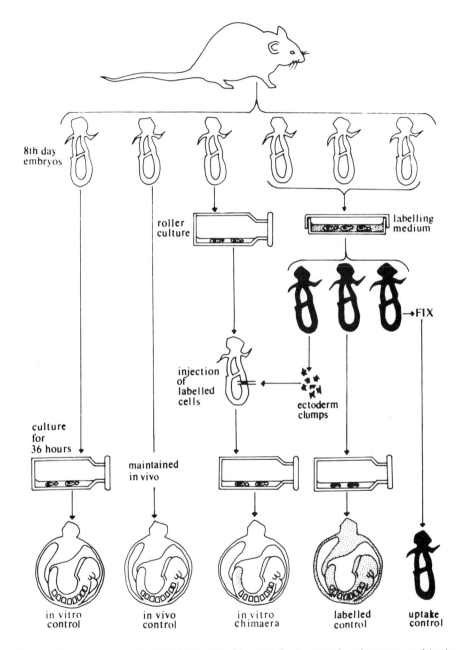

8th day embryos

roller culture

labelling medium

→FIX

injection of labelled cells

ectoderm clumps

culture for 36 hours

maintained in vivo

in vitro control

in vivo control

in vitro chimaera

labelled control

uptake control

Figure 7. General strategy for producing *in vitro* chimaeras. *In vivo* control: embryo recovered *in vivo*, of the same age as embryos at the end of the culture period, providing standard for the expected degree of development. *In vitro* control: non-experimental embryo grown in culture providing standard for expected development in culture. Labelled control: embryo labelled to the same extent as donor tissue with [^{3}H]thymidine; this controls for non-toxicity of label, ubiquitous distribution of label and for the expected dilution of label during the culture period. *In vitro* chimaera: injected embryo. Uptake control: embryo fixed immediately after labelling providing measure of percentage of cells labelled.

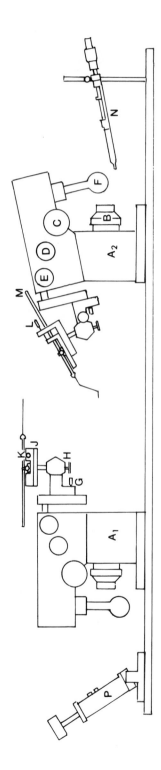

Figure 8. Essential components of the micromanipulator assembly required for injecting cells or label into the embryo. A_1 and A_2: right and left micromanipulator units clamped to base plate (Leitz). A_1 is set up for use with a Puliv micromanipulator chamber on a fixed-stage microscope with image-erected optics. A_2 is set up for manipulation in a Petri dish using a dissecting microscope. **B**: vertical control, **C**: tilt control, **D**: lateral control, **E** forwards control, **F** joystick, **G**: height control, **H**: instrument head angle control, **J**: instrument head, **K**: fine lateral control, **L**: fine forward control, **M**: instrument holder, **N**: Agla micrometer syringe (Wellcome, Derbyshire, UK), **P**: de Fonbrune suction and force pump (Beaudouin, Paris, France).

complicated manipulations may require the introduction of extra instruments, such as straight or recurved solid needles, a description of which can be found elsewhere (27). Alternatively, instruments incorporating electrodes may be required for the iontophoretic injection of tracers, such as horseradish peroxidase (HRP), into individual cells *in situ*. Again, specific details of this technique are provided elsewhere (28,29). In practice, the best combination of instruments for a particular procedure is largely a matter of trial and error but once a simple technique, such as the two-instrument injection method, has been mastered the principles of making and setting up extra instruments will be self-evident.

7.1 Equipment

Controlled injections of either solutions, cells or solid tissue require a micromanipulator assembly (Leitz, Wetzlar, FRG). The essential components of this assembly are shown in *Figure 8*. Primitive streak stage or younger embryos can be manipulated in hanging drops of media, suspended from a siliconized coverslip, mounted on a Puliv manipulation chamber (Leitz) filled with paraffin oil (for further details, see Chapter 12, Section 2.3.1). In this case a fixed-stage microscope with image-erected optics and long distance working objectives is most appropriate (e.g. Microtec M2, MicroInstruments, Oxford). Older embryos are best manipulated in a drop of medium on the bottom of a liquid paraffin-filled bacteriological Petri dish or Petri dish lid. Here a standard dissecting microscope is more suitable. Primitive streak stage embryos can also be manipulated using a dissecting microscope but the lower resolution makes such manipulations less precise.

The most critical ingredient for successful micromanipulation is good instruments (see Section 9), which are correctly aligned for use.

7.1.1 Setting up the instruments

The most important aspects of setting up instruments are:

(i) there should be no air bubbles in the columns of paraffin oil;
(ii) the instruments should be absolutely straight, bisecting the microscope field horizontally, and lying exactly parallel to each other.

Procedures are given in *Table 9* (see also Chapter 12, Section 3.2).

Where a dissecting microscope is used and injections are conducted in a Petri dish, it is important that the tips of the instruments are inclined at a slight angle towards the bottom of the dish so that the shoulder left after pulling the capillary does not prevent the mouth of the pipette reaching the bottom of the dish (see Section 9). This is achieved by introducing two bends into the instruments (*Figure 11D,b*) and by tilting the instrument head and manipulator as shown in *Figure 8*. For manipulations in hanging drops it is essential that the two instruments should run parallel to the microscope stage.

7.2 Preparation of label or cells for injection

7.2.1 Label

Recently, it has been shown that the injection of wheat germ agglutinin (WGA) coupl-

Table 9. Setting up instruments.

1.	Ensure that the tubing attached to the Agla syringe and to the de Fonbrune suction pump is full of paraffin oil, that there are no air bubbles and that there is an adequate reservoir of oil in the syringe and pump.
2.	Insert an instrument holder, having removed the cap and brass washer, into the end of each length of tubing and fill the holder with oil so that the oil meniscus emerges from the mouth of the holder.
3.	Put the caps and washers, in the appropriate orientation, onto the injection and holding pipettes. Insert the injection pipette into the instrument holder attached to the de Fonbrune pump making sure that the capillary is firmly held by the rubber tubing washer. Do the same for the holding pipette, attaching it to the Agla syringe. Completely fill both instruments with paraffin oil.
4.	Focus the microscope on the drop of medium in which the injections will be carried out.
5.	Mount the instrument holders on the manipulator and bring the instruments into focus in the microscope field. Align the instruments so that they are directly opposite one another and bisect the microscope field horizontally. Ensure that both instruments, as they are finally set up, can be moved anywhere in the microscope field with the joy stick and still remain in focus.
6.	Withdraw the instruments. Where a manipulation chamber is being used the instruments can simply be withdrawn to the limit of the horizontal control. Where a Petri dish is to be used it is more convenient to lift the instruments out of the drop of medium using the 'tilt' control on the manipulator.

Table 10. Preparation of [^{3}H]thymidine-labelled cells for injection.

1.	Add 10 μCi/ml of [^{3}H]thymidine made up to a specific activity of about 10.5 Ci/mM, to the α modification of Eagles medium (Flow Laboratories). Pre-equilibrate the medium in 5% CO_2 in air at 37°C.
2.	Incubate intact embryos, with Reichert's membrane reflected, in the medium for at least 2 h. Two hours is sufficient to label over 90% of cells in the late primitive streak stage embryo but older embryos may require longer. This must be checked by doing an initial time course and examining serial sections after autoradiography of embryos fixed immediately after labelling (uptake controls; *Figure 9a*).
3.	Wash the embryos in three changes of PB1[a] + 10% FCS + 10 μl/ml of 10^{-3} M thymidine. Unlabelled thymidine, at this concentration (10^{-5} M), should be present in all media used for subsequent dissection, manipulation and culture.
4.	Place two or three labelled embryos into roller culture to serve as labelled controls (*Figure 7*). It is important to include labelled controls in *every* experiment since they provide an internal control for any variation in labelling, for dilution of label during the experiment and for variations in autoradiographic processing (*Figure 9b*).
5.	Isolate the particular tissue or region for injection as described in Section 4.
6.	Cut the tissue into appropriate sized fractions for injection with glass needles, and transfer the pieces of tissue to the manipulation drop ready for injection into a recipient embryo.

[a]See Chaper 13, Section 3.1.1 for composition of PB1.

ed to labels such as colloidal gold or HRP produces effective labelling of cells lining the amniotic cavity (24). It is likely that other high molecular weight molecules which will bind to epiblast or ectoderm cells, such as other lectins or perhaps antibodies (see Chapter 2, Section 4), may prove suitable labels for selectively marking these cells or cells lining other cavities in the embryo. If the object is to label only cells lining a cavity it is crucial to avoid increasing the hydrostatic pressure of the cavity to such an extent that label is forced into internal tissues. Therefore, always prepare a concentrated solution of the label and inject a minimum quantity. For example, approximately 2×10^{-3} μl of a 1 mg/ml (w/v) solution of WGA − HRP is sufficient to label all the epiblast cells in a late primitive streak stage embryo, the label still being detectable

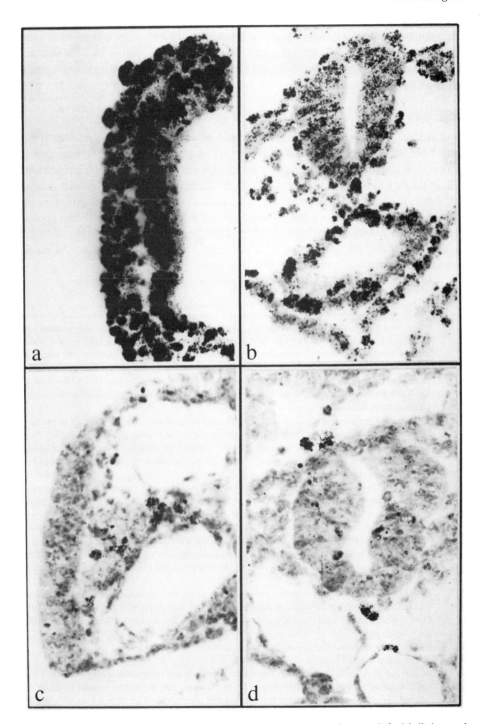

Figure 9. Autoradiographs of control and experimental embryos. **a**, uptake control; **b**, labelled control; **c**, injected embryo chimaeric in the primitive streak region; **d**, injected embryo chimaeric in the notochord and gut.

after culture for 24 h. Having prepared the solution of label, place a drop of it in the manipulation chamber or dish, sufficiently far away from the drop of medium in which the injections will be carried out, to prevent the two coalescing when the injection pipette is passed from one to the other.

7.2.2 Cells

In principle, single cell genetic markers could be used to distinguish injected donor cells from surrounding host tissue. However, at present, apart from interspecific markers, no ubiquitous genetic marker exists which can be used to distinguish individual cells in the early embryonic tissues of the post-implantation conceptus. The most satisfactory extrinsic marker is [³H]thymidine which, due to the rapid cell cycle of early post-implantation embryonic cells, is quickly taken up by the vast majority of cells and appears not to be deleterious to development (25). Labelling of cells with [³H]thymidine is described in *Table 10*, and *Figure 9* shows representative autoradiograms of uptake and labelled controls together with two chimaeras.

Table 11. Injection of cells into the epiblast of the 8th day embryo.

1.	Place one or more embryos to be injected, together with the donor cells, in a manipulation drop of PB1[a] + 10% FCS + 10^{-5} M 'cold' thymidine. Bring the previously aligned injection and holding pipettes (Section 7.1.1) into the same field and suck a small quantity of medium into each.
2.	Focus on the clump of donor tissue and bring the tip of the injection pipette into the same focal plane. Very gently suck the cells into the pipette until they lie approximately $10-20$ μm from the tip. Violent suction will result in the medium dispersing the oil column into droplets with concomitant loss of hydraulic control.
3.	Focus on the embryo and rotate it so that the desired point of injection lies adjacent to the tip of the injection pipette. Gently suck the embryo onto the end of the holding pipette (*Figure 10a*). Bring the injection pipette into exactly the same plane of focus as the proposed site of injection (*Figure 10a*). Confirm that they are truly aligned, with the joy stick centred, by gently nudging the embryo with the tip of the pipette.
4.	Using only the joy stick move the injection pipette to one side of the embryo and advance the joy stick as far forwards as it will go. Using the forwards control bring the tip of the injection pipette in line with the furthest extreme of the amniotic cavity (i.e. so that when it is pushed to its furthermost limit with the joy stick it will not penetrate the opposite side of the embryo; *Figure 10b*).
5.	Using the joy stick bring the tip of the pipette back in line with the proposed site of injection (*Figure 10c*). Withdraw it slightly and then push the joy stick rapidly and forcefully to its forwardmost limit. The pipette should penetrate the embryo at the intended site and emerge into the amniotic cavity (*Figure 10d*).
6.	Gently withdraw the pipette with the joy stick until its tip lies just behind the intended region of injection. Slowly release the cells into the space created by the retraction of the pipette (*Figure 10e*). Once the cells are released, withdraw the injection pipette altogether and release the embryo from the holding pipette.
7.	Rotate the embryo and check that the graft is in the correct position.
8.	Transfer the injected embryo to roller culture.
9.	At the end of the culture period, fix the injected embryos and labelled controls in Carnoy's fluid (60% absolute alcohol; 30% chloroform; 10% acetic acid) and process for routine histology and autoradiography. It is recommended that one labelled control should be included in every block of injected embryos prepared for sectioning.

[a]See Chapter 13, Section 3.1.1 for composition of PB1.

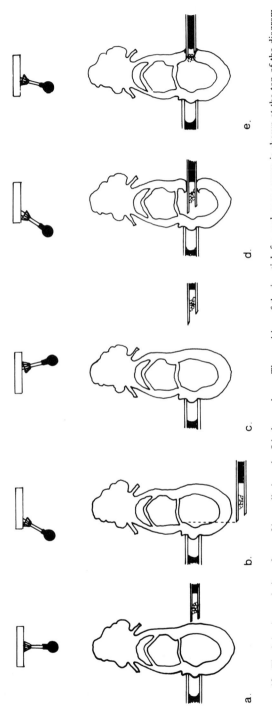

Figure 10. The injection technique for grafting cells into the 8th day embryo. The position of the joystick for each manoeuvre is shown at the top of the diagram. The detailed manipulations are described in *Table 11*.

7.3 **Injection procedure**

Essentially the injection procedure is the same for both solutions and cells. Therefore, only the injection of cells is described here (*Table 11* and *Figure 10*). An important point to remember when injecting solutions is never to put the holding pipette into the drop of label. Each manipulation drop should contain only one embryo and a fresh drop should be used for each injection to minimize any possible contamination due to label diffusing out of the injection pipette. Furthermore, injections into the mesoderm or ectoderm require going through the endoderm layer. While this generally does not matter when cells are injected, solutions may leak into tissues other than that intended for labelling. This problem may be avoided by careful choice of the injection route, for example, by injecting into the amniotic cavity via the extra-embryonic region.

8. INTRODUCTION OF CELLS OR LABEL INTO EMBRYOS IN VIVO

Over the last 10 years several successful attempts have been made to introduce cells or virus into the embryo *in utero* (30 − 32). Obviously, this has the advantage that host embryos can be recovered at any subsequent stage and that adult chimaeras can be obtained. Perhaps the most dramatic example of initiating chimaerism *in vivo* is the recent report of viable mice showing unequivocal melanocyte chimaerism produced by injecting neural crest cells, from a pigmented donor, into albino embryos *in utero* (13). The disadvantages of *in vivo* injections are that precisely localized injections are impossible and that there may be only a restricted number of cell types which can integrate into the early post-implantation embryo following injection into the visceral yolk sac or amniotic cavity. Furthermore, chimaeras have not been obtained from embryos injected before the middle of the 9th day of gestation, presumably because younger embryos are physically too small to be successfully penetrated, or to tolerate penetration.

Before attempting to produce chimaeras it is advisable to practice the injection procedures, given in *Table 12* , using readily identified solutions or suspensions, such as pontamine sky blue or India ink, or cells labelled with a fluorescein dye, to establish that the injection pipette is consistently reaching the target area. It is possible to inject 'free-hand' but, particularly for younger embryos, holding the injection pipette in a Singer manipulator produces a more reliable and steadier motion.

9. PREPARATION OF MICROSURGICAL GLASS INSTRUMENTS
9.1 **Glass**

All the instruments are made from glass capillary tubing. Injection and holding pipettes are produced from 'thin-walled' tubing (o.d. 1.0 mm; i.d. 0.8 mm) which can be obtained from Drummond Scientific Company, Pennsylvania. This is available either in soda glass or Pyrex, both of which produce suitable instruments although it is easier to clean the Pyrex glass. Solid microneedles are made from 'thick-walled' tubing (o.d. 1.0 mm; i.d. 0.5 mm) supplied by Leitz. All the capillary tubing should be pre-cleaned: soak the glass in chromic acid for 24 h and rinse very thoroughly with repeated washings in distilled water. Dry the tubing in a hot oven and store in a dust-free container.

9.2 **Equipment**

All instruments are made by first pulling gradually tapered needles which are subsequently modified to produce the requisite implement. Therefore, the two basic requirements are an electrode puller and a microforge. Electrode pullers may be either electromagnetically or gravity-operated and are commercially available from a number of suppliers. Minor modification may be necessary in order to fit tubing of 1.0 mm diameter because most pullers are designed to accommodate the slightly larger, standard microelectrode capillary tubing.

A suitable microforge is that designed by de Fonbrune and available from Beaudouin, Paris. A V-shaped filament of platinum wire (diameter 0.3 mm) with a small globule of glass fused to its apex should be fitted to the filament holder. The microscope should be fitted with a calibrated ocular micrometer and should give a magnification of approximately ×100.

9.3 **Instruments**

The most comprehensive description of the different methods for making various glass

Table 12. Injections *in utero*.

1.	In order to monitor the volume of fluid injected it is advisable to calibrate the injection pipette (Section 9) into convenient volume increments (e.g. 0.5 μl). Knowing the internal diameter of the glass tubing and treating it as a standard cylinder, it is simple to calculate the appropriate length for a given volume along the unpulled shaft of the pipette, and to make a series of calibration marks proximal to the tapered end.
2.	Mount the injection pipette in an instrument holder attached to a de Fonbrune suction and force pump and fill it with paraffin oil.
3.	Anaesthetize the pregnant mouse (e.g. Avertin: 0.1 ml/5 g body weight). Place it on the stage of a dissecting microscope and illuminate its lower abdomen, preferably using fibre optic illumination.
4.	Make a mid-ventral incision (~ 1.0 − 1.5 cm long) through the skin and the body wall of the lower abdomen. To avoid excessive bleeding, cauterize the wound by using very hot scissors to make the incision.
5.	Gently pull the uterus through the opening.
6.	Pick up the donor cells or label in the injection pipette, sucking up sufficient medium or label so that the meniscus of the paraffin oil lies within the calibrated region of the pipette. Apply *gentle* suction so that the oil does not form small droplets in the medium. Mount the instrument holder on a Singer manipulator.
7.	Hold the uterus, adjacent to a decidual swelling, with fine forceps. Advance the injection pipette so that its tip lies opposite the mid-line of the decidual swelling, approximately a third of the way up from the antimesometrial end.
8.	Make a hole through the uterine muscle with the tip of a syringe needle. Insert the injection pipette through this hole into the middle of the decidual swelling.
9.	Observing the movement of the paraffin oil meniscus, inject about 0.5 μl into the embryo.
10.	Withdraw the injection pipette.
11.	Repeat this procedure (steps 7 − 10) on the remaining decidual swellings, or on as many as it is convenient to inject. It may be necessary to change the injection pipette if it gets clogged with decidual tissue.
12.	Replace the uterus in the peritoneal cavity and sew up the body wall. The skin can either be sewn or closed with wound clips.

microinstruments is provided by de Fonbrune himself (33). Several basic procedures are included in the instruction manual supplied with the de Fonbrune microforge. The procedures used in this laboratory for making solid dissecting needles and holding pipettes are given in *Table 13* and those for *in vivo* injection pipettes in *Table 14* . *Figure 11* illustrates these procedures. (The reader is also referred to Chapter 12, Section 2.4).

Needles can be produced by simply pulling Leitz capillary on an electrode puller but such hollow needles can cause damage to tissue if they break because of the ensuing capillary action. Solid needles are preferable both for this reason and also because they have greater mechanical strength.

Injection pipettes are made in exactly the same way as holding pipettes except that they are broken off at a smaller internal diameter. The final diameter of the injection pipette will depend on what is to be injected (ranging from 1 μm for intracellular injection to 60 μm for large pieces of tissue). If relatively large pieces of tissue are to be transferred, requiring an internal diameter greater than about 35 μm, it is useful to have a bevelled end to the pipette. While bevelling machines are available commercially satisfactory instruments can be prepared by breaking the capillary tip against a cold filament. If the pipette is repeatedly driven head-on into the globule of cold glass a break can be produced at approximately the correct internal diameter. Usually these breaks are not perpendicular but at an angle, and so long as the end is carefully flame-polished, minor irregularities in the line of the break do not matter.

Table 13. Production of solid needles and holding pipettes.

Solid dissection needles

1.	Fuse the central portion of the capillary tubing into a solid glass rod over a Bunsen flame, rotating the glass throughout to prevent reduction in the tube diameter. Remove the tubing from the heat and, pausing for about 1 sec to allow it to cool down, pull the tubing straight without decreasing the outer diameter of the fused portion.
2.	Place the capillary in the electrode puller with the fused region in the heated filament and pull to produce a needle with a tapered length of at least 5 mm.
3.	In order that the needle should run parallel to the bottom of the dish for dissections it is convenient to introduce two bends in the tubing as shown in *Figure 11D,a*. This is done over a micro-Bunsen burner, which can be made by attaching a syringe needle to the end of the gas outlet tubing. If during fusion of the capillary a prominent shoulder of solid glass has been produced on one side of the instrument make the first bend towards this shoulder so that in the final instrument, it will face away from the bottom of the dish.
4.	Finally, siliconize the needle by dipping its tip into Repelcote (Hopkin and Williams).

Holding pipette

1.	Using a 'thin-walled' capillary, pull a needle on the electrode puller. Insert the needle into a Leitz instrument holder and mount it horizontally on the microforge.
2.	Bring the filament and the horizontal shaft of the needle, at the point where its internal diameter is 80 μm, into the same focal plane in the microscope field.
3.	Heat the filament so that the glass at its tip is just molten (*Figure 11A,a*) and bring it up to fuse with the underside of the capillary (*Figure 11A,b*). This should be achieved without distorting or bending the capillary.
4.	Turn off the heating circuit of the filament. The contraction of the filament on cooling should result in a clean break in the capillary perpendicular to its long axis (*Figure 11A,c*).
5.	Flame-polish the end of the pipette by re-heating the filament and bringing it end-on towards the pipette until the broken surface begins to thicken (*Figure 11A,d*). Turn off the filament.

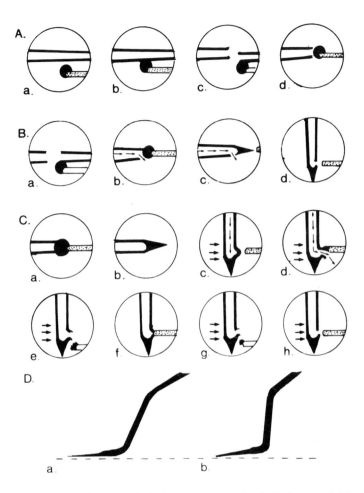

Figure 11. The production of microsurgical glass instruments (after de Fonbrune; ref. 33). **(A)** *In vitro* holding and injection pipettes; **(B)** *in vivo* injection pipette (Method 1); **(C)** *in vivo* injection pipette (Method 2); **(D)** requisite bends in **(a)** solid needles, **(b)** injection or holding pipettes to be used in a Petri dish. Descriptions of the procedures shown are given in *Table 13* and *14*.

Straight holding and *in vitro* injection pipettes are appropriate for manipulations carried out in a manipulation chamber. For manipulations on 9th day or older embryos, which have to be done in a Petri dish it is necessary to introduce two bends over a microburner, similar to those described for solid needles (*Table 13*), although the exact angle of the bends will depend on how the micromanipulator is set up (*Figure 8*). As noted earlier, it is important that the tips of the holding and injection pipettes are inclined downwards at a very slight angle (*Figure 11D,b*) otherwise the shoulder of the pulled glass may prevent the pipettes making contact with the embryo or donor tissue.

There are two basic requirements for an *in vivo* injection pipette. The first is that the instrument should penetrate the tissue easily and the second is that the mouth of the pipette should not become obstructed by adherent tissue during penetration. To meet these requirements de Fonbrune recommends that pipettes with a solid pointed end should

Table 14. Production of *in vivo* injection pipettes.

Method 1

1. Pull a hollow needle, using thin walled capillary, on the electrode puller.
2. Insert the needle into an instrument holder attached by tubing either to a source of compressed air or to a 20 ml syringe full of air.
3. Mount the instrument horizontally on the microforge. Break the pipette at an internal diameter of 15 μm (*Figure 11B,a*).
4. While blowing air through the pipette, fuse the heated filament to its tip such that a small outlet for the air remains at one point just behind the fused glass (*Figure 11B,b*).
5. Maintaining the air flow, rapidly retract the filament producing a straight, solid pointed end (*Figure 11B,c*). If the pointed tip is too large, or crooked, break off the end and repeat the fusion and retraction.
6. Orientate the instrument vertically with its mouth directed towards the filament. Bring the heated filament adjacent to the aperture in order to flame-polish its edges (*Figure 11B,d*).

Method 2.

1. Break off a hollow needle at the required diameter (e.g. 50 μm).
2. Fuse the heated filament to the end of the pipette and quickly retract it to produce a solid, pointed tip (*Figure 11C,a and b*).
3. Orientate the instrument vertically. Focus the microforge air jets onto the pipette just behind the solid tip (this helps to localize the heating effect of the filament). Blow air through the instrument while bringing the heated filament adjacent to, but not touching, the point where the pipette mouth is required. Increase the temperature of the filament until a small bubble of molten glass forms (*Figure 11C,c*). When the bubble touches the filament immediately turn off the heat (*Figure 11C,d*). The bubble will break off leaving a relatively large, jagged opening (*Figure 11C,e*).
4. Fuse the filament to the edges of the aperture, which will decrease in size (*Figure 11C,f*).
5. When the aperture is reduced sufficiently, turn off the filament and move it vertically downwards. The protruding glass should break off flush with the main shaft of the pipette (*Figure 11C,g*).
6. Flame-polish the edges of the aperture (*Figure 11C,h*).

be used, the mouth of the pipette being located laterally just behind the tip (*Figure 11B and C*). Two methods for achieving this are given in *Table 14*. Method 1 is suitable for small diameter pipettes whereas method 2 is for larger ones. If the injection pipette is renewed between operations, solutions can be injected using the simpler *in vitro* injection pipette.

10. CONCLUDING REMARKS

Implantation of the mouse embryo presents a serious obstacle to the analysis of tissue diversification during gastrulation and early organogenesis. However, recently, new techniques have been devised for studying the behaviour of embryonic cells in their normal environment in the conceptus during these stages of development. Some of these techniques have been described in this chapter. In many respects they may appear somewhat crude and certainly there is much room for improvement, particularly in the areas of whole embryo culture and precise manipulations of the embryo, both *in vitro* and *in vivo*. Nevertheless, over the past 10 years or so, the predominantly histological and morphological descriptions of early post-implantation embryogenesis have been augmented by valuable information relating to the more dynamic aspects of development, such as the normal fate and developmental potency of different tissues or regions of the embryo. Such information, derived from experimental perturbation of the embryo, may lead eventually to a clearer understanding of the processes governing the initial organization and differentiation of the fetus.

11. ACKNOWLEDGEMENTS

I would like to thank Professor R.L.Gardner, Dr D.L.Cockroft and Dr P.P.L.Tam for their discussion of the manuscript. The author is a Lister Institute for Preventive Medicine Research Fellow.

12. REFERENCES

1. Gardner,R.L. (1983) *Int. Rev. Exp. Pathol.*, **24**, 63.
2. Beddington,R.S.P. (1987) In *Experimental Approaches to Mammalian Development*. Pedersen,R.A. and Rossant,J. (eds), Cambridge University Press, New York, in press.
3. Jolly,J. and Ferester-Tadie,M. (1936) *Arch. Anat. Microsc.*, **32**, 323.
4. Rugh,R. (1968) *The Mouse*. Burgess Publishing Company, USA.
5. Snell,G.D. and Stevens,L.C. (1966) In *Biology of the Laboratory Mouse*. Green,L. (ed.), McGraw-Hill, New York, p. 205.
6. Whittingham,D.G. and Wales,R.G. (1969) *Aust. J. Biol. Sci.*, **22**, 1065.
7. Theiler,K. (1972) *The House Mouse*. Springer-Verlag, Berlin.
8. Levak-Svajger,B., Svajger,A. and Skreb,N. (1969) *Experientia*, **25**, 1311.
9. Snow,M.H.L. (1978) In *Methods in Mammalian Reproduction*. Daniel,J.C.,Jr (ed.), Academic Press, New York, p. 167.
10. Gardner,R.L. and Rossant,J. (1979) *J. Embryol. Exp. Morphol.*, **52**, 141.
11. Kruse,P.F. and Patterson,M.K. (eds) (1973) *Tissue Culture: Methods and Applications*. Academic Press, New York.
12. Ito,K. and Takeuchi,T. (1984) *J. Embryol. Exp. Morphol.*, **84**, 49.
13. Jaenisch,R. (1985) *Nature*, **318**, 181.
14. Snow,M.H.L. (1981) *J. Embryol. Exp. Morphol.*, **65** (Suppl.), 269.
15. Rossant,J. and Ofer,L. (1977) *J. Embryol. Exp. Morphol.*, **39**, 183
16. Diwan,S.B. and Stevens,L.C. (1976) *J. Natl. Cancer Inst.*, **57**, 937.
17. Damjanov,I., Solter,D. and Skreb,N. (1971) *Wilhelm Roux' Arch. Entwicklungsmech. Org.*, **173**, 228.
18. New,D.A.T. (1978) *Biol. Rev.*, **53**, 81.
19. Sadler,T.W. and New,D.A.T. (1981) *J. Embryol. Exp. Morphol.*, **66**, 109.
20. Buckley,S.K.L., Steele,C.E. and New,D.A.T. (1978) *Dev. Biol.*, **65**, 390.
21. Steele,C.E. (1972) *Nature, New Biol.*, **237**, 150.
22. Tam,P.P.L. and Snow,M.H.L. (1980) *J. Embryol. Exp. Morphol.*, **59**, 131.
23. Lowry,O.H., Rosebrough,N.J., Farr,A.L. and Randall,R.J. (1951) *J. Biol. Chem.*, **193**, 265.
24. Smits-Van Prooije,A.E., Poelmann,R.E., Dubbeldam,J.A., Mentink,M.T. and Vermeij-Keers,C. (1986) *Stain Technol.*, **61**, 97.
25. Beddington,R.S.P. (1981) *J. Embryol. Exp. Morphol.*, **64**, 87.
26. Beddington,R.S.P. (1982) *J. Embryol. Exp. Morphol.*, **69**, 265.
27. Gardner,R.L. (1978) In *Methods in Mammalian Reproduction*. Daniel,J.C.,Jr (ed.), Academic Press, New York, p. 137.
28. Lawson,K., Meneses,J.J. and Pedersen,R.A. (1986) *Dev. Biol.*, **115**, 325.
29. Lo,C.W. and Gilula,N.B. (1979) *Cell*, **18**, 399.
30. Weissman,I.L., Papaioannou,V.E. and Gardner,R.L. (1978) In *Differentiation of Normal and Neoplastic Haemopoetic Cells*. Clarkson,B., Marks,P.A. and Till,J.E. (eds), Cold Spring Harbor Laboratory Press, New York, p. 33.
31. Jaenisch,R. (1980) *Cell*, **19**, 181.
32. Fleischman,R.A. and Mintz,B. (1979) *Proc. Natl. Acad. Sci. USA*, **76**, 5736.
33. De Fonbrune,P. (1949) *Technique de Micromanipulation (Monographies d'Institut Pasteur)*. Masson et Cie., Paris.

CHAPTER 4

Meiotic analysis in germ cells of man and the mouse

ANN C.CHANDLEY

1. INTRODUCTION

Mammalian meiotic chromosome preparation has progressed considerably since the early days when a good squash preparation was the best that could be achieved. The vastly superior and more rapid method for metaphase chromosome preparation, of air-drying of germ cells following a short hypotonic pre-treatment, has long since superceded squashing. In addition, a new and powerful tool in the form of 'spreading' for meiotic prophase analysis, has been introduced over recent years.

The aim of this chapter will be to describe in detail the currently available techniques for air-drying and spreading of oocytes and spermatocytes in man and the mouse.

2. PREPARATION OF MALE MEIOTIC CHROMOSOMES BY AIR-DRYING

While gonadal material can easily be obtained from mice, testicular samples for the study of human meiosis are not so easy to come by. Nevertheless, they should be possible to obtain from elderly patients undergoing orchidectomy in the course of treatment for prostatic cancer, or from young men undergoing infertility investigations.

Meiosis begins at puberty in the male, and is an integral part of spermatogenesis, a process which begins with a series of mitotic divisions in the spermatogonia. Following the pre-meiotic S-phase interval, the cells proceed as spermatocytes through a long meiotic prophase and then through the two cell divisions leading to the production of haploid gametes. When air-dried suspensions of cells from the testis of a sexually mature male with normal fertility are prepared, three types of dividing cells are distinguishable on the slides, spermatogonial metaphases, and first and second meiotic metaphases. If an infertility problem is diagnosed, then spermatogenesis may well be depressed or arrested, and in these cases, the later germ cell stages will be absent or deficient on the slides.

Metaphase I (MI) is the stage of meiosis used in routine analysis for the establishment of bivalent number, chiasma frequency or position, and for the detection of a chromosomal anomaly such as a reciprocal translocation which can be recognized as a multivalent configuration in the complement. At MI in man, the X and Y pair in an end-to-end fashion, by the two short arms (p), and in the mouse, by the two long arms (q). In a small proportion of cells in both species, the sex chromosomes are seen as univalents at MI, and such cells may occur with much higher frequencies in sterile cases. Cells in the second meiotic division (MII) are often difficult to analyse on account of the twisting and overlapping of chromatid arms, but when well spread, they can, and have, been used to assess aneuploidies arising from a segregational error at the first meiotic division (1,2).

The air-drying method in regular use for the preparation of meiotic divisions in man and the mouse is that of Evans *et al.* (3). The technique as described originally is applied to the mouse testis below (Section 2.1), and modified in our laboratory for use with human testicular material (Section 2.2).

2.1 Air-drying method for use in the mouse

(i)   Remove the testis from the tunica albuginea and place in isotonic sodium citrate solution (2.2% w/v) at room temperature.

(ii)   Transfer the testis to fresh 2.2% sodium citrate solution contained in a small glass Petri dish and gently pull out the tubules. Hold the mass of tubules with fine, straight forceps and thoroughly tease out their contents with fine, curved forceps. When the tubules appear 'flat' and opaque allow them to settle, and carefully transfer the supernatant fluid into a 15 ml centrifuge tube.

(iii)   Centrifuge the cell suspension obtained at 50 g for 5 min. This generally leaves the majority of sperm in suspension and sediments the larger cells including the spermatocytes. Discard the supernatant fluid and resuspend the sedimented cells in approximately 3 ml of 1% sodium citrate solution.

(iv)   Leave the cells in hypotonic solution for 12 min at room temperature. Centrifuge at 50 g for 5 min with slow acceleration.

(v)   Remove as much of the supernatant fluid as possible. Resuspend the cells in the remainder by flicking the tube so that a thin film of suspension adheres to the wall of the tube. Add about 0.25 ml of fixative (3:1 methanol:glacial acetic acid) rapidly to the suspended cells. Thoroughly mix using a Pasteur pipette. Add more fixative down the side of the tube, maintaining the mixing process, until the tube is about one-third full. After 5 min, sediment the cells again by centrifuging and resuspend in fresh fixative. Repeat the change of fixative after a further 10 min, controlling the volume of the final suspension until it appears slightly cloudy with cells.

(vi)   Make a trial preparation by dropping one drop of suspension onto a grease-free (acid/alcohol washed) slide at room temperature. If the slide is thoroughly clean and the fixation satisfactory, the droplet should expand evenly and, by gently blowing on the slide, the final evaporation will be hastened. Examine by phase contrast. If the cell suspension is too thin, centrifuge again and add a smaller volume of fresh fixative. If too thick, add more fixative.

2.2 Modification of the technique for use in man

(i)   Collect testicular biopsy in 1% hypotonic sodium citrate solution.

(ii)   After about 30 min chop the tubules very finely with scissors in a glass Petri dish tilted slightly, allowing large pieces of tubule to settle to the bottom.

(iii)   Draw off the cell suspension with a pipette and spin down for 8 min at 450 *g* (it is our experience that thinner suspensions make better fixed preparations).

(iv)   Discard most of the supernatant, leaving behind just enough citrate covering the

cells to flick the cells into a suspension by gently tapping the sides of the tube.
(v) Add fixative (3:1 methanol:glacial acetic acid) slowly down the sides of the tube until the volume of cell suspension is about trebled. Then pipette, firmly but gently, breaking up any clumps of cells which may be formed. Add more fixative to a volume of about 5 ml. Spin down for 8 min at 450 *g*.
(vi) Discard the supernatant. Add about 5 ml of fixative and allow the suspension to stand at room temperature for about 1 h.
(vii) Spin down for 8 min at 450 *g*.
(viii) Add about 1 ml of fixative (freshly prepared) until a slightly cloudy suspension is obtained.
(ix) Allow clumps of cells to settle to the bottom and then make a trial preparation by allowing one drop to evaporate onto a microscope slide. Examine under the phase microscope and if it is too thin, spin down again and remove some of the fixative. Resuspend.

2.3 Conventional staining

A recommended conventional stain is freshly prepared Giemsa (Gurrs Giemsa 'R 66' 1 ml to 40 ml of buffer pH 6.8 made with Gurr's buffer tablets). Stain for 5 min in a Coplin jar.

Alternatively, in our laboratory we have made good use of carbol fuchsin stain (4).

Solution A:	basic fuchsin	3 g
	70% ethanol	100 ml
Solution B:	solution A	10 ml
	5% phenol	90 ml
Staining solution:	solution B	45 ml
	glacial acetic acid	6 ml
	37% formaldehyde	6 ml

When freshly prepared, a period of 30 min or more may be required to stain the slides, but with maturity, this stain 'ripens' and staining times can be progressively shortened. Carbol fuchsin is especially useful for autoradiography as it will not interfere with emulsion and will not wash out in developer. Cells can therefore be stained prior to filming, for example in *in situ* hybridization experiments.

Cells in spermatogonial mitosis, MI, and MII, from a human testicular biopsy prepared by air-drying and stained with carbon fuchsin, are shown in *Figure 1a − c*. At MI, the XY bivalent is readily recognizable by its chain-like appearance. Autosomal bivalents can be arranged in order of size, but unambiguous identification of individual bivalents is not possible. At MII, the No. 9 chromosome can sometimes be identified by the stretched region on the long arm corresponding to the secondary constriction. An MII cell from the male mouse is shown in *Figure 1d*.

2.4 C-banding

For further and more accurate identification of meiotic bivalents, or in the interpretation of abnormal configurations, C-banding will be found helpful. In the mouse, a C-band is located adjacent to the centromere in all but the Y chromosome (see also Chapter

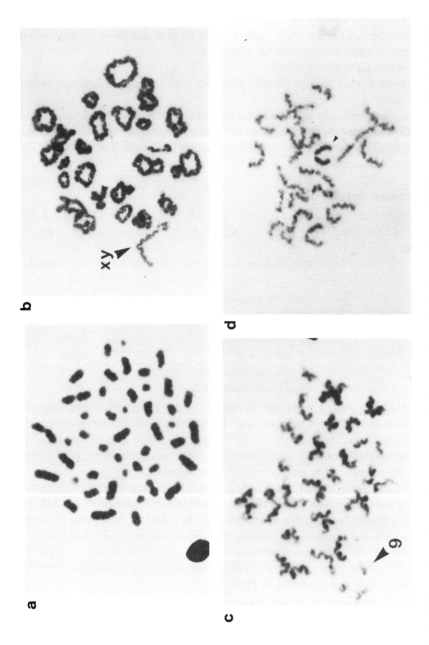

Figure 1. Air-dried metaphases from human and mouse testicular suspensions stained with carbol fuchsin. (**a**) Human spermatogonial mitosis; (**b**) human meiotic MI (the XY bivalent is arrowed); (**c**) human meiotic MII (the No. 9 chromosome with its stretched secondary constriction, is arrowed); (**d**) mouse MII (the X chromosome is arrowed).

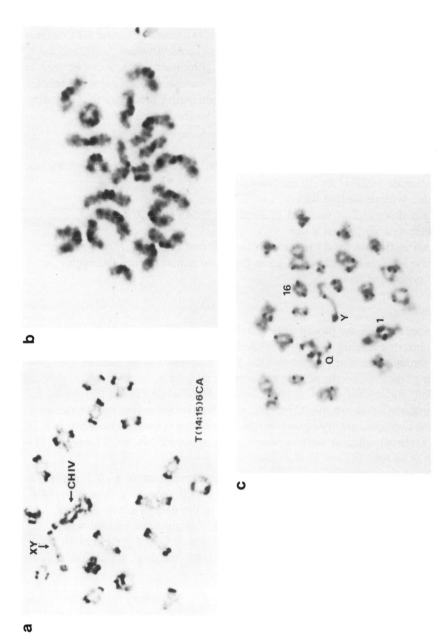

Figure 2. C-banded spermatocytes. (**a**) MI from the mouse translocation T(14;15)6 Ca. Association between the quadrivalent of the reciprocal translocation (CHIV) and XY bivalent is shown. [From (5) reproduced with permission.] (**b**) MII from a normal male mouse. (**c**) MI from a human reciprocal translocation t(9;22) heterozygote. The quadrivalent (Q) shows the prominent C-bands of the No. 9 chromosomes. Bivalents 1 and 16 can also be identified by their large C-bands as well as the Y chromosome.

5, Section 5.2. *Figure 2a* shows a C-banded MI from the mouse translocation T(14;15)6 Ca. The C-band heterochromatin on the X and all autosomes is prominent. In this particular male-sterile strain, associations between the X centromere and translocation quadrivalent are common (5). The Y does not display a prominent C-band. A C-banded MII mouse spermatocyte is shown in *Figure 2b*. *Figure 2c* shows a human MI C-banded cell from a t(9;22) reciprocal translocation heterozygote. Bivalents 1, 16 and the Y are readily identifiable and the large C-bands of chromosome No. 9 can be seen in the quadrivalents (Q). The method used to produce C-bands at meiosis is that of Chandley and Fletcher (6) modified from Sumner (7). (An alternative procedure for C-banding of mitotic chromosomes is described in Chapter 5, Section 5.2.)

(i)   Place the slides in 0.2 M HCl at room temperature for 1 h.
(ii)   Rinse with de-ionized water.
(iii)   Place the slides in 5% barium hydroxide at 50°C for 30 sec (human), 4% barium hydroxide at 37°C for 30 sec (mouse).
(iv)   Rinse with de-ionized water.
(v)   Place slides in 2 × standard saline citrate (SSC) at 60°C for 1 h.
(vi)   Rinse with de-ionized water.
(vii)   Stain in Giemsa for 45 min to 1 h (see Section 2.3).
(viii) Rinse with de-ionized water, leave for a few minutes to dry thoroughly, soak in xylene and mount.

2.5 Q-banding

Fluorescence staining has proved of little value in meiotic analysis of the mouse, but in man quinacrine staining originally revealed that it was the non-fluorescent short arm of the Y chromosome which paired with the X at metaphase I of meiosis (8). The more recent use of the AT-specific peptide antibiotic distamycin A, in combination with the AT-specific fluorescent dye 4′,6′-diamidino-2-phenylindole, DAPI (9), on human meiotic bivalents has shown, like C-banding, that chromosome regions containing constitutive heterochromatin are highlighted. Bright fluorescence is found on pairs No. 1, 9, 16 and the Y chromosome as well as, occasionally, bivalent No. 15 (*Figure 3*). The technique is as follows.

(i)   Flood the slide with distamycin A (Sigma) solution (0.1−0.2 mg/ml in McIlvaine's citric acid−Na_2HPO_4 buffer, pH 7.0), enough to float a coverslip. Incubate at room temperature for 15 min in the dark in a wet chamber.
(ii)   Wash the coverslip off using a wash bottle containing de-ionized water.
(iii)   Apply one large drop of DAPI (Sigma) solution (0.2−0.4 g/ml in McIlvaine's buffer pH 7.0) and lower a fresh coverslip onto the drop. Return the slide to the damp chamber in the dark for 30 min.
(iv)   Wash off again using de-ionized water.
(v)   Mount the slide in two drops of McIlvaine's buffer.
(vi)   Blot gently with filter paper until no more buffer comes out.
(vii)   Seal around the coverslip with rubber solution (e.g. Holdtite).

Distamycin A tends to lack stability in aqueous solution and it is not recommended to store it in solution. Stained preparations may fade rapidly when first examined but usually stabilize after a day or so stored in the dark at 4°C.

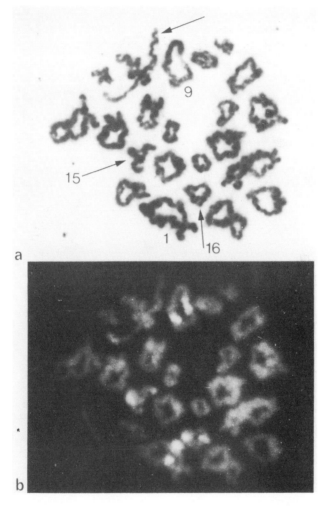

Figure 3. Human MI stained sequentially with (**a**) carbol fuchsin and (**b**) DA/DAPI. Fluorescence enables bivalents 1, 9, 15, 16 and the Y chromosome to be identified. The XY bivalent is arrowed in (**a**).

2.6 Pachytene mapping

Special air-drying techniques using extended room temperature hypotonic treatments, or shorter hypotonic treatments at higher temperatures, have been devised for the production of extended pachytene chromosomes which can be used for mapping of individual bivalents (10–13). At late pachytene, each bivalent exhibits a linear sequence of compacted regions of chromatin or 'chromomeres' which vary in number, size and sequence for each chromosome in the complement. A striking correspondence between chromomeres and mitotic G-bands has been observed, and complete pachytene karyotypes in man (12) and the mouse (13) have been produced. C-banding can aid in the localization of centromere positions (*Figures 4a* and *b*). The technique used in the preparation of these human pachytene cells is that of Luciani *et al.* (12).

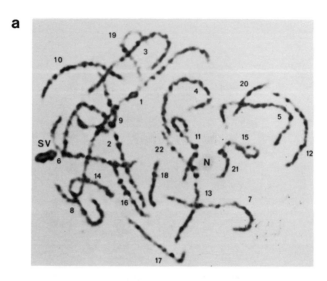

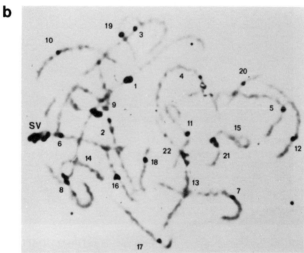

Figure 4. Human pachytene spermatocyte in which each of the 22 autosomal bivalents has been identified on the basis of its chromomere pattern (**a**) and centromere position (**b**). N, nucleolus; SV, sex vesicle. [From (12) reproduced with permission.]

(i)   Immerse testicular fragments in 10 ml of 0.88% KCl and keep at room temperature for 8−10 h.

(ii)   Transfer to fixative (3:1 methanol:glacial acetic acid) and leave overnight at room temperature.

(iii) The next day, shred the fragments in the fixative.

(iv)   Pipette the cell suspension into a conical vial and centrifuge at 150 *g* for 7 min.

(v)   Resuspend the pellet in 5 ml of 45% glacial acetic acid, then immediately centrifuge at 150 *g* for 5 min.

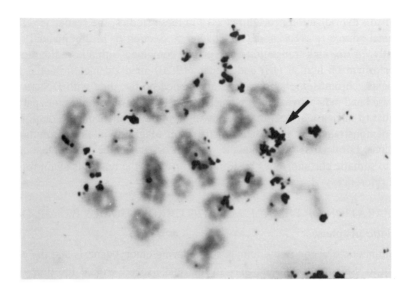

Figure 5. Human MI probed with the No. 9 sequence Xb₁ by *in situ* hybridization. The No. 9 bivalent is arrowed. [From (15) with permission.]

(vi)   Make preparations by dropping suspension from a Pasteur pipette on clean pre-cooled slides and gently dry over a low gas flame.

(vii)   Stain with phosphate-buffered Giemsa solution (pH 6.8) (see Section 2.3).

(viii) To visualize centromeres, remove the Giemsa using methanol, and place the slides in 1 M HCl for 5 min at room temperature. Wash in water and treat for 3 min in 5% barium hydroxide solution at 58°C. After washing, place the slides for 20 min in 2 × SSC at 58°C and adjust the pH to 7.0. Rinse the slides again and stain with Giemsa.

The same cells photographed for chromomere patterns at the end of step (vii) are re-photographed after step (viii).

An alternative method for preparing pachytene karyotypes using a 1 h hypotonic KCl treatment at 37−38°C, has been described for man by Hungerford (10) and for the mouse by Fang and Jagiello (13). Pachytene preparations made in this way allow the visualization of 'parameres' on chromosome No. 9 in man (10,14,15) when Giemsa staining is used. These are highly refractile bodies seen only at the pachytene stage, and which correspond to the No. 9 C-band region seen at metaphase.

2.7 The use of air-dried male meiotic preparations in molecular analyses

Extended pachytene chromosomes prepared by the techniques described in Section 2.6 have been used for the localization of genes by *in situ* hybridization techniques (16,17). Th resolution with pachytene chromosomes is claimed to be better than that obtained with high resolution banded somatic chromosomes (16,17).

In situ localization of chromosome 9 sequences in man have been made using MI and MII air-dried metaphases prepared as in Section 2.2 (15). An MI cell probed with the No. 9 secondary constriction sequence Xb₁ is shown in *Figure 5*. For full details of the *in situ* hybridization procedures used in these studies, the reader is referred to the original source references and to Chapter 5, Section 6, in this manual.

Recently, *in situ* nick translation experiments combined with autoradiography have been performed on human (18) and mouse (19,20) air-dried meiotic cells. In these experiments, chromosome preparations are treated on the slide with DNase I, DNA polymerase I and radioactive deoxyribonucleotides. Regions of chromosomes particularly susceptible to nuclease attack, that is showing a more open chromatin conformation, can be demonstrated by a localized distribution of grains in the autoradiographic preparations. For details of the technique, which is essentially the same as that originally applied to somatic chromosomes by Kerem *et al.* (21), the reader is again referred to original references.

3. PREPARATION OF FEMALE MEIOTIC CHROMOSOMES BY AIR-DRYING

3.1 Meiotic prophase

The first meiotic division in the female starts during embryogenesis and the prophase stages up to diplotene are found only in the fetal ovary. Around the time of birth, oocytes enter the resting diplotene or dictyotene stage, where they remain until shortly before ovulation. Early cytogenetic studies of meiotic prophase in females employing the use of air-drying techniques in the mouse, include those of Röhrborn and Hansmann (22), and in man, those of Luciani and Stahl (23) and Kurilo (24).

Meiotic prophase in the mouse commences on day 13 of gestation with maximum numbers of pachytenes being seen over days 16−18. In the human female, meiosis commences towards the end of the first trimester of fetal development and, with increasing gestational age, more and more cells enter the prophase stages of leptotene, zygotene and pachytene. Ovaries from aborted human fetuses have been shown to have cells in the pachytene stage during weeks 16−23 of gestation (25).

In recent years, significant developments in the preparation of fetal oocytes in man and the mouse by 'microspreading', have taken place, and because of the superior quality of the preparations and the vast improvement in detail which can be discerned in such spreads, all discussion regarding chromosome preparation at meiotic prophase in oogenesis will be deferred until later (Section 4.3).

3.2 MI and MII oocyte preparation by in vitro culture

Oocytes in MI are obtained by *in vitro* culture from the germinal vesicle stage of the arrested dictyate stage, a technique pioneered by Pincus and Enzmann (26) with oocytes of the rabbit. The stimulation to resume meiosis appears to be the liberation of the oocyte from its follicle (27). An early published method for the culture of mouse and human oocytes is that of Edwards (28). Continuation of the culture period will also yield oocytes in MII, but an alternative technique for this stage of oocyte maturation is to collect freshly ovulated oocytes from the fallopian tubes, a procedure described in Section 3.3.

For the preparation of *in vitro* matured mouse eggs the method of Speed (29) has been used successfully in our own laboratory.

(i)   Excise ovaries and place them in physiological saline (0.9%). Using a dissecting microscope for visualization of follicles, puncture large follicles with a sterile scalpel.

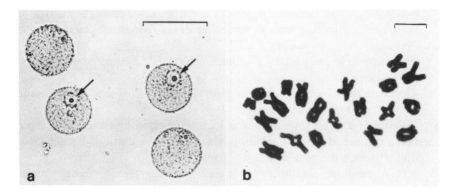

Figure 6. Mouse oocytes (**a**) on release from follicle showing germinal vesicle (arrowed) (**b**) after culture to MI. (Bar in **a** = 100 μm; in **b** = 10 μm.) [From (29) reproduced with permission.]

(ii)   Pool eggs showing a clear germinal vesicle (gv) in culture medium (*Figure 6a*), using a sterile micropipette. The medium used is fetal calf serum (FCS, 9.5 ml) and tryptose phosphate broth (Oxoid, 0.5 ml), containing 0.25 mM sodium pyruvate, 1.0 mM glutamine and 5.56 mM glucose.

(iii) Mechanically remove any follicular cells that remain attached to the eggs by pipetting them in and out of a narrow bore micropipette.

(iv)   Place 20−40 eggs in 1 ml of medium in the central well of a 60 × 15 mm organ tissue culture dish (Falcon 3010), and culture under 1 ml of sterile paraffin oil (BDH lightweight 0.840−0.860 g) at 37°C in a high humidity gassed incubator (5% CO_2 in air). After 5 h, eggs reach MI (*Figure 6b*) and after 17 h, MII.

(v)   Wash the eggs twice and leave for 20 min in fresh 1% sodium citrate at room temperature. Place a microdrop of 1% sodium citrate with an egg on a grease-free slide using a fine pipette. Mark the reverse side with a circle.

(vi)   The fixation procedure is that of Tarkowski (30), also described in Chapter 5, Section 4, of this manual. Expel a few drops of 3:1 fixative (methanol:glacial acetic acid) from another fine pipette whose tip is brought just over the microdrop containing the egg. The optimal number of drops of fixative is three.

(vii) Stain for 5 min with 2% Giemsa or any other conventional stain (see Section 2.3), or C-band as described in Section 2.4.

An alternative method for egg culture to the MI stage and fixation, is that of Kamiguchi *et al.* (31), modified for use in the mouse by Sugawara and Mikamo (32). This technique employs a more gradual fixation procedure and is less likely to produce broken preparations showing artefactual losses of chomosomes.

(i)   Excise ovaries between 12.00 h and 18.00 h on the day of dioestrum, that is the mid-point of the oestrous cycle, as verified by vaginal smearing. Place them in a watch glass containing Hank's solution and collect the eggs from large follicles by pricking with a sharp needle.

(ii)   Remove the adhering cumulus cells mechanically with a fine drawn pipette.

(iii) Culture those eggs having a germinal vesicle in 1 ml of medium for 3.5−6 h at 37°C with 5% CO_2 in air. The culture medium is 85% TC199 (Gibco Biocult) and 15% FCS. About 72% of eggs reach MI in this medium.

(iv)   For MII eggs, a longer culture period is required (see earlier). Alternatively, these can be collected from the ampullae of the oviducts following ovulation (Section 3.3).

(v)   Treat mouse eggs with 0.02% pronase (Sigma) for 15−20 sec to soften the zona pellucida.

(vi)   Treat with two hypotonic solutions, first 40% FCS for 10 min, then 1% sodium citrate for 40 min, both treatments at 37°C.

(vii)   Place the eggs with a small amount of 1% sodium citrate hypotonic solution gently into a mixture of absolute methanol:glacial acetic acid:distilled water (5:1:4). This dissolves the zona and gently fixes the eggs.

         Up to this stage, the dish or watch glass containing the eggs must be replaced for each change of solution. Observation of all steps is carried out under the dissecting microscope.

(viii)   Five minutes after placing eggs in the above solution, suck up a single egg, together with a small drop of fixative, into a Spemann pipette with a fine point, then expel within a circle (marked on the reverse side) on a grease-free slide. While the fixative spreads, the egg sticks to the slide near the tip of the pipette.

(ix)   Re-fix the egg by an immediate application of Carnoy's solution.

(x)   Put the slide at once into a Coplin jar filled with the same fixative and leave for at least 20 min.

(xi)   Dip the slide into a mixture of absolute methanol:glacial acetic acid: distilled water (3:3:1) for 1 min to help the egg flatten during the drying process.

(xii)   Remove the slide very slowly and air-dry using air which is moisturized by passing through water at 37°C.

(xiii)   Stain routinely with Giemsa (Section 2.3), or by C-banding (Section 2.4).

Slides prepared with this method appear to be more reliable than those made using any previous method, including Tarkowski's (30), in respect of the spreading and maintenance of chromosomes (32). Its use for cultured human MI oocytes has been described by Wramsby and Liedholm (33).

3.3 MII oocyte preparation following ovulation

Stimulation of meiosis by hormonal injection into the female (Chapter 1, Section 5.4) can be performed to collect eggs in MI and MII. Oocytes in MI must be collected from ovarian follicles after the required time interval following injection of follicle-stimulating hormone [pregnant mare's serum gonadotropin (PMSG)] and human chorionic gonadotrophin (hCG). Oocytes in MII can be collected from the ampullae of the fallopian tubes following ovulation. In the mouse, a suitable method has been described by Röhrborn and Hansmann (22).

(i)   Inject 2.5 IU of PMSG and 2.0 IU of hCG 48 h later, intraperitoneally.

(ii)   Collect MI oocytes from ripe follicles 5 h after hCG, as described in Section 3.2.

(iii)   Collect ovulated MII eggs 15 h after hCG from the ampullae of the fallopian tubes (see Chapter 2, Section 2.1). Use pre-warmed medium at 37°C to collect the eggs.

(iv)   Dissolve away the cumulus cells surrounding MII eggs in a solution of 100 IU/ml hyaluronidase (Type III Sigma) for 2−5 min (see Chapter 2, Section 2.1).

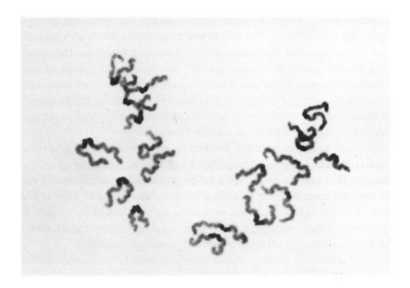

Figure 7. C-banded MII of the female mouse.

Proceed as in Section 3.2 from step (v) onwards (methods of Speed, ref. 29), for chromosome preparation and staining.

A C-banded MII egg from the mouse is shown in *Figure 7.* Application of C-banding can aid greatly in the accurate counting of MII chromosomes, for example when aneuploidy studies are made.

In man, a superovulation procedure for the collection of eggs in MI and MII has been described by Jagiello *et al.* (34) and, more recently, the collection of eggs for human *in vitro* fertilization following hCG administration has been used by Angell *et al.* (35).

It should be pointed out, however, that high quality preparations of MI human eggs in which reliable chiasma counts could be made, have never really been obtained despite numerous attempts on the part of many workers over a number of years. Little problem has ever been encountered with other mammalian species and why human eggs should give problems, with both the *in vitro* and *in vivo* methods, is not at all clear.

4. MICROSPREADING FOR PROPHASE ANALYSIS

4.1 Surface spreading for electron microscope analysis of spermatocytes

The introduction of a simple and rapid surface microspreading technique for the visualization of the synaptonemal complex at meiotic prophase in insect spermatocytes (36) has revolutionized our understanding of the behaviour of chromosomes during this important stage of meiosis. Application of the technique to mammalian spermatocytes, pioneered by Moses (37−39), has demonstrated its potential for the analysis of the behaviour of chromosome abnormalities at meiotic prophase as well as the analysis of normal meiotic development. A large body of 'spreading' data is now accumulating in the cytogenetic literature, spanning all eukaryote species from plants to man.

Spread preparations of synaptonemal complexes (SC) are obtained through surface tension on an aqueous solution. The SC, to which chromatin fibrils are attached, appears prior to or at synapsis, and disappears at or following desynapsis. The appearance of the pairing events in the SC reflect exactly the pairing behaviour of the meiotic chromosomes throughout prophase. The spreading technique is not applicable to cells once they reach the metaphase stage. A wide variety of methods for preparing spreads has been described and it would not be possible here to give precise details of them all. Nevertheless, some generalizations can be made.

Basically, there are two principle techniques and two levels of observation, light microscope (LM) and electron microscope (EM). One technique involves the picking up of spread cells from the surface of a saline or sucrose solution onto a slide and is generally used for spermatocyte spreading when an abundance of cells is available in suspension. The second involves the settling of cells onto a slide surface through a sucrose, saline or other solution, and has been used principally for oocytes, but may be useful for spermatocytes if only a thin suspension is available.

A method of surface spreading of spermatocytes originally devised by Moses (37), based on Counce and Meyer (36), is as follows.

(i)   Chop the testis or testicular biopsy in Eagles Minimum Essential Medium (Gibco Biocult, MEM without glutamine), gently macerating the tubules with curved blunt forceps. Take up the cell suspension in a 1 ml syringe without a needle. Add a drop or two of MEM to the syringe, making the volume up to 1 ml. Stand the syringe upright for 1 min to allow larger pieces of debris to settle into the tip. Then expel and discard these. Expel the remaining clear suspension and draw it back into the syringe several times to break up clumps of cells.

(ii)   Again make up the volume to 1 ml with MEM, expel into a conical glass centrifuge tube, and centrifuge for 5 min at about 150 g at room temperature.

(iii) Remove the supernatant, except for about twice the volume of the pellet. Resuspend the pellet. Keep the suspension on ice.

(iv)   With a micropipette, carefully expel a drop of suspension and gently touch to the surface of the spreading solution. The latter consists of a filtered 0.5% solution of NaCl filling a black embryological dissecting dish. Prior to adding the cell suspension, sweep the surface clean with either lens paper or a hard surface toilet paper, and adjust the liquid level so that it is slightly convex. Allow the spread to stabilize for 1 or 2 min.

(v)   Prepare copper girds (Pelco:GC-100) as follows. Film the grids with 0.3% formvar, and then coat with carbon. Before use, make the grids hydrophilic by ionizing in a slow discharge (2 min, 1.3 kV, 0.1 Torr). Keep the forceps used for handling grids scrupulously clean and wipe them after each step by punching the tips through No. 1 Whatman filter paper, or bibulous paper.

(vi)   Touch the grids one at a time, to different regions of the surface of the spread. Use approximately 10 grids.

(vii) Fixation. The fixative is 4% paraformaldehyde in 0.1 M sucrose; heat to $60-80°C$ with six drops of NaOH and stir till clear. Finally adjust the pH to 8.5 with pH 9 borate buffer. Filter and store at 4°C for up to 1 week.

(viii) Quickly float the grids on the clean surface of the fixative in a small dish and leave there, covered, for at least 5, and no more than 10 min, five grids per dish. Alternatively, dispense 10 double drops of fixative on a wax plate (dental wax, or a sheet of parafilm stuck down on a glass surface), and float each grid on the surface of a drop (1 grid per drop − use each drop once only). Small disposable beakers, inverted, are used as covers during fixing.

(ix) After fixation, remove each grid with clamping, anti-capillary forceps, drain excess fixative by touching the edge of the grid to the surface of the fixative, and hold the grid face down on the surface of 0.4% Photoflo (Eastman Kodak, freshly prepared in borate buffer, pH 8 and filtered before use). Hold for 20−30 sec, moving gently to agitate. If floated, grids tend to sink. Drain the grid by touching the edge to filter paper, allow to dry partially under an incandescent lamp, and then place on filter paper in a Petri dish to complete drying. If Photoflo creeps onto the reverse side of the grid, it may leave drying artefacts. Grids may be stained at any point after drying (some grids have been stained a year or more after preparation).

(x) Staining. The stain is a 1:3 dilution in 95% ethanol of a 4% solution of phosphotungstic acid (PTA) freshly made up and filtered just before use.
Using the anti-capillary forceps, hold the dried grid face-down on the surface of the ethanolic PTA for 1 min with gentle agitation, drain on the surface, transfer to the surface of 95% ethanol to rinse for 15−30 sec, drain on filter paper and allow to air-dry.

N.B. Care must be taken not to allow either stain or ethanol to flow onto the non-specimen-bearing surface, as this produces surface tensions during drying that may rupture the carbon−formvar film.

(xi) Before or after staining, align the grids face-up on a microscope slide, cover with a No. 1 coverslip, and examine with dry $20\times$ or $40\times$ phase objectives. Suitably spread spermatocytes can be recognized, and SCs are clearly visible. Entire SCs, attachment plaques and kinetochore regions are resolvable. Desirable grids can thus be selected quickly.

(xii) An EM operating at low magnification, is best suited for karyotype study $(250-3000\times)$.

The above method is specifically designed for EM analysis of SC preparations. Nevertheless, because SCs and other structures such as nucleolar organizing regions (NORs) and attachment plaques are visible at the light microscope level when silver staining is used, a number of authors have devised methods for LM analysis (40), or for sequential LM and EM investigation (41). LM observation can be carried out rapidly and can be used to scan slides and select good cells for further analysis in the EM. One such method is described in the following section.

4.2 Spreading for light microscope and electron microscope sequential analysis

The procedure currently used in our own laboratory is an extension of the LM method of Fletcher (40).

4.2.1 *Preparation of cell suspension*

(i) *For the mouse.* Remove the testis from tunica albuginea and place in medium F10 supplemented with 20% FCS (Gibco Biocult). Chop with fine scissors in a tilted glass Petri dish. Allow tubule fragments to settle for about 20 min. Draw off the suspension with a Pasteur pipette leaving debris behind, and transfer to a conical centrifuge tube. Centrifuge at 150 *g* for 10 min. Discard most of the supernatant, leaving behind just enough to resuspend the cells.

(ii) *For human spermatocyte preparation.* Collect testicular biopsies in Dulbecco's phosphate-buffered saline (PBS, Dulbecco A tablets from Oxoid), and chop immediately with fine scissors. Agitate the cells and tubule fragments using a stirring bar and magnetic stirrer for about 1 h. Draw off the cell suspension and centrifuge at 150 *g* for 10 min in a conical centrifuge tube. Discard the supernatant and resuspend the cells in fresh PBS. Wash the cells three times and leave in a small volume of PBS.

4.2.2 *Spreading the cells*

(i)   Prior to making spreads, coat pre-cleaned alcohol-washed slides in an 0.5% solution of Optilux (Falcon) plastic in chloroform (42). Use a Coplin jar for dipping, and then stand the slides on end in a rack to dry in a dust-free atmosphere. When dry, seal the plastic coating to the edges of the slides with Holdtite rubber adhesive.

(ii)   Spray 50 mm watch glasses with black 'DEXION' paint to provide a water-repellant surface. Any paint which will give a dark background and a water-repellant surface can be used.

(iii) Fill the black watch glass with the spreading solution (0.2 M sucrose, made fresh and filtered); the sucrose should sit in a nice 'bubble', and the level of the solution should be higher than the edge of the watch glass (6−7 ml).

(iv)   Draw up a small amount of the cell suspension into a long form Pasteur pipette and allow one drop (~ 0.02 ml) to hang from its tip. Gently touch this drop onto the surface of the spreading solution. The process can be watched under a low power dissecting microscope, but this is not absolutely necessary as the cell spreading can be readily observed by eye.

(v)   Allow the cells to spread over the surface for about 30 sec then pick them up by touching the surface of the solution with the plastic-coated slide.

(vi)   Leave the slide lying flat on the bench for about 10 min to allow more cells to settle, then fix the cells in paraformaldehyde.

(vii) After 5−10 min, drain off all excess fixative and wash the slide for 30 sec in 0.4% Photoflo from a wash bottle.

(viii) Allow the slide to dry on the bench at room temperature.

4.2.3 *Staining*

Staining can be carried out most conveniently by the rapid colloidal silver method of Howell and Black (43), a technique originally developed for staining of NORs. The method requires the use of two solutions.

(i)   Prepare a colloidal developer solution by dissolving 2 g of powdered gelatin in

100 ml of de-ionized water and 1 ml of pure formic acid. Stir constantly for 10 min in order to dissolve the gelatin. The solution is stable for 2 weeks.

(ii)   Prepare an aqueous silver nitrate solution by dissolving 4 g of $AgNO_3$ in 8 ml of de-ionized water. This solution is stable. Store both the colloidal developer and silver solution in capped, amber-glass bottles.

(iii)   Pipette two drops of the colloidal developer and four drops of the aqueous silver nitrate onto the surface of the microscope slide containing spread cells. Mix the solutions and cover with a coverslip.

(iv)   Place the slide on the surface of a slide warmer (hot plate) which has been stabilized at 70°C.

(v)   Within 30 sec, the silver-staining mixture will turn yellow, and within 2 min, it will become golden brown. Remove the slide and the coverslip and rinse off the staining mixture using de-ionized water.

(vi)   Blot the slide dry and examine immediately.

4.2.4 *Preparation of grids for electron microscopy*

(i)   Locate good silver-stained spreads under low power (×25) with the LM. Using a Leitz diamond slide marker, score a circle in the plastic coating around the cells required for EM analysis.

(ii)   When all desired cells on a slide are located, float off the round discs of plastic onto the surface of distilled water contained in a large square glass staining dish. This is achieved by gently lowering the slide at an angle, into the water, when the surface tension will pull the discs of plastic away from the slide.

(iii)   Pick up discs one by one as follows. Using watchmakers forceps carefully lower an EM grid [G200 HS Cu (Gilder)], to a position under the disc. Gently hold the disc in place on the surface of a grid by means of a strong eyelash mounted to the end of a thin wooden (orange) stick, and pick up the disc by bringing the grid up under the disc and out of the water. Allow the grids to dry before EM examination.

Spread diplotene spermatocytes prepared by this method from the mouse (*Figure 8a*) and human (*Figure 8b*) are shown. In *Figure 8a*, the preparation is photographed in the light microscope, in *Figure 8b*, at the EM level. It should be noted that visualization of centromeres and the central element of the SC is not usually possible in silver-stained preparations. For these, staining should be carried out in PTA (Section 4.1).

The application of this spreading method in our laboratory to the spermatocytes of a translocation heterozygote in man (44), to several inversion heterozygtes in the mouse (45), and in the study of X-Y pairing in the human male (46), can be found in the literature in addition to many other published accounts where spreading techniques have been employed in male meiotic prophase analysis. The technique has indeed opened up a whole new area of meiotic investigation in the male.

4.3 Spreading of oocytes

The material (mouse or human ovaries) must be obtained as fresh as possible. After 24 h the morphology of the pachytene cells is greatly reduced. In the human female

a

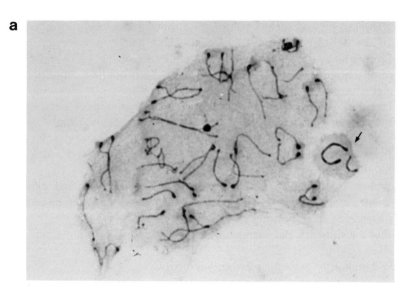

b

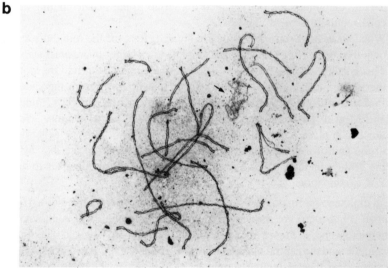

Figure 8. Diplotene spermatocytes prepared by surface spreading and silver staining. The XY axes are arrowed. (**a**) mouse (LM), (**b**) human (EM). In (**a**), the X and Y are unattached.

fetal ovary the maximum number of pachytenes occurs at about 26 weeks, although they appear as early as 14 weeks and continue to about 30 weeks of gestation. Maximum numbers of pachytenes occur on about day 18 post-coitum in mouse, although this will vary with the strain.

A simple spreading method for oocytes, pioneered by Speed (47), is as follows.

(i)   Place the ovary in Dulbecco's PBS and, in the case of human, cut into small pieces about 2 mm³ using a scalpel (for mouse one whole ovary is used to prepare one slide).

a

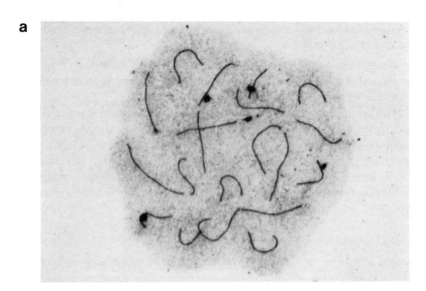

b

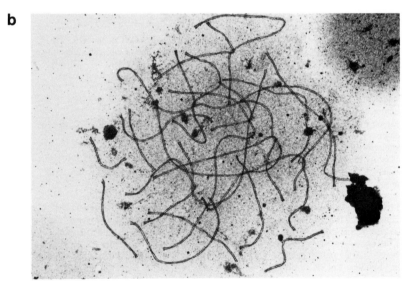

Figure 9. Pachytene oocytes prepared by spreading and silver staining. (**a**) mouse (LM), (**b**) human (EM).

(ii)   Place one piece of ovarian tissue (or whole ovary in the case of mouse) onto a clean slide in 2−3 drops of 0.2 M (4.5%) sucrose (made up in distilled water).

(iii) Tease the material apart using the blunt edge of the scalpel and a dissecting needle. Remove large debris and disperse the cells in the sucrose by gently stirring with the needle. Oocytes will sink through the sucrose and adhere to the slide.

(iv)   If cells are required for examination at LM level only, leave the slide to dry for a minimum of 30 min and even overnight. If analysis at the EM is necessary transfer the cells in sucrose to a plastic-coated slide (see Section 4.2.2) and carefully spread over an area 1−1.5 cm² without touching the coating.

(v) Fixation and staining and the preparation of EM grids is as described in Sections 4.2.2, 4.2.3 and 4.2.4.

Spread pachytene oocytes prepared by this method from the mouse (*Figure 9a*) and human (*Figure 9b*) are shown. The mouse pachytene spread is photographed at LM level: the human spread is taken in the EM.

The application of the spreading technique to oocytes has been of enormous benefit, just as in spermatocytes, in revealing intimate detail of pairing and other configurations at meiotic prophase. Detail of normal oocytes, and oocytes showing a variety of anomalies, are given in papers for mouse by Speed (47) and Speed and Chandley (48) and for the human by Speed (25). In the study of human trisomy 21 oocytes (49) and mouse XO oocytes (50), the technique has proved invaluable in revealing the synaptic configurations present in these abnormal situations.

5. ACKNOWLEDGEMENTS

The author is grateful to R.M.Speed for advice in the preparation of Section 3 and for providing *Figures 6* and *9*. J.D.Brook is thanked for *Figures 1d* and *7*.

6. REFERENCES

1. Beatty,R.A., Lim,M.-C. and Coulter,V.J. (1975) *Cytogenet. Cell Genet.*, **15**, 256.
2. Laurie,D.A., Firkett,C.L. and Hulten,M.A. (1985) *Ann. Hum. Gent.*, **49**, 23.
3. Evans,E.P., Breckon,G. and Ford,C.E. (1964) *Cytogenetics*, **3**, 289.
4. Carr,D.H. and Walker,J.E. (1961) *Stain Technol.*, **36**, 233.
5. Forejt,J. and Gregorová,S. (1977) *Cytogenet. Cell Genet.*, **19**, 159.
6. Chandley,A.C. and Fletcher,J. (1973) *Humangenetik*, **18**, 247.
7. Sumner,A.T. (1972) *Exp. Cell Res.*, **75**, 304.
8. Pearson,P.L. and Bobrow,M. (1970) *Nature*, **226**, 959.
9. Schweizer,D., Ambros,P. and Andrle,M. (1978) *Exp. Cell Res.*, **111**, 327.
10. Hungerford,D.A. (1971) *Cytogenetics*, **10**, 23.
11. Luciani,J.M., Morazzani,M.R. and Stahl,A. (1975) *Chromosoma*, **52**, 275.
12. Luciani,J.M., Guichaoua,M.R. and Morazzani,M.R. (1984) *Hum. Genet.*, **66**, 267.
13. Fang,J.-S. and Jagiello,G. (1981) *Chromosoma*, **82**, 437.
14. Page,B.M. (1973) *Cytogenet. Cell Genet.*, **12**, 254.
15. Mitchell,A.R., Ambros,P., McBeath,S. and Chandley,A.C. (1986) *Cytogenet. Cell. Genet.*, **41**, 89.
16. Jhanwar,S.C., Neal,B.G., Hayward,W.S. and Chaganti,R.S.K. (1984) *Cytogenet. Cell Genet.*, **38**, 73.
17. Chaganti,R.S.K., Jhanwar,S.C., Antonarakis,S.E. and Hayward,W.S. (1985) *Somat. Cell Mol. Genet.*, **11**, 197.
18. Chandley,A.C. and McBeath,S. (1986) In *Chromosomes Today*. Stahl,A. (ed.), Geo. Allen and Unwin, London, Vol. **9**, in press.
19. Sperling,K. and Marcus,M. (1984) In *Chromosomes Today*. Bennet,M.D., Gropp,A. and Wolf,U. (eds), Geo. Allen and Unwin, London, Vol., **8**, p. 169.
20. Richler,C., Teitelboim,E. and Wahrman,J. (1986) In *Chromosomes Today*. Stahl,A. (ed.), Geo Allen and Unwin, London, Vol. 9, in press.
21. Kerem,B., Goitein,R., Richler,C., Marcus,M. and Cedar,H. (1983) *Nature*, **304**, 88.
22. Röhrborn,G. and Hansmann,I. (1971) *Humangenetik*, **13**, 184.
23. Luciani,J.M. and Stahl,A. (1971) *Bull. Assoc. Anat. (Nancy)*, **15**, 445.
24. Kurilo,L.F. (1981) *Hum. Genet.*, **57**, 86.
25. Speed,R.M. (1985) *Hum. Genet.*, **69**, 69.
26. Pincus,G. and Enzmann,E.V. (1935) *Anat. Rec.*, **75**, 537.
27. Edwards,R.G. (1962) *Nature*, **196**, 446.
28. Edwards,R.G. (1965) *Lancet*, **i**, 926.
29. Speed,R.M. (1977) *Chromosoma*, **64**, 241.
30. Tarkowski,A.K. (1966) *Cytogenetics*, **5**, 394.
31. Kamiguchi,Y., Funaki,K. and Mikamo,K. (1976) *Proc. Jap. Acad.*, **52**, 316.

32. Sugawara,S. and Mikamo,K. (1986) *Chromosoma*, **93**, 321.
33. Wramsby,H. and Liedholm,P. (1984) *Fertil. Steril.*, **41**, 736.
34. Jagiello,G., Karnicki,J. and Ryan,R.J. (1968) *Lancet*, **i**, 178.
35. Angell,R.R., Templeton,A.A. and Aitken,R.J. (1986) *Hum. Genet.*, **72**, 333.
36. Counce,S.J. and Meyer,G.F. (1973) *Chromosoma*, **44**, 231.
37. Moses,M.J. (1977) *Chromosoma*, **60**, 99.
38. Moses,M.J. (1977) In *Chromosomes Today*. de la Chapelle,A. and Sorsa,M. (eds), Elsevier/North Holland, Amsterdam, Vol. 6, p. 71.
39. Moses,M.J. (1980) In *Animal Models in Human Reproduction*. Serio,M. and Martini,L. (eds), Raven Press, New York, p. 169.
40. Fletcher,J.M. (1979) *Chromosoma*, **72**, 241.
41. Dresser,M.E. and Moses,M.J. (1979) *Exp. Cell Res.*, **121**, 416.
42. Felluga,B. and Martinucci,G.B. (1976) *J. Submicrosc. Cytol.*, **8**, 347.
43. Howell,W.M. and Black,D.A. (1980) *Experientia*, **36**, 1014.
44. Chandley,A.C., Speed,R.M., McBeath,S. and Hargreave,T.B. (1986) *Cytogenet. Cell Genet.*, **41**, 145.
45. Chandley,A.C. (1982) *Chromosoma*, **85**, 127.
46. Chandley,A.C., Goetz,P., Hargreave,T.B., Joseph,A.M. and Speed,R.M. (1984) *Cytogenet. Cell Genet.*, **38**, 241.
47. Speed,R.M. (1982) *Chromosoma*, **85**, 427.
48. Speed,R.M. and Chandley,A.C. (1983) *Chromosoma*, **88**, 184.
49. Speed,R.M. (1984) *Hum. Genet.*, **66**, 176.
50. Speed,R.M. (1986) *Chromosoma*, **94**, 115

Karyotyping and sexing of gametes, embryos and fetuses and *in situ* hybridization to chromosomes

E.P.EVANS

1. INTRODUCTION

The importance of cytogenetic techniques to studies of mammalian development is well established, for example, in the identification of chromosomally unbalanced zygotes, sexing of embryos, recognition of the inactive X chromosome, analysis of chimaerism, mapping of native or introduced genes by *in situ* hybridization and the study of chromatin structure. In a field where biological material is limited, experimental cytogenetic techniques often require only a few cells.

Mammalian embryogenesis provides an abundant source of mitotic divisions. However, the yield of metaphase plates of sufficient quality for cytogenetic study is determined by the accessibility and response of these divisions to established chromosome preparatory techniques. The purpose of this chapter is to describe some of the available techniques and the ways of utilizing the preparations obtained by them in a variety of applications which range from the simple determination of chromosome counts to the *in situ* hybridization of gene probes to banded chromosomes. Most of the experience described has been gained from the mouse but many of the techniques and applications are equally suitable for other mammals, including man.

2. BASIC TECHNIQUES

The one requirement for the cytogenetic study of mitotic cells is to produce a well spread, flattened metaphase plate as free from overlapping chromosomes as possible. In attempting to achieve this goal, most of the techniques currently used involve the same four essential steps which are:

(i)   initial recovery of the mitotic cells which may be aided by earlier exposure to a mitotic arrestant,

(ii)   treatment with a hypotonic solution so that they swell,

(iii)   fixation to preserve chromosome morphology, and

(iv)   spreading and flattening onto glass slides.

The methods of applying these four steps will vary according to the stage of embryogenesis being examined and they will be described in turn. Once preparations are obtained they can be treated to serve a number of purposes and some of these will be described.

3. EQUIPMENT

The equipment required for the recovery of whole and parts of pre- and post-implantation embryos is described in Chapters 2 and 3 and can be used for recovering suitable specimens to make chromosome preparations; however, for this purpose, requirements can often be less exacting. Any reasonable dissecting microscope with incident and transmitted light, a good working distance and an optical magnification of up to ×50 will suffice. Moreover, specimens can be handled with sufficiently fine hand drawn micropipettes and, providing the specimens are handled quickly, they can be left in unsophisticated media without serum or gassing. If possible, siliconizing of equipment should be avoided since any carry over of silicon can severely inhibit the spreading of mitotic cells.

A centrifuge with slow acceleration/deceleration and providing 200 g is desirable for handling cells made into suspensions. Since slides are treated at a number of different temperatures in the protocols, it is convenient if a sufficient number of water baths and ovens can be maintained at these temperatures without the need for continuous readjustment. Examination of chromosomes, particularly if they are banded and carry silver grains, is optically demanding and the best possible microscope should be obtained. Ideally, this should have a phase-contrast facility for checking unstained slides, good quality ×16 and ×100 objectives and a built-in 35 mm camera. It should always be set up to give maximum resolution. Filters are useful accessories for viewing and photographing stained chromosomes; interference filters such as the Balzer No.8 both sharpen the image and enhance the contrast. A number of 35 mm films are available and are used to photograph chromosomes. Each has its attributes but for all round performance Kodak technical Pan film 2415 is recommended, with subsequent development in Kodak HC110 developer for about the suggested time at the manufacturer's dilution D.

All slides used for making chromosome preparations should be clean. Most commercially available pre-cleaned slides are contaminated, as is evident from the interference rings which are visible between the slides when they are removed from the box. Pre-cleaned slides are best further cleaned by leaving in a 1:1 mixture of absolute alcohol:concentrated HCl overnight, washed thoroughly the next day in running tap water, rinsed in de-ionized water and stored in a 1:1 mixture of absolute alcohol:diethyl ether. Immediately prior to using, take out the slides with forceps and, avoiding direct finger contact, wipe dry on coarse, grease-free tissues such as Kimwipes.

4. CHROMOSOME PREPARATIONS FROM EMBRYOS AND FETUSES

4.1 Preparations from pronuclear stages

The mitotic divisions recovered from the pronuclear stages are unique because they reveal gametic chromosome complements; thus, in a fertilized egg, male and female contributions can be distinguished, and in unfertilized eggs, the female chromosomes can be revealed by parthenogenetic activation with a variety of treatments (ref. 1, and see Chapter 12). Studies of pronuclear division in fertilized eggs are useful for assessing the primary level of chromosome abnormalities, some of which may act as zygotic lethals or become disguised in subsequent cleavages. The single pronuclear haploid eggs produced by activation serve as a starting point for parthenogenetic studies in mammals (2).

Due to the asynchrony of fertilization and the time of appearance of pronuclear chromosomes, natural ovulations and matings are unrewarding and various other procedures have had to be introduced to improve recovery. Superovulation (Chapter 1, Section 5.4) increases both the number of eggs recovered and the synchrony, and, if combined with the use of a mitotic arrestant to accumulate eggs in metaphase, the success rate is greatly increased. However, over-exposure to the arrestant is deterimental in that male and female complements lose their morphological identity due to chromosome contraction. Initially, the male can be distinguished from the female complement by the more elongated chromosomes and staining differences (3) but, as development progresses, the two become more equivalent and less easy to separate. If it is required

Table 1. Chromosome preparations from the pronuclear stage.

1.	Superovulate the female mice with an intraperitoneal (i.p.) injection of approximately 5 IU of pregnant mare's serum followed 40−48 h later by injection of approximately 5 IU of human chorionic gonadotropin (hCG) (see Chapter 1, Section 5.4 for details). After hCG, mate the females overnight.
2.	Kill the females in the early afternoon of the next day and release the eggs from the ampulla of the oviduct (Chapter 2, Section 2.1) into culture medium (M2, Chapter 2, Section 5.2).
3.	The pronuclear mitosis should occur in the course of the next night between 28 and 32 h after injecting hCG. Normally a mitotic arrestant is added to the medium a few hours before nuclear division but in this case, unless there is a particular need to recover uncondensed chromosomes, it is more feasible to add the arrestant (0.1 μg/ml Sigma Colcemid or Lilly Velbe) at the time of setting up the culture or later and to leave the eggs to be exposed overnight.
4.	Next day, remove the cumulus cells with hyaluronidase (Chapter 2, Section 2.1) and wash the eggs in fresh medium.
5.	Handle the eggs in batches of three until experience has been gained. Place the first three into a deep well slide containing 1 ml of hypotonic aqueous solution of 1% tri-sodium citrate or 0.56% potassium chloride. If the former is used, leave for 7−10 min at room temperature; if the latter is used, 6−8 min should suffice.
6.	Pick up the eggs in a mouth-controlled micropipette with a bore slightly larger than the diameter of an egg after hypotonic treatment (it should be noted that the eggs swell in hypotonic solution and may stick in a pipette which accommodated them before treatment) and place, with minimum carry over of residual hypotonic solution, in the middle of a clean microscope slide. Tarkowski (5), the originator of the method, suggests a drop of residual hypotonic solution not greater than 2 mm in diameter. An excess causes turbulence with breakage and loss of chromosomes, an insufficient amount results in unspread chromosomes.
7.	With as much accuracy as possible, drop onto the eggs one drop (~0.02 ml) of a freshly prepared fixative mixture of 3:1 methanol:glacial acetic acid so that the drop engulfs the residuum in a concentric manner. Do not leave more than 1−2 sec between placing the eggs on the slide and fixing, otherwise the eggs begin to flatten prematurely and do not spread. Expel drops of fixative from a height of at least 2 cm, since there is a possibility that the eggs will be prematurely fixed by vapours and will fail to spread.
8.	Leave the fixative drop to spread to its maximum periphery and aid drying by blowing with the mouth and holding the slide up to the heat from a 60 W bench-lamp bulb.
9.	Proceed with further egg batches. Batches may be increased to more than three but avoid eggs spreading in close proximity when their contents may become mixed.
10.	The slides obtained can be conventionally stained (Section 5.1), C-banded (Section 5.2) or G-banded (Section 5.3).

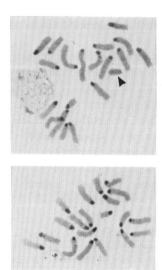

Figure 1. Pronuclear chromosomes with the more elongated male complement uppermost next to the polar body. The cell is C-banded to identify the Y chromosome (arrowhead) as a further aid to sexing. Because of the wide spreading, the two components were photographed separately.

to positively identify the components, it is recommended that one of the parents is made homozygous for a marker chromosome such as T6 (Section 5.1).

A number of methods have been described (3,1,4) to obtain pronuclear chromosomes and they include the use of superovulation with fertilization and mitotic arrest *in vivo* and also activation or fertilization and mitotic arrest *in vitro*. Many of the methods involve both similar and different steps but most, if not all, produce the final preparation by the Tarkowski method (5) of fixing and spreading. In *Table 1*, an *in vivo/in vitro* method is described, although it should be noted that there are a number of alternative methods available. A preparation obtained by the method described in *Table 1* is illustrated in *Figure 1*.

4.2 Preparations from pre-implantation stages

Chromosome preparations can be obtained from all the recoverable stages of pre-implantation embryos described in Chapter 2. Since, after normal ovulation, between 6 and 12 eggs are recovered, success is measured by the number which are undergoing mitotic division and the number of divisions which can be analysed. The numbers ovulated can be considerably increased by superovulation (Chapter 1, Section 5.4). Embryos from both induced and normal ovulations show asynchrony in development from an early stage of cleavage and preparatory techniques have to be devised to attempt to recover as many as possible with analysable chromosomes. Injection of the mother with a mitotic arrestant many hours before embryo recovery will increase the proportion containing mitotic divisions but long exposure also leads to excessive chromosome contraction which makes analysis difficult. A balance can be achieved by exposing the eggs for a moderate time; for example, a 4 h *in vivo* exposure accumulates analysable

Table 2. Preparations from pre-implantation stages.

---

1. Obtain time mated females (Chapter 1, Section 5.2) for the required stage of pre-implantation and inject i.p. with 0.25 ml of a 0.04% aqueous solution of Colcemid or Velbe. Leave for 4 h and recover the embryos as described in Chapter 2, Section 2. Alternatively, culture embryos with the mitotic arrestant added to the medium (0.1 μg/ml).
2. Transfer the eggs into a hypotonic solution of either aqueous 1% tri-sodium citrate or 0.56% potassium chloride. Since the latter is the more efficient hypotonic agent and may quickly cause metaphase breakage in sensitive material such as early eggs, it is suggested that, until experience is gained, it may be more feasible to use the former solution.
3. Vary the exposure time to hypotonic according to the stage of development; two-cell eggs are extremely sensitive and 2 min in tri-sodium citrate may suffice, late blastocysts are resistant and may require up to 15 min. Time can be saved by using several well slides of hypotonic and starting the treatment of single embryos at 1 min intervals. Where there is asynchrony in developmental stage in a batch of embryos, start with the earliest so that subsequent embryos receive increasingly longer times in hypotonic.
4. Make slides in the same way as for the pronuclear stage (*Table 1*), handling embryos one at a time to avoid overlap of spreads. One drop of fixative is usually sufficient to fix and spread early morulae and blastocysts, later blastocysts may, however, require several drops to spread adequately.
5. The slides can be conventionally stained (Section 5.1), C-banded (Section 5.2) or G-banded (Section 5.3).

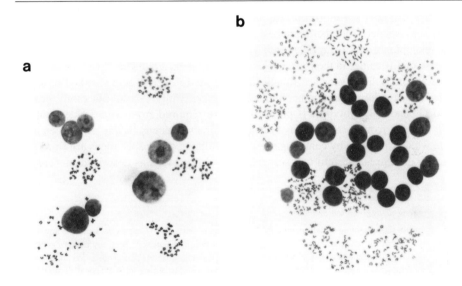

Figure 2. (a) Morula with five mitotic cells. Since the father was heterozygous for seven different Robertsonian translocations, the cells contain metacentric chromosomes. **(b)** Early, triploid *in vitro* with nine mitotic cells which also contain metacentric chromosomes.

metaphases in 50−80% of embryos at the early morula to early blastocyst stage. A thorough knowledge of the cleavage rate and expected fluctuations in the mouse strain being used provides a useful guide to the timing of exposure to the mitotic arrestant.

Pre-implantation development can proceed *in vitro* in culture (Chapter 2) and the arrestant added to the medium at the stage of development required. Preparations are then made in the same way as for the embryos recovered directly from the mother.

It should be noted that chromosomally abnormal eggs frequently lag in development as compared with their normal counterparts (6) and if they become amitotic, it is possible they may either be misrepresented or excluded from a scored sample. If samples are scored for aneuploidy, a significant chromosome count identifying aneuploidy is regarded as consisting of minimally two mitotic cells with the same count.

The main method which has been used for obtaining chromosome preparations from pre-implantation embryos is again that of Tarkowski (5) and it is described in *Table 2*. If conditions of hypotonic treatment and fixation are controlled, the method can give satisfactory preparations from most pre-implantation stages (*Figure 2a* and *b*) except possibly from the very late blastocyst stage. The alternative is either to process these by the 'double fixation and softening' method of Dyban and Baranov (7) or to disaggregate them on a slide in 60% acetic acid as described in *Table 3a*.

4.3 Preparations from post-implantation stages

The technical approach is determined by the composition and dimension of the specimen and also by the nature of the investigation. These are factors which decide whether the specimen is initially handled as a solid and disaggregated after fixation, or is immediately disaggregated and handled from the outset as a cell suspension. Because of their dimensions (≤ 5 mm), the majority of the embryos described in Chapter 3, and the tissues isolated from them, are best treated as solids to conserve the small specimen. The fetal membranes from older embryos are also best handled initially as solid tissue because of difficulties in their mechanical disaggregation. Following hypotonic treatment and fixation, solid tissues are finally disaggregated in acetic acid (method described in *Table 3a*). The preparations obtained will successfully C-band but rarely produce good G-banded karyotypes. If preparations are required from whole older embryos (≥ 5 mm) or from larger organs such as livers and spleens of day $12-20$ embryos from which some loss by centrifugation can be tolerated, the specimens are first mechanically disaggregated into cell suspensions and processed as described in *Table 3b*.

Table 3. Chromosome preparations from post-implantation stages.

(a) Acetic acid disaggregation method

1. Obtain tissue sample which may be the whole fetus or parts of it (Chapter 3, Section 3 and 4) and, if required, incubate at $36-37°C$ for about 1 h in medium (Chapter 3, Sections 5 and 6) containing a mitotic arrestant such as Colcemid or Velbe at a concentration of 0.1 $\mu g/ml$. Glass containers are preferable to plastic since some plastics become part soluble during the procedure. If samples are large (5 mm) cut them into smaller pieces to increase surface exposure and solution penetration. Large fetal membranes can be left intact.

2. Place the tissues into an aqueous solution of either 1% tri-sodium citrate or 0.56% potassium chloride. Optimum times should be established by experimentation, e.g. young fetal material of day 6 is very sensitive and only requires $3-4$ min exposure to 0.56% potassium chloride or $4-6$ min to 1% tri-sodium citrate.

3. Fix the tissue in freshly prepared 3:1 mixture of methanol:glacial acetic acid (the fixative can be used for up to 3 h after preparation). Transfer the tissue from hypotonic either with watchmaker's forceps or with a wide bore pipette into a 10-fold excess of fixative. Alternatively, remove the hypotonic with a pipette and add an excess of fixative to the specimen. The mixing of the solutions creates turbulence and small specimens often become transparent on fixing, so take care that the specimen is not lost. Fixed material can be stored in an excess of fixative for up to 3 months at $4°C$ and still yield usable preparations.

4. Disaggregate in a 5-fold excess of aqueous 60% glacial acetic acid at room temperature. Observe disaggregation under a dissecting microscope.

5. After about 5 min make a test slide by pipetting off 0.5 ml of the suspension and placing on a slide on a hot plate at $40-60°C$. Keep the suspension moving, either by tilting the slide so that the drop moves across the surface, or by pipetting in and out and moving the drop to fresh areas of the slide. As the drop moves and retracts, cells spread and dry on the surface. The higher temperature of 60°C leads to rapid drying and slide making but the lower temperature of 40°C frequently yields better preparations.

6. Monitor the slide under ×128 phase contrast optics and, if satisfactory, make further slides if required.

7. If the test slide is unsatisfactory, make another slide from the cell suspension, and, if disaggregation is slow, it can be aided by teasing the specimen with watchmaker's forceps or by pipetting. Prolonged exposure to the 60% acetic acid, however, leads to oversoftening and breakage of mitotic cells.

8. Small (≤ 1 mm) tissue samples can be directly disaggregated on a slide in a drop of 60% acetic acid. If this is done, care should be taken to ensure that the sample is initially kept in the centre of the drop to prevent premature drying of the whole mass before cells become detached.

9. The slides can be conventionally stained (Section 5.1) or C-banded (Section 5.2).

(b) Cell suspension method

1. Obtain samples of whole fetuses ($5-10$ mm dimension) or whole organs such as the livers or spleens from day $12-20$ embryos and convert them into cell suspensions by breaking them up with forceps and pipetting in a suitable volume of medium ($10×$) such as TC199 or MEM (Flow).

2. Tilt the dish to allow the larger tissue fragments to settle before removing the supernatant fluid containing the freed cells. If a mitotic arrestant (Colcemid or Velbe at 0.1 μg/ml) is used, incubate the cell suspension for about 1 h at $36-37°C$.

3. Centrifuge the cell suspensions in a 10 ml conical glass centrifuge tube at 200 g for 5 min, discard the supernatant fluid and resuspend the cell pellet in a 10-fold volume of hypotonic solution of either 1% tri-sodium citrate or of 0.56% potassium chloride at room temperature. If embryonic livers or spleens are sampled, the latter will lyse red cells giving a cleaner final preparation. Cells from young fetuses require $6-8$ min in 0.56% potassium chloride or $8-10$ min in 1% tri-sodium citrate, cells from older fetuses need longer. Optimum times should be determined by experimentation.

4. Centrifuge the cell suspension at 200 g for 5 min, remove the supernatant fluid and fix the cell pellet by carefully pipetting in about 5 ml of a freshly prepared 3:1 mixture of methanol:glacial acetic acid. Proper fixation is a critical step and is best achieved by adding the fixative without disturbing the pellet, removing the fixative and adding a further quantity twice before the pellet is vigorously dispersed by flicking the centrifuge tube with the forefinger whilst simultaneously adding further fixative to resuspend the cells. Penetration of the pellet by the fixative can be observed by a colour change from cream to white and, depending on the size of the pellet, the whole fixing procedure should take between 2 and 5 min. Overlong fixation of the pellet is detrimental since it overhardens and becomes difficult to disperse.

5. Centrifuge at 200 g for 5 min and resuspend in a small volume of further fixative.

6. Place three drops of suspension in a row on a clean slide. Leave the drops to spread to their maximum periphery and as they begin to dry (at this point interference rings will be visible) blow onto the slide with the mouth whilst holding up to the gentle heat from a 60 W bench-lamp bulb. The combination of a stream of moist air and gentle heat promotes good metaphase spreading.

7. Monitor the slide under ×128 phase contrast optics and, if required, add further drops of cell suspension and repeat the drying procedure.

8. If the mitotic cells are not spreading adequately, due either to insufficient earlier hypotonic treatment or the chromosomes being embedded in cytoplasm, preparations can be greatly improved by adding three drops of fixative at a point when the drops of cell suspension are beginning to dry, leaving the fixative to spread to its maximum periphery, and blowing dry as described above.

9. The slides can be conventionally stained (Section 5.1), C-banded (Section 5.2), G-banded (Section 5.3) or used for *in situ* hybridization (Section 6).

Although the technique is more time consuming, the mitotic cells obtained are of better quality and are useful for C- and G-banding (*Figure 5*) and can also be used for *in situ* hybridization (Section 6).

As is apparent from Chapter 3 (Section 2), the growth and development taking place in the early fetus provides a rich source of mitotic division. Later, the processes slow down with the shift to increased differentiation with a rapid fall in mitotic index in both embryo and fetal membranes after day 14, thus it becomes difficult to obtain chromosome preparations from the later stages, particularly shortly preceding birth. Following birth, there is a resurgence of mitotic activity and excellent preparations can be obtained from the liver and spleen up to 5 days later. In the young fetus, the high mitotic index obviates the use of a mitotic arrestant such as Colcemid (Sigma) or Velbe (Lilly), but if the mitotic index is low, as in older specimens, or if large numbers of mitotic cells are required, mitotic arrest can be used to advantage. Injection into the mother is not recommended since the placental barrier poses a problem in the transmission to the fetus. The arrestants can be directly injected into the later fetuses to greatly increase metaphase numbers but the procedure requires operative skill and can be traumatic to the mother. A much simpler way of introducing an arrestant is to remove the tissue and incubate briefly in culture medium (see Chapter 3, Sections 5 and 6) containing 0.1 μg/ml of either Colcemid or Velbe. If the incubation period is for less than 1 h, which should be sufficient to produce some metaphase accumulation, there is probably no need to include serum or to gas the medium.

The vast majority of chromosomally unbalanced mouse fetuses die between day 6 and 15 (8) and mitotic activity often decreases several hours or even days prior to the event. Nevertheless, even if the embryo is in a fairly advanced stage of resorption, some mitotic preparations can be obtained from the fetal membranes and used to establish whether death was chromosomally related. The facility offered by the membranes for karyotyping lends itself to studies in which there is a need to preserve the abnormal or normal embryo intact.

5. CHROMOSOME ANALYSIS BY STAINING

5.1 Mouse chromosomes after conventional staining

The normal mouse has 40 chromosomes which are all acrocentric with centromeres close to the terminus and almost invisible short arms. In mid-metaphase, mouse chromosomes range in length from about 2 to 5 μm. This morphology does not lend itself to analysis after conventional staining since only a few of the autosomes show distinguishing features in the form of secondary constrictions (*Figure 3a*) and, of the sex chromosomes, only the Y of some strains can be recognized by virtue of occasionally visible unique features (*Figure 3a*). Conventional staining is, however, useful for chromosome counting in situations in which aneuploidy is suspected. Some of the methods are given in *Table 4*.

The problem presented by the uniform morphology of conventionally stained mouse chromosomes in the recognition of chimaeric or mosaic conditions can be overcome by using marker chromosomes to label cells. The markers can be in the form of the

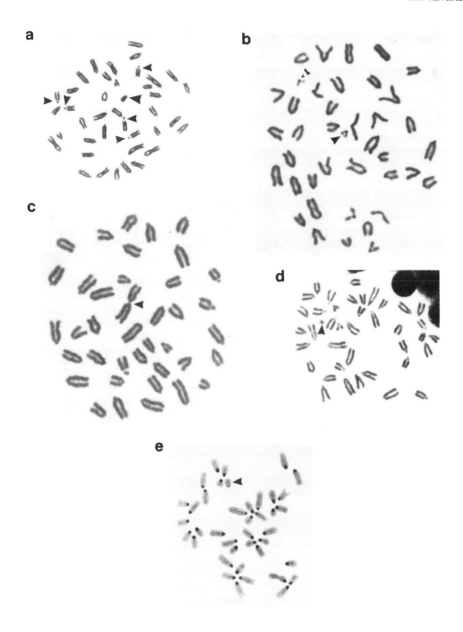

Figure 3. (a) A male metaphase stained with Giemsa and showing secondary constrictions (small arrowheads) and the Y chromosome (large arrowhead). The Y is recognizable by the closely paralleled chromatids and the small short arm. **(b)** The T6 marker chromosome (arrowheads) in a homozygous condition. It is the product of the unequal reciprocal translocation, T(14;15)6Ca (17), but the reciprocal product does not show up as a long marker in a conventionally stained cell. **(c)** One of the Robertsonian or metacentric chromosomes, Rb(11.13)4Bnr (17), which can also be used as a marker (arrowhead). **(d)** A metaphase which is easily sexed by the inclusion of a small, submetacentric Y chromosome (arrowhead). **(e)** The Y chromosome virtually lacking a C-band shows up in sharp contrast to the other chromosomes.

Table 4. Conventional staining of chromosomes.

There are innumerable chromosome stains and it is only appropriate to mention two of the commonly used ones here, Giemsa and toluidene blue. Frequently, the cytoplasm of a mitotic cell also takes up stain which occludes the chromosomes. This can be prevented by hydrolysis before staining.

1. Place the slides in a Coplin jar containing 5 M HCl at room temperature, leave for 5 min. Remove the slides and wash well in running tap water, drain off excess water and stain in either Giemsa or toluidene blue.
2. Staining with Giemsa
    (a) Commercially available Giemsa stains vary in concentration. Merck Giemsa stain solution is concentrated and a 1% solution in phosphate buffer at pH 6.8 (Gurr buffer tablets) will stain satisfactorily in about 15 min. For other Giemsa solutions monitor the wet slides under ×128 optics bearing in mind that mitotic cells will be darker stained when dry.
    (b) After staining, quickly rinse in pH 6.8 buffer, drain by touching onto filter paper and dry the remaining buffer solution with air puffed from a large rubber bulb.
    (c) To make permanent preparations, mount with a coverslip in a xylene-based mountant such as DePex (Gurr) or Eukitt (Riedel-de Haen). If preparations are for temporary examination, apply immersion oil directly to the slide. Note that modern immersion oils can rapidly leach out the stain.
3. Staining with toluidene blue
    Toluidene blue provides an alternative stain to Giemsa and can be obtained in the form of a ready to use stain mountant (ASCO). Add a drop of this to a slide under a coverslip and gently press out the excess stain. This will give instant staining which will not intensify. Such preparations will remain usable for 4−6 weeks if stored away from light. If the preparation fades, wash off the coverslip in water at 60°C and re-stain in fresh stain mountant.

Table 5. C-banding of chromosomes.

A number of C-banding techniques are described in the literature (and see also Chapter 4, Section 2.4), many are effective but some can also distort chromosome morphology. The method described here is effective, simple and conserves morphology. Slides for C-banding can be aged from 1 to 30 days.

1. Place slides in 5 M HCl at room temperature for 5 min.
2. Wash free of acid under running tap water and place in 2 × SSC (0.3 M sodium chloride/0.03 M tri-sodium citrate) in a lidded Coplin jar at 65°C for 15 min.
3. Place in pH 6.8 buffer solution (Gurr's tablets) at room temperature for 5 min.
4. Stain in Giemsa (e.g. 2−3% Merck Giemsa) in pH 6.8 buffer solution. Monitor wet slides under ×128 optics.
5. If the contrast between the C-bands and the rest of the chromosome is considered sufficient, rinse the slides in buffer, drain and dry the slide with air blown from a large rubber bulb.
6. Permanently mount the dry slide under a coverslip in DePex or Eukitt, or observe directly under immersion oil.
7. If C-band contrast is poor and the remainder of the chromosome is darkly stained, the time spent in the 2 × SSC is probably insufficient. If unoiled, slides can be returned for further treatment and re-stained. If contrast is poor because both C-bands and the remainder of the chromosomes are pale-staining in spite of using a more concentrated stain, the time spent in 5 M HCl and/or in the 2 × SSC at 65°C was too long and the preparations cannot be rescued. It is suggested that short times should be used initially with a return of the slides to the hot 2 × SSC if further differentiation is required.

shorter and/or longer chromosomes produced by induced, unequal reciprocal translocations, or spontaneously arising metacentric Robertsonian fusions in laboratory or feral populations. Examples of these markers, including the much used T6 chromosome, are shown in *Figure 3b* and *c*.

Y chromosome recognition after conventional staining can be greatly simplified by

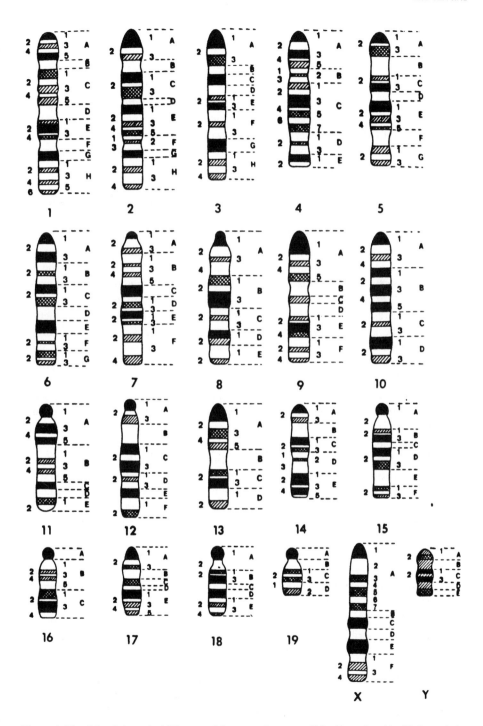

Figure 4. The G-banded standard idiogram of the mouse karyotype (26). (Reproduced by kind permission of the editor and publisher.)

using variants which have arisen spontaneously. Two are available (refs. 9 and 10) and both appear as short, sub-metacentric chromosomes (*Figure 3d*).

5.2 C-banded preparations and their use for Y chromosome recognition

C-banding methods demonstrate constitutive heterochromatin which, in the mouse, is normally composed of centromere-associated satellite DNA. Inter-strain differences in the amount of C-banded material occurs between certain autosomes (11) and have been used as cell markers in chimaeras (12). In contrast, the Y chromosome in all known strains is virtually without a C-band and this absence has proved an easy and positive method of identifying cellular sex. The C-banding method used in this laboratory (see also Chapter 4, Section 2.4) is described in *Table 5* and its use illustrated in *Figure 3e*.

5.3 G-banded preparations and their analysis

In common with other mammalian species, the individual chromosomes of the mouse can be recognized by using a variety of different banding methods but G-banding has found the greatest favour and also provides the basis for the standard idiogram which is illustrated in *Figure 4*. One method of achieving this differentiation is described in *Table 6* and the result illustrated in *Figure 5*. The latter represents the traditional method of analysing the karyotype with the homologous chromosomes arranged in pairs in decreasing order of linear length so that any disparities should become obvious. With experience, much of the analysis can be done at the microscope and cut-out photographs need only be used to detect subtle changes.

The standard idiogram lists 312 distinct regions within the mouse karyotype but rarely, if ever, have they all been observed in a single metaphase plate. The nomenclature adopted (13) has the advantage of being able to accommodate further band information if this should materialize, either through the development of higher resolution techniques or from more astute observations. Some additional information has emerged but, unfortunately, mouse chromosomes have not responded well to the high resolution techniques developed for human chromosomes and it is only recently that a higher resolution idiogram has been published (14). Currently, the standard idiogram has remained unchanged since its adoption in 1974, although an updated version is planned for the near future.

The G-banded karyotype has many uses which include, for example, the identification of absent, additional or aberrant chromosomes, the identification of balanced and unbalanced chromosome segments after meiotic segregation and for the linear placement of loci on the chromosome either following *in situ* hybridization (5,6), or by traditional genetic methods using marker genes, linkage and translocation breakpoints. From the point of view of describing abnormal karyotypes, the recommended nomenclature rules (15) should be followed.

The only consistent variation found within the G-banded karyotypes from different mouse strains has been in the amount of A1 band material (regions which are synonymous with C-bands) observed in certain chromosomes. A1 band sizes are frequently a strain characteristic and inter-strain differences can be used, for example, to monitor for genetic contamination (16). Some of the A1 band differences are illustrated in *Figure 5b*.

Table 6. G-banding of chromosomes.

Freshly prepared slides band poorly and although methods have been described for improving their potential, a greater improvement can be achieved by ageing the slides in boxes at room temperature for between 3 and 30 days. Optimum banding quality is reached after about 10 days.

1. Place suitably aged slides in a lidded Coplin jar in aqueous 2 × SSC at 60−65°C. Leave for 1.5−2 h. This time interval is not critical and can be extended to 3 h.

2. Make up a 0.025% trypsin (Difco 1:250) solution in 0.85% normal saline solution (Oxoid saline tablets) at room temperature.

3. Cool the Coplin jar with slides to room temperature with running cold tap water. Transfer the slides to 0.85% saline and leave for 5 min at room temperature.

4. Remove one slide, drain off the excess saline by touching onto filter paper, place the slide in a horizontal plane (e.g. across two lengths of glass rod) and flood uniformly with the 0.025% trypsin solution from a Pasteur pipette. The exposure time to the trypsin is critical, underexposure gives poor band resolution whereas overexposure destroys chromosome morphology. Several factors determine chromosome sensitivity to trypsin:
   (a) contracted chromosomes are more sensitive than elongated ones;
   (b) recently made (2−7 day old) slides are more sensitive than older slides (7−30 day old);
   (c) 'softly fixed' chromosomes (which, unstained, appear less optically dense under phase contrast) are more sensitive than 'hard fixed' chromosomes (optically dense).
   In this laboratory, for example, a 10 day old slide carrying a mixed population of contracted and elongated 'hard fixed' chromosomes would require a trypsin exposure time of about 20 sec to yield a satisfactory result. It is recommended that one slide should be put through the whole trypsin/staining procedure before the other slides are committed to the trypsin. The trypsin solution remains effective for up to 2 h after making.

5. Following the trypsin treatment, quickly drain the slide by tipping, and plunge it into 0.85% saline solution to dilute out the tryptic activity.

6. Rinse the slide in pH 6.8 buffer solution (Gurr's tablets) and transfer it to Giemsa stain diluted with the same buffer (e.g. 1−2% solution of Merck Giemsa). Monitor the wet slide under ×128 optics.

7. Rinse the slide in pH 6.8 buffer, drain off the excess buffer and dry the slide with air puffed from a large rubber bulb.

8. Directly examine the dry slide under immersion oil which gives the highest optical resolution.

9. Remember, modern oils leach out the stain. Unless an analysis can be made, or a photograph taken, within 24 h of oiling, it is advisable to make the slide permanent by mounting with a coverslip in a mountant such as DePex or Eukitt.

The success of the G-banding technique largely depends on having reasonably fresh solutions and clean glassware. Make up solutions in glass-distilled water and do not keep for longer than 2 weeks. Keep Coplin jars specifically for G-banding and wash simply with hot tap water immediately after using. Carry over of detergent or strong acids and alkalis can have a disastrous effect on banding.

5.4 Staining for the inactive X chromosome

If it is required to cytogenetically demonstrate the activity or otherwise of paternally or maternally derived X chromosomes in cells of fetal origin, the first requirement is to mark one of them so that it can be distinguished from the other one. Suitable markers include X-autosomal reciprocal translocations such as T(X;4)37H or the X-autosomal insertion Is(In7;X)1Ct, both of which present a marker chromosome longer than the normal longest mouse chromosome (see ref. 17 for a list of such markers). The second requirement is to examine the mitotic cells under conditions where the inactive can be distinguished from the active X. Numerous methods are available which, for example, include autoradiography and systems of 5'-bromodeoxyuridine (BrdU) substitution, but

105

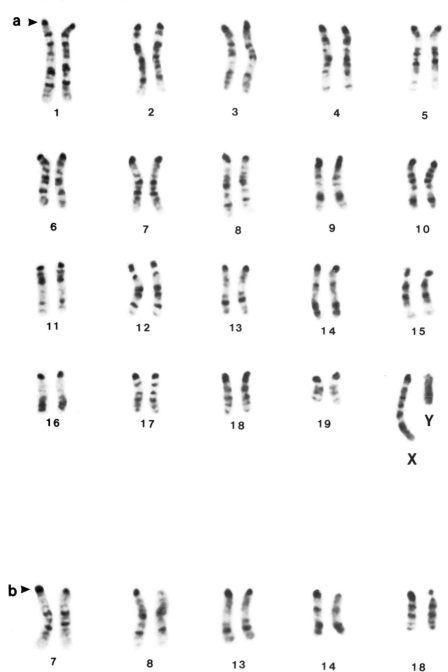

Figure 5. (a) G-banded karyotype of an inbred mouse (C57BL/601a). The A1 band regions, which are synonymous with C-bands (as an example, arrowhead shows the A1 band in chromosome 1), are of equal size, or homozygous, for each homologous pair of chromosomes. **(b)** In contrast, some of the different sized, or heterozygous, A1 band variants (as an example, arrowhead shows the A1 band region in chromosome 7) which can be used as strain markers. The examples were obtained from different strains (see ref. 11 for listings).

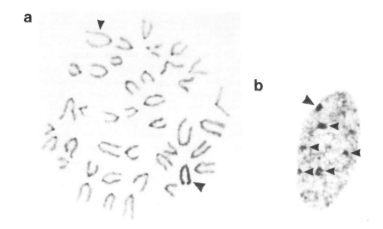

Figure 6. (a) A metaphase showing the inactive, normal X (large arrowhead) and the marked (Is1Ct) active X chromosome (small arrowhead). (b) Interphase nucleus from the amnion showing the sex chromatin body (large arrowhead) and the chromocentres (small arrowheads).

here, the simpler Kanda (18) method will be described (*Table 7*) since it can be done quickly using the minimum of equipment.

Unfortunately, the Kanda method is destructive and many of the mitotic cells become unscorable but a sufficient number are conserved to give a clear indication of X-chromosome activity. *Figure 6a* illustrates such a mitotic cell in which the normal X is inactive and is seen as being shorter than normal (the X is normally the fifth longest chromosome) and darker staining than the other chromosomes. The other X, marked with Is(In7;X)1Ct, remains unchanged although it should be noted that the anticipated longer length is not always obvious. Should the marked chromosome constitute the inactive X, however, it is reasonably easy to observe as the darkest and longest chromosome in the cell. The differential staining reflects a different response to chromatin uncoiling brought about by the extended hypotonic treatment (*Table 7*). Reasons for assuming that the dark staining does demonstrate inactivity are well founded since, in a number of comparative tests, cytogenetic determination has been found to concur with enzymatic determination.

5.5 Staining for sex chromatin

If an investigation requires embryos to be sexed and chromosome preparations are considered unhelpful due, for example, to the Y chromosome not being distinctive, or efforts to C-band are not considered worthwhile, it is possible to separate females from males by the presence or absence of sex chromatin. The stained sex chromatin body in mouse nuclei is not as distinctive as in the nuclei of some other mammalian species because the chromocentres (regions in which centromeric heterochromatin collect) are localized and stain up almost as intensely. Nevertheless, it can be used for sexing and the mouse amnion is a favourable source of nuclei (19). Preparations can be made by isolating the amnion, fixing it in 3:1 methanol:glacial acetic acid and spreading the nuclei in aqueous 60% acetic acid as described in *Table 3a*. The spread nuclei can be simply stained in acetic orcein (Gurr) under a coverslip and if the staining is slow,

Table 7. Staining for the inactive X chromosome.

1.	Obtain samples from fetuses and produce cell suspensions in medium as described in *Table 3a*. Centrifuge the suspensions at 200 *g* for 5 min and discard the supernatant fluid.
2.	Resuspend the cell pellet in an aqueous solution of 0.56% potassium chloride at 50°C for periods of up to 30 min. The hypotonic treatment at this temperature and for the maximum time severely distorts chromosome morphology and it is suggested that shorter periods should be tried since they may give an adequate differentiation without excessive distortion.
3.	Centrifuge at 200 *g* for 5 min, fix and make slides as described in *Table 3b*.
4.	Stain the slides in a 2−5% Giemsa solution in pH 6.8 buffer (Gurr's tablets). Monitor the wet slides under ×128 optics, it should be possible to observe the emergence of the dark staining X chromosome.
5.	If it is adequately stained, rinse the slide in pH 6.8 buffer, drain and blow the slide dry with air.
6.	Either examine directly under oil immersion or make permanent by mounting with a coverslip in DePex or Eukitt.

or the slides are to be kept for a short time, the coverslips can be ringed with rubber solution (Weldtite) to prevent the stain drying out. If the nuclei are from a female fetus, the sex chromatin body becomes visible in a proportion of the nuclei as the dark staining body illustrated in *Figure 6b*. The proportion will vary between 40 and 80% in nuclei from different female fetuses, most of the variation resulting from technical reasons such as, for example, the number of nuclei which flatten on the slide with the body in an easily observed peripheral position. In these nuclei the body can readily be distinguished from the chromocentres since it is appreciably darker staining and lies in a small 'notch' in the nuclear membrane. The nuclei from some male fetuses may present false-positive sex chromatin bodies due to the occasional peripheral location of a large chromocentre, but the total from any one rarely exceeds 10%.

6. IN SITU HYBRIDIZATION TO CHROMOSOMES

If the requirement for *in situ* hybridization to chromosomes is to probe for the completely unknown localization of unique sequences in the mouse genome, existing techniques are limited to tissue samples which give a plentiful supply of both slides and mitotic cells. This is due to the likelihood of the technical loss of material in proceeding through the protocol, the possibility of a weak hybridization signal from small weakly labelled sequences and the need to be able to score a statistically significant sample of silver grain counts. A further problem arises from the featureless morphology of mouse chromosomes and the need to introduce some form of banding technique to identify the individual chromosomes and the fine regions within these. If repetitive sequences are to be probed, because of the stronger expected hybridization signal, it may be possible to work with fewer slides and mitotic cells but the need for banding remains. On the other hand, with both unique and repetitive sequences, if they are known to map to a certain chromosome, it may be possible to locate them approximately using unbanded marker chromosomes to identify that chromosome. Moreover, if the approximate position on a chromosome is also known, it may be possible to split that region if an appropriate unequal reciprocal translocation is available. Since the number of suitable translocations, reciprocal and Robertsonian, which can be used as easily recognized markers is limited (see 17 for listings and linkage information), it is proposed here to describe a more general approach to *in situ* hybridization assuming

Table 8. *In situ* hybridization to chromosomes[a].

The technique described is adapted from one described for human (24).

1. Obtain a flask of third subculture mouse embryonic fibroblasts in growth phase (21), add BrdU (Sigma) at a concentration of 200 μg/ml and continue to culture for 16 h. At this concentration of BrdU many of the cells are blocked and synchronized in the middle of S phase. After adding BrdU, and subsequently, protect the culture, the cell suspensions and the slides obtained from them from light, otherwise the DNA will be nicked. A simple method of light protection is to wrap all containers in aluminium foil.

2. After 16 h, wash the cells four times with medium to remove the BrdU totally. Incubate in fresh medium containing 10 μM thymidine and harvest the cells after $6-6.5$ h.

3. Harvesting the cells. Reduce the medium to 10 ml and preferentially dislodge the rounded mitotic cells by vigorously tapping the flask on the heel of the hand.

4. If step 3 fails, dislodge the mitotic cells by gentle trypsinization using 0.25% trypsin in PBS. Remove the medium and retain, gently pour in 10 ml of trypsin solution and gently swirl over the cells for 1 min to loosen them, quickly pour off the trypsin and return the medium to the flask. A gentle tap of the flask will now dislodge the mitotic cells.

5. Centrifuge the cell suspension at 200 g for 5 min. Remove the supernatant fluid and resuspend the cell pellet in an aqueous solution of 0.56% potassium chloride. Leave for 8 min at room temperature.

6. Centrifuge at 200 g for 5 min and fix the cells and make slides as described in *Table 3b*. For replication bands, slides are best made after the cell suspension has been stored overnight in fixative at 4°C, the fixative changed and the cells resuspended in fresh fixative. Suspensions can be stored in fixative for up to 1 week and remain usable.

7. For *in situ* hybridization, from this point in the protocol, slides, whether they are replication banded, pre-banded or destined to be post-banded, are equivalent. If slides are pre-banded, photograph suitably G-banded mitotic cells (Section 5.3) and locate their positions either by vernier or graticule slide. Modern immersion oils are difficult to remove completely from slide surfaces, thus, for photography, slides should be mounted under a coverslip in an aqueous mountant such as Hydramount (Gurr) which can be removed in running water. The purpose of the investigation will determine the number of photographs required. The mapping of repeat sequences with a probe labelled to a high specific activity will need the least (≤ 50), and mapping a small unique sequence with a probe labelled to a low specific activity the most (≥ 100) photographs. It is convenient to work with 10 slides at a time, and solutions, if not required freshly made up, should be prepared in advance.

8. Before *in situ* hybridization, thoroughly de-stain the slides in fixative (3 methanol:1 glacial acetic acid) since stains of the Giemsa type are positively chemographic with nuclear research emulsions.

9. Boil a 100 μg/ml RNase (Sigma) solution made up in 2 $\times$ SSC for 10 min to remove any contaminating DNase. Cool and place 200 μl per slide under a coverslip (No.1, 64 $\times$ 22 mm). Place slides in a chamber moistened with 2 $\times$ SSC for 1 h at 37°C. Suitable chambers can be made from plastic sandwich boxes (Stewart). Unused RNase can be stored frozen for at least a year.

10. Remove the coverglasses by tipping and place the slides in a ridged glass carrier, wash in four changes of 2 $\times$ SSC then dehydrate by passing the slides through a 10, 50, 75, 96 and 100% aqueous/ethanol series. Air-dry the slides and store in a desiccator. Slides are best kept in the carriers throughout the procedures and the carriers moved gently through the solutions to minimize the possibility of cell loss. Glass slide carriers, wire handles and 'Continental' pattern staining troughs which accommodate them, are available as sets (Horwell).

11. Denature the DNA on the slides to single strands in 70% formamide and 0.1 mM EDTA in 2 $\times$ SSC at pH 7 for 4 min at 65°C. The pH is adjusted by adding 5 M HCl. Wash through four changes of 2 $\times$ SSC, dehydrate through the alcohol series and store in a desiccator.

12. If the intended probe is labelled with ^{125}I, acetylate the slides in 0.25% acetic anhydride in 0.1 M triethanolamine at pH 8 at room temperature for 10 min. Acetylation decreases background but is unnecessary if other labels are used. Wash through four changes of fresh 2 $\times$ SSC, dehydrate through the alcohol series and store in a desiccator.

13. Details of nick translation, labelling and counting of the probe are given in Chapter 10. Since an isotope is used from this point onwards, safety precautions should be observed.

14.   For 10 slides, use $100-150$ ng of ^{32}P-labelled DNA probe (see Chapter 10, Section 2) dissolved in 300 μl of hybridization buffer consisting of 50% formamide, 5 × Denhardt's solution, 5 × SSPE pH 7.2 (0.9 M sodium chloride/50 mM sodium di-hydrogen phosphate/5 mM EDTA), 10% dextran sulphate and 200 μg/ml of salmon sperm DNA (Sigma). The dextran sulphate is difficult to dissolve and can either be heated in boiling water and shaken vigorously or briefly microwaved.

15.   Before use, denature the hybridization mix containing the probe by boiling for $2-3$ min and then plunge onto ice. Use a closed Eppendorf tube pierced by a 21 gauge 1.5 inch hypodermic needle which serves both as a handle for holding the tube in the boiling water and as a vent to prevent the mix from excessive bubbling as it boils.

16.   Pipette 30 μl of hybridization mix onto each slide under a 64mm × 22 mm coverslip. This volume should be just sufficient to cover the area providing it is placed in three equidistant drops and the coverslips carefully lowered without producing air bubbles. Seal the coverslip edges with rubber solution (Weldtite) and place the slides in a moist chamber at 42°C overnight. Plastic sandwich boxes (Stewart) with tight fitting lids should be used and the slides supported on glass rods above the humidifying solution of 2 × SSC.

17.   Peel off the rubber solution with forceps and carefully remove the coverslips in 5 × SSC (0.75 M sodium chloride/0.075 M tri-sodium citrate) by slide immersion and agitation. Wash in 2 × SSC at room temperature for $1-2$ h.

18.   The stringency of subsequent washes will depend on the type of probe used and the wash procedures outlined below can only serve as a guideline and may need to be modified.

19.   For unique sequences, wash the slides in 2 × SSC at 65°C for 1 h with one change of hot 2 × SSC during this period. Follow with one wash in 0.2 × SSC at room temperature for 30 min and one wash in 0.1 × SSC at the same temperature and for the same time.

20.   For repetitive sequences, wash in SSC of a strength range of $0.5-0.1$ times at 60°C for 1 h, with one change of hot SSC during the period. Follow by a wash in 0.1 × SSC at room temperature.

21.   After the stringency washing, dehydrate the slides through the alcohol series and store in a desiccator.

22.   Prepare the emulsion for slide dipping. Many detailed reviews are available on the subject of auto-radiography (e.g. 25) and only experience with Ilford Nuclear Emulsion L4 will be briefly described here. The emulsion is delivered in 50 ml volumes, contained in a dark glass bottle, and in the form of semi-solid 'strings'. Store at 4°C and away from any radioactive materials. In a darkroom with a deep red safelight (e.g. Kodak No.6B) deflected away from the bench, divide the emulsion approximately into halves, replace one half in the dark bottle for future use and dissolve the other half for 30 min in 25 ml of distilled water at 55°C in a water bath. The dissolving can be done in a 50 ml plastic centrifuge tube (Nunc) which can also be used as a slide dipping container. Before dipping, carefully mix the contents to ensure homogeneity by slowly turning it with a microscope slide held with plastic forceps (metal instruments should not be used as they can produce background grains).

23.   Dip the slides slowly, withdraw, wipe the backs with paper tissue and drain the slides by resting them in an almost vertical position. Leave the slides to dry and harden for 30 min and, if available, this can be assisted by placing in a drying cabinet. The diluted emulsion, if stored in a light-proof bottle at 4°C, can be used $3-4$ times without an appreciable loss of sensitivity or an increase in background.

24.   Load the slides into light-proof slide boxes with desiccant, tape up the boxes with black plastic tape and store at 4°C to expose.

25.   It is prudent also to include some control slides for subsequent analysis of any technical problems which may arise. Controls should consist of a slide from a previously successful probing experiment as a positive control, an unprobed slide from the current series as a negative control and a blank slide to test for background in the emulsion.

26.   Exposure time will depend on several factors such as the nature of the sequence and the activity of the probe. Repetitive sequences with high activity probe require short periods, unique sequences and low activity require longer periods. For ^{125}I it could be from 2 to 9 days, for ^{3}H, $10-30$ days. Decide on an exploratory exposure time and use the worst slides, for example those with a low mitotic index or, if previously G-banded, those of which the fewest photographs were taken.

27.   Warm the slides to room temperature for 1 h, develop in total darkness in D19 (Kodak) at 20°C for 5 min without agitation. Rinse them in a 1% acetic acid stop bath and fix in Amfix (May and

Baker) diluted 1:4 with distilled water with added hardener (Ilford). Wash the slides in gently running water for 1 h in the dark, rinse in distilled water and air-dry by leaving the slides to drain in a near vertical position. If the backs of the slides have not been wiped free of emulsion after dipping, they can be wiped at this point or after staining.

28. Re-stain previously G-banded slides in a 2−4% Giemsa solution in pH 6.8 buffer, monitor the wet slides for staining intensity and, when satisfactory, rinse them in buffer and air-dry by standing them in a near vertical position. Locate previously photographed mitotic cells and either compare with the G-banded photographic prints and re-photograph if they are informative, or routinely re-photograph each one for an eventual overall analysis. If slides have not been pre-banded, attempt to induce post *in situ* hybridization banding using Wright's stain (22).

29. Protect replication banded slides from direct light during and after drying. Stain the dry slides in Hoechst 33258 (Sigma) at a concentration of 10 μg/ml in 2 × SSC for 30 min using an excess of stain under a coverslip. Rinse off coverslips in 2 × SSC, place the slides horizontally in 2 × SSC (e.g. in a flat plastic tray) and expose, uncovered, for 1 h at a distance of 15 cm away from a long wave u.v. light source (Sylvania). Rinse the slides in pH 6.8 buffer and stain them in 5−10% Giemsa in buffer. Monitor the wet slides until satisfactory replication bands appear, rinse them in buffer and air-dry by standing them in a near vertical position. Since modern oils may subsequently destain the chromosomes, it may be preferable to make the slides permanent by mounting under a coverslip in mountants such as DePex or Eukitt.

30. If the developed slides are satisfactory from the point of view of hybridization signal and background, further slides can be developed. If unsatisfactory, slides can be left to expose for longer, although it should be noted that this may eventually lead to an unaccceptable background level. The vagaries of silver grains and exposure times are well described elsewhere (25).

[a]See also refs 27, 28, 29.

no prior knowledge of the position of the unique sequence in the genome. It should be borne in mind, however, that if repeat sequences are to be probed, or if it is known to which chromosome or approximate chromosome region the repetitive or the unique sequences map, procedures can be simplified.

6.1 Embryo fibroblast culture and replication banding

The problem of introducing a banding regime which survives the *in situ* hybridization protocol has been largely overcome in man through the introduction of replication G-banding by BrdU substitution (20). The use of this method in the mouse is currently in its infancy and since the procedure requires cells to pass through a round of replication in the presence of BrdU, it can only be used in cultured cells, which, because of the numbers required, are limited to third passage embryonic fibroblasts from pooled day 12−15 embryos. A number of methods for initiating embryonic fibroblast culture are described in the literature (e.g. 21) and it is not proposed to review them here but only to mention the introduction of BrdU and subsequent harvest of the cells (*Table 8*). Embryonic fibroblast culture and replication banding has the advantage of providing a large number of slides and mitotic cells which could be used for several different probes but the disadvantages of having to pool a considerable number of embryos for the initiation of the culture and a lack of knowledge as to the source and lineage of the cells which become established in culture. It is also a time-consuming procedure.

6.2 Pre- and post-banding methods

The alternatives to providing morphologically recognizable chromosomes without resorting to replication banding are either to 'pre-band' or to 'post-band' after hybridization.

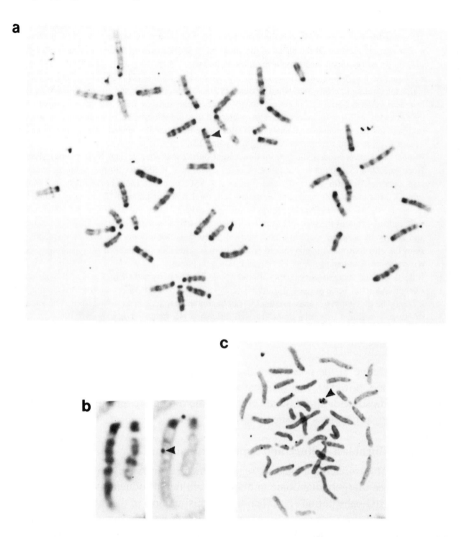

Figure 7. (a) Replication banded metaphase and *in situ* hybridization with a probe for a human 2'-5' oligo A synthetase gene. A significant number of grains were scored overlying band A4 of mouse chromosome 11 (arrowhead). **(b)** Part of a mitotic cell pre-banded for G-bands and photographed before *in situ* hybridization with a human Hprt probe. The grain (arrowhead) on the post *in situ* photograph locates the sequence to band A6 on the X chromosome. **(c)** A metaphase heterozygous for T(1;17)19OH, an unequal reciprocal translocation which splits chromosome 17 in the B band and presents a long and short marker chromosome. After *in situ* hybridization with a Tcp-1 (T complex protein) probe, a majority of the grains were observed to overlie the small marker (arrowhead), thus showing that Tcp-1 is more proximal than distal in the 17B band.

Pre-banding, which, for example, can be done using the G-banding procedure described here (*Table 6*), does not survive the *in situ* hybridization protocol so mitotic cells have to be photographed, their positions located by either vernier readings or a graticule slide, and re-photographed after hybridization in order to compare bands and silver grain positions.

Since, for most purposes, at least 50 G-banded cells have to be photographed, this procedure is tedious and time consuming. Post-banding after hybridization can be achieved using Wright's stain (Gurr or Merck) (22) but often, band definition is not detailed and, from hearsay, a number of researchers have experienced problems with the method and failed to obtain bands. Pre- or post-banding avoids the need to culture cells since adequate numbers of mitotic cells can be obtained directly from mid-term fetuses (*Table 3b*).

In summary, replication banding of embryonic fibroblast cells provides the best means of locating unique sequences in an unmarked mouse genome and a protocol for this method is described in *Table 8* and the results illustrated in *Figure 7a*. References to pre- and post-banding are also made in *Table 8* and pre-banding illustrated in *Figure 7b*. The use of an unequal reciprocal translocation unbanded, to both mark and split a chromosome region, is illustrated in *Figure 7c*.

6.3 Scoring

Scoring procedures will be determined by the nature of the investigation. If the location of the sequence, or sequences, is completely unknown and requires fine mapping to a G-band (replication bands are equivalent) on a chromosome, the practice is to identify each chromosome and to score the position of each silver grain overlying it. Before commencing, a decision should be made as to whether grains touching chromosome edges are to be included or not. It is convenient to prepare scoring sheets which include a G-banded copy of each mouse chromosome (*Figure 4*) and to enter grain positions directly on to them. In this way any 'hot spots' will become apparent as scoring proceeds. The aim should be to score between 50 and 100 mitotic cells. Results can be analysed by reference to the total number of grains overlying the chromosomes in all the mitotic cells scored, the total number of grains overlying each individual chromosome and the significance of the latter in relation to the number expected to overlie that chromosome assuming that grains are randomly distributed per unit length of chromosome. A table of the haploid percentage length is published (23). For example, if 265 grains are scored overlying the chromosomes in 100 mitotic cells, and since chromosome 1 represents 7.20% of the genome, then approximately 19 grains (7.20% of 265 = 19.08) would be expected to overlie the homologous pair. By estimating the expected number of each chromosome and comparing it with the observed number, any significant departures can be recorded. If a significant departure appears for a certain chromosome, the chromosome can be examined in more detail and the position of the peak grain distribution related to the G-bands.

If marker chromosomes are used, scoring can be simplifed since the expected to the observed grain counts need only be related to the total percentage length of the markers within the genome. It should be noted, however, that if markers are used in the heterozygous state, the expected grain count is halved since the other half is expected to overlie the unmarked, and therefore unscored normal homologues.

7. ACKNOWLEDGEMENTS

Within the space available, an attempt has been made to describe some of the techniques available for obtaining and utilizing chromosome preparations from different stages of embryogenesis. Some of the innovators of techniques have been acknowledged in the bibliography, through lack of space others have not and to these I apologize. I particularly wish to record my gratitude to Dr Veronica Buckle and to Mike Burtenshaw for their help in establishing the *in situ* hybridization methods.

8. REFERENCES

1. Kaufman,M.H. (1982) *J. Embryol. Exp. Morphol.,* **71**, 139.
2. Robertson,E.J., Kaufman,M.H., Bradley,A. and Evans,M.J. (1983) In *Teratocarcinoma Stem Cells. Cold Spring Harbor Conference on Cell Proliferation.* Silver,L.M., Martin,G.R. and Strickland,S. (eds), Cold Spring Harbor Laboratory Press, New York, Vol. **10**, p. 647.
3. Donahue,R.P. (1982) *Proc. Natl. Acad. Sci. USA,* **69**, 74.
4. Maudlin,I. and Fraser,L.R. (1977) *J. Reprod. Fertil.,* **50**, 275.
5. Tarkowski,A.K. (1966) *Cytogenetics,* **5**, 394.
6. Baranov,V.S. (1983) *Genetica,* **61**, 165.
7. Dyban,A.P. and Baranov,V.S. (1978) *'Nauka'.* Moscow, Russia.
8. Gropp,A. (1982) *Virchows Arch. (Pathol. Anat.),* **395**, 117.
9. Winking,H. (1978) *Mouse News Lett.,* **58**, 53.
10. Burgoyne,P.S. (1987) In *International Index of Laboratory Animals.* 5th Edition, Festing,M.F.W. (ed.), Laboratory Animals Ltd, PO Box 101, Newbury, Berkshire, UK.
11. Davisson,M.T. (1981) In *Genetic Variants and Strains of the Laboratory Mouse.* Green,M.C. (ed.), Gustav Fischer Verlag, Stuttgart and New York, p. 357.
12. McLaren,A. (1975) *J. Embryol. Exp. Morphol.,* **33**, 205.
13. Nesbitt,M.N. and Francke,U. (1973) *Chromosoma,* **41**, 145.
14. Sawyer,J.R. and Hozier,J.C. (1986) *Science,* **232**, 1632.
15. Rules for Nomenclature of Chromosome Anomalies. (1981) In *Genetic Variants and Strains of the Laboratory Mouse.* Green,M.C. (ed.), Gustav Fischer Verlag, Stuttgart and New York, p. 314.
16. Evans,E.P., Burtenshaw,M.D. and Adler,I.-D. (1985) *Genet. Res. Camb.,* **46**, 353.
17. Searle,A.G. (1981) In *Genetic Variants and Strains of the Laboratory Mouse.* Green,M.C. (ed.), Gustav Fischer Verlag, Stuttgart and New York, p. 324.
18. Kanda,N. (1973) *Exp. Cell Res.,* **80**, 463.
19. Monk,M. and McLaren,A. (1981) *J. Embryol. Exp. Morphol.,* **63**, 75.
20. Zabel,B.U., Naylor,S.L., Sagkaguchi,A.Y., Bell,G.I. and Shows,T.B. (1983) *Proc. Natl. Acad. Sci. USA,* **80**, 6932.
21. Freshney,R.I. (1983) *Culture of Animal Cells.* Alan R.Liss, Inc., New York.
22. Cannizzaro,L.A. and Emanuel,B.E. (1984) *Cytogenet. Cell Genet.,* **38**, 308.
23. Roderick,T.H. (1981) In *Genetic Variants and Strains of the Laboratory Mouse.* Green,M.C. (ed.), Gustav Fischer Verlag, Stuttgart and New York, p. 279.
24. Buckle,V.J., Mondello,V., Darling,C., Craig,I.W. and Goodfellow,P.N. (1985) *Nature,* **317**, 739.
25. Rogers,A.W. (1979) *Techniques of Autoradiography.* Elsevier/North-Holland Biomedical Press, Amsterdam/New York/Oxford.
26. Davisson,M.T. (1981) In *Genetic Variants and Strains of the Laboratory Mouse.* Green,M.C. (ed.), Gustav Fischer Verlag, Stuttgart and New York, p. 317.
27. Buckle,V.J. and Craig,I.W. (1986) In *Human Genetic Disease—A Practical Approach.* Davies,K.E. (ed.), IRL Press, Oxford, p. 85.
28. Malcolm,S., Cowell,J.K. and Young,B.D. (1986) In *Human Cytogenetics—A Practical Approach.* Rooney,D.E. and Czepulkowski,B.H. (eds), IRL Press, Oxford, p. 197.
29. Pardue,M.L. (1985) In *Nucleic Acid Hybridisation—A Practical Approach.* Hames,B.D. and Higgins,S.J. (eds), IRL Press, Oxford, p. 179.

CHAPTER 6

Cell marking techniques and their application

B.A.J.PONDER

1. INTRODUCTION

This chapter describes techniques for marking one cell or a group of cells and their clonal descendants in an animal. The techniques can be used to address two types of question. First the marked cells may identify particular cell lineages and cellular contributions to organs or tissues during normal or abnormal development. Thus, studies on the fate of marked cells introduced at morula and blastocyst stages have shown the progressive restriction in developmental potency of cells in different regions of the pre-implantation mouse embryo (1); introduction of quail neuroectodermal cells into chick embryos has provided the most complete picture of the cell lineages descended from this tissue (2); and analysis of the mosaic composition of adult tissues of chimaeras formed by aggregation of early mouse embryos has given information on clonal histories in terms of cell proliferation and mingling during development (3). Pathological processes can also be examined: for example, the mosaic composition of tumours may indicate whether they have originated from one single cell or from more than one (4). For all these studies it is preferable that the marked cells should be identical in their behaviour to the unmarked cells in the same tissue. In other words, one of the criteria for choosing a marker is that it should perturb the behaviour of the cells as little as possible.

The second type of question which can be addressed with cellular markers concerns the cellular site of action of mutant genes. Thus, the defect associated with the *pcd* gene which causes cerebellar ataxia was expressed in *pcd/pcd*←→wild-type chimaeras (see below) only where the Purkinje cells were of mutant genotype (5). The cellular basis of any phenotypic difference between mouse strains can, in principle, be explored in this way: for example, whether the high or low tumour incidence characteristic of some strains is intrinsic to the target cells, or the result of modifying factors in another tissue (e.g. the immune system). For these kinds of question it is necessary that the marked cell populations differ in the expression of the gene or other behaviour which is under examination but once again the cell marker phenotype should not introduce additional complicating differences.

The ideal cellular marker should be cell-autonomous (that is, its expression should be limited strictly to individual cells of the appropriate genotype and not communicated to neighbouring cells); it should not affect behaviour of the marked cells in any way which is relevant to the experiment; it should be stably expressed in all of the cells of the tissues under examination and under different physiological or pathological con-

115

ditions; and it should be technically easy to demonstrate. In general terms, an ideal marker will be ubiquitous − that is, expressed in all tissues at all stages of development − but this is not important for a specific experiment limited to a specific tissue. It is also ideal that both cell populations in the chimaeric tissue be differentially marked and identifiable. Otherwise one must infer that all unmarked cells belong to the other population, which may not be true. Most cell marking techniques are not 100% efficient, and if the identification of a cell population which is present in a small minority is crucial to the experiment, it is better that this population should be the one marked for positive identification.

No ideal marker system has yet been developed. The advantages and disadvantages of those which are available will be discussed below. The experimental systems generating differentially marked cell populations fall into five broad groups:

(i)   experimental chimaeras;
(ii)   X-inactivation mosaics;
(iii)   marking of single cells by injection of marker substances;
(iv)   marked cell populations produced by integration of cloned genes in some cells of the early embryos;
(v)   markers induced by somatic cell mutation.

Each will be considered in turn.

2. CHIMAERAS

2.1 Definition

A chimaera is an animal containing cells derived from more than one zygote. For reviews of mammalian chimaeras, see (6,7). In principle, the different cell populations can be combined at any stage of the life of the animal. Thus, although the chimaeras familiar to developmental biologists are constructed by combining cells from two or more pre-implantation embryos, they may also be produced by introduction of cells of different genetic origin into post-implantation embryos (see Chapter 3, Sections 7 and 8), or into adult animals. A patient with a kidney transplant is a chimaera. 'Radiation chimaeras' constructed by irradiation of a host animal to ablate the bone marrow and lymphoid tissues, followed by engraftment of bone marrow from a genetically distinct donor, have been extensively used in immunology and radiobiology. Whatever the origin of the chimaerism, the principles of the use of cell markers are similar. The discussion which follows will concentrate on mouse embryo chimaeras and their application to developmental biology.

2.2 Types of chimaera and their production

2.2.1 Embryo aggregation chimaeras

These are produced by the aggregation, in culture, of 4- to 8-cell embryos of mice of different strains, followed by transfer of the aggregated embryos into the uterus of a pseudopregnant 'incubator mother', where they develop to term. The procedure is given in *Table 1*.

Table 1. Making embryo aggregation chimaeras[a].

1.	Arrange for 4-cell embryos of the two strains to be used to make the chimaera to be available on the same day and the pseudopregnant foster mother to receive the embryos to be available on the following morning.
2.	Prepare culture drops of M16 + BSA medium (Chapter 2, Section 5.2) under oil in a Falcon plastic dish. Equilibrate in a 5% CO_2 incubator at 37°C.
3.	Prepare two glass cavity blocks each with 1 ml of M16 + BSA covered with a layer of oil for washing the embryos, one wash glass for each strain. Place in an incubator.
4.	Kill pregnant female (3rd day of pregnancy) of one strain of mouse by cervical dislocation. Follow the steps in Chapter 2, Section 2.2, to obtain 4- to 8-cell embryos. Wash in M2 + BSA medium (Chapter 2, Section 5.2).
5.	Digest the zona pellucida by placing the embryos in pronase (Chapter 2, Section 3.2). Between 6 and 15 min when the membrane appears fragile and hazy (shortly afterwards it will disappear) transfer the embryos back to M2 + BSA. Pipette up and down gently to remove any fragments of zona. Do not treat too many embryos at once — no more than 15. (Acid tyrodes solution — see Chapter 2, Section 3.2 — is faster but therefore less easy to control and more likely to damage the embryos. Also the embryos seem, in our hands, to aggregate better after pronase).
6.	Transfer the embryos to fresh M2 + BSA medium.
7.	Wash embryos in one wash glass of M16 + BSA (use the other for the second strain).
8.	Place one embryo in each culture drop of M16 + BSA in Falcon dishes and replace in the incubator. (Try not to keep the M16 + BSA, even under oil, out of the incubator too long or the pH will alter.)
9.	Repeat for the second strain of mouse. Place one of these embryos in each of the same culture drops containing an embryo of the first strain.
10.	Nudge the two embryos in the culture drops together with a drawn out pipette. Repeat 2−3 times during the course of the day.
11.	Incubate overnight.
12.	Check the embryos next day and note how many have aggregated. Transfer to pseudopregnant recipients as described in Chapter 13, Section 6.4.

[a]For further details on the isolation and handling of embryos, removal of zonae, culture medium, preparation of pseudopregnant foster mothers, equipment and materials see Chapters 1 and 2. For embryo transfer see Chapter 13, Section 6.4.

2.2.2 *Blastocyst injection chimaeras*

These chimaeras are produced by injection of one or a small number of cells into the blastocyst, rather than by aggregation of two morulae. (For method, see refs 8, 9.) This technique is more difficult, but it has some advantages. Removal of the zona pellucida is not necessary, so the technique is suitable for species such as the rabbit in which the zona is necessary for subsequent development. The trophoblast is entirely of host type, which may permit successful development of some interspecific combinations which would otherwise fail to implant. Because of the later stage of chimaera formation, blastocyst chimaeras allow the analysis of the developmental fate of later embryonic tissue than can be achieved with morula aggregation. Finally, blastocyst injection offers one route by which marked embryonic stem cells can be introduced into the germ line.

3. CELLULAR MARKERS IN CHIMAERIC SYSTEMS

The range of markers which have been used and their characteristics are summarized in *Table 2*. (For a more complete description of potential markers which have not been

Table 2. Markers for use in chimaeric mice[a].

Marker	Method	Tissues	Choice of strains[b] (see ref 10)	References
A. Yielding spatial information				
Pigment				
Melanin	direct inspection	retinal pigment epithelium	wide: (albino/ non-albino)	6,11
		membrane labyrinth inner ear		
Melanin/hair follicle effects of coat colour genes	direct inspection	hair	wide: (range of genes acting on melano- cytes or hair follicle)	6
Enzyme polymorphisms				
β-Glucuronidase	histochemical[c]	liver miscellaneous others claimed	wide	11,12
β-Galactosidase	histochemical[c]	Purkinje cells of cerebellum cranial nerve nuclei mammary epithelium	wide	11,13
Malic enzyme	histochemical	extraembryonic membranes, ?others		1
Glucose phosphate isomerase (Gpi)	histochemical and immunohisto- chemical	many	wide	11,14
Polymorphic cell surface markers				
H-2 antigens	immunohisto- chemical	several; not in embryo or neonate	wide	15
Ia antigens	immunohisto- chemical	lymphomyeloid	wide	not reported
Thy-1 antigens	immunohisto- chemical	? thymocytes, fibroblasts, CNS, ?mammary myoepithelial cells ? may not be cell- autonomous	restricted (θ1.1 uncommon: AKR and few others)	16
Carbohydrate poly- morphisms recognized by lectins	lectin binding histochemistry (DBA)	intestinal epi- thelium and vascular endo- thelium after 12 days gestation	restricted (see *Table 7*)	17,18,19
Morphological markers				
Beige	histochemical intracellular granules	granulocytes, mast cells, osteoclasts, ? others	restricted	11,20,21
Specific cellular degeneration				
Mouse retinal degeneration (rd)	morphology in tissue sections	retina	restricted	22

Table 2 continued.

Marker	Method	Tissues	Choice of strains[b] (see ref 10)	References
Purkinje cell degeneration (pcd)	morphology in tissue sections	Purkinje cells of cerebellum	restricted	5
Species difference in DNA				
Satellite DNA differences between mouse species	*In situ* hybridization in tissue sections	any	restricted at present to *Mus musculus/ Mus caroli*	23

B. Not yielding spatial information

Electrophoretic polymorphisms

Marker	Method	Tissues	Choice of strains	References
Glucose phosphate isomerase (Gpi)	electrophoresis of tissue homogenate	ubiquitous, including early embryo	wide	1,10,24
Isocitric dehydrogenase	electrophoresis of tissue homogenate	liver, kidney, spleen, muscle and others	wide	25
Major urinary protein + potentially others (see refs 10, 11)	urine	liver in live mice	wide	26

Chromosomal markers

Marker	Method	Tissues	Choice of strains	References
T6	morphology in cell cultures from tissue	in principle, any mitotic tissue; short-term culture may be required	restricted	10,11,27 Chapter 5
Y chromosome	morphology in cell cultures from tissue	in principle, any mitotic tissue; short-term culture may be required		11,28
Tetraploidy	morphology in cell cultures from tissue	in principle, any mitotic tissue; short-term culture may be required (development abnormal)		29
Others	morphology in cell cultures from tissue			10

[a]For further general discussions of cell markers in chimaeras, see refs 1, 6, 7 and 11.
[b]The 'choice of strains' refers to those available. Of course, any of these markers could in principle be bred onto the strain background of one's choice. Useful listings of genetic variants in inbred strains of mice are to be found in ref 10, and in the quarterly issues of Mouse News Letter. It is useful to incorporate a coat colour marker of chimaerism into any pair of strains which will be used to make chimaeras, so as to identify easily the offspring which are chimaeric from those which are probably not.
[c]These enzymes may not always give a consistent positive stain in all cells of a normal (non-chimaeric) tissue. Suitable positive and negative controls of staining must be rigorously applied (preferably with appropriate sections flanking the chimaeric section on the same slide) before the results in chimaeric tissue are interpreted.

developed to practical use, and markers in non-mammalian species, see West, ref. 11.)

The most important practical distinction is between markers which can be demonstrated visually in tissue sections or whole mounts, and markers such as the electrophoretic

Table 3. Immunohistochemical localization of H-2 antigens in mice.

Consistently positive	Variable sometimes positive	Consistently negative
Epithelia		
small intestine	bladder	squamous epithelia of
colon	bronchus	tongue
uterus	glandular stomach	oesophagus
breast	bile duct	cervix
skin	pancreatic duct	vagina
trachea	genital tract (other	hepatocytes
transitional epithelium of	than uterus)	
renal pelvis	thyroid follicles	
kidney collecting tubules	salivary gland acini	
salivary gland ducts	(other than parotid)	
parotid acini		
Other tissues		
thymic medulla	renal glomeruli	adrenal gland
spleen	interstitial	pancreatic islet
lymph nodes	cells of testis	cells
Peyer's patch		ovarian follicle
cells of monocyte-		brain, spinal cord
phagocytic series		peripheral nerve
vascular endothelium		skeletal and
sinusoidal lining cells		cardiac muscle
of liver		
lung parenchyma[a]		
adipocytes		

[a]Diffuse staining of alveolar walls that could not be assigned to one cell type.

polymorphisms which rely on the estimation of the proportions of cells of each genotype in homogenates of small tissue samples. The advantages and disadvantages of each type of marker in the analysis of chimaeric tissues are discussed in Section 5. Detailed methods will only be given here for the use of H-2 antigens and the carbohydrate polymorphism recognized by *Dolichos biflorus* agglutinin (DBA). For methods for other markers see references and brief comments in *Table 2*.

3.1 The use of H-2 antigens as markers

H-2 antigens can be detected immunocytochemically using anti-H-2 monoclonal antibodies on a wide variety of tissues of mice more than about 7 days of age (30) (see *Table 3*). Although H-2 antigen expression in mouse embryos can be detected serologically, only in late-gestation thymus are the amounts sufficient to be detectable by the immunohistochemical technique described here (see Chapter 2, Section 4 and Chapter 3, Sections 7 and 8 for labelling of embryonic cells). There is no evidence that the H-2 difference in chimaeras between strains of differing H-2 type has any effects on development, but this cannot be taken as proven.

The advantages of H-2 antigenic markers are the large number of tissues and wide range of mouse strains to which the techniques can be applied, the ability to stain the

Table 4. Some technical points when making enzyme conjugates with antibody.

1. For alkaline phosphatase conjugates, use high specific activity enzyme from calf intestine (e.g. from Boehringer, or Sigma). These conjugates can be used with 1 mM levamisole (Sigma) as an inhibitor of endogenous alkaline phosphatase in tissues other than intestine (ref. 33; see *Table 5*, step 18).
2. Apparently similar enzymes from different suppliers may differ in their 'stickiness' and give variable background problems on tissue sections. It may be worth trying different sources if you have a problem.
3. Make sure that all buffers (e.g. Tris) containing free amino groups have been dialysed out of enzyme or antibody preparations before adding protein cross-linking reagents.
4. It may be necessary to try a range of antibody:enzyme ratios to see which gives the best results.
5. Test the newly prepared conjugate at a range of 2- or 3-fold increasing dilutions and choose the highest intensity which also gives clean background on control tissue. If background staining is a problem, fractionation of the conjugate on a gel filtration column (e.g. Sephadex G-200 or Sephacryl) may be helpful, especially with peroxidase conjugates.
6. Do not freeze the conjugate. Aliquot and store at 4°C in buffer at pH 7.5−8.0 with 5 mg/ml bovine serum albumin and 0.1% sodium azide.

two genotypes in the same section using antibodies to the two different H-2 specificities, and the possibility of staining whole mounts of some tissues (e.g. epidermis — see Section 6.4).

The disadvantages of H-2 antigenic markers are the technical difficulties involved. Firstly, the epitopes recognized by currently available antibodies are labile to most fixatives and standard wax embedding procedures, so that lightly fixed cryostat sections must be used (see below). Secondly, H-2 antigens are expressed at the cell surface, and, with the problems with tissue processing, the resolution of the staining at the single-cell level in a mixed tissue is not always good.

The following points should be taken into account when setting up the system. Note that the same general principles will apply to any system based on immunohistochemistry with monoclonal antibodies.

(i) Choose the antibodies (sources:commercial; American Type Culture collection, gifts from other laboratories) and the mouse strains to be used. Mouse H-2 haplotypes are detailed in Green (10).

(ii) Check the specificity and strength of the antibodies for immunohistochemistry in the chosen mouse strains on cryostat sections of thymus from young adult mice. Sections of each strain, adjacent on the same slide, unfixed or fixed with PLP (see below, *Table 5*), are exposed to serial dilutions of monoclonal antibody (e.g. tripling dilutions 1:10 → 1:810). The monoclonal antibody may be directly labelled (see below) or it may be visualized by a second antibody which is anti-mouse immunoglobulin conjugated with peroxidase or alkaline phosphatase. If the optimum titre of the second antibody is not already known, this can be determined by using a series of tripling dilutions of the second antibody with each dilution of first antibody.

Choose conditions where the thymic medulla of one strain stains strongly with a dilution of its specific monoclonal anti-H-2 antibody which gives no stain on the thymus of the other strain. Some staining may occur with the second (anti-mouse) immunoglobulin antibody: include a control without anti-H-2 antibody.

(iii) If simultaneous staining of both phenotypes in the same section of chimaeric tissue

121

Cell marking techniques and their application

Table 5. Double H-2 antigen staining of tissue sections[a,b].

Time for a single staining run: 5−6 h.

1. Use 4 μm cryostat sections of unfixed tissue on gelatin-coated slides.
2. Store the sections dry at −20°C until needed (up to several weeks).
3. Plan your staining run. Number the slides (pencil on frosted slide ends), and write yourself a schedule of the antibody conjugates and staining to be applied to each. It is very easy to get confused. Do not do too many at once — 30 slides if the combination of antibodies/stains is not complicated, 15−20 if it is.
4. Place the slides in a plastic staining tray (*Figure 1*). Check the tray is level (use a spirit level). Allow slides to dry at room temperature for at least 20 min.
5. Rinse with PBS. Pick the slides up one by one, flood them with a *gentle* stream from a plastic wash bottle, aiming the jet well away from the sections. Coax the PBS over non-wettable parts of the sections, or flood these directly (and gently) using a Pasteur pipette. Replace the slides, each flooded with PBS, in the tray.
6. Fix with PLP (0.5% paraformaldehyde−0.15 M lysine−10 mM sodium periodate) for 2 min (this timing is critical) at room temperature. Drain off the PBS by tipping the slide, and flood with PLP using a Pasteur pipette. PLP is freshly made up: 1 part of 4% paraformaldehyde in distilled water, 3 parts of 0.2 M lysine, sodium periodate crystals to 10 mM. The lysine and paraformaldehyde stock solutions can be stored at +4°C. The periodate crystals may take a few minutes to dissolve.
7. Wash the PLP off by flooding the slide with PBS.
8. Block endogenous peroxidase. We use 0.1% phenylhydrazine hydrochloride (PHD) in PBS because it does not cause bubble formation (as the peroxide-based methods do) which can destroy frozen sections. Tip off the PBS and flood the slides with PHD. Leave for 5 min; then drain the slides and rinse with PBS as above.
9. Wash the slides with PBS containing 0.5% bovine serum albumin (PBS + BSA).
10. Provided you are not using a method which involves an anti-mouse Ig antibody conjugate, incubate the slides for 15 min at room temperature with PBS containing 10% mouse serum (any strain) to block non-specific binding sites for mouse immunoglobulins. Meanwhile carry out the following step.
11. Make up the dilutions of the first antibody conjugate to be applied in PBS containing 10% mouse serum. Allow 100 μl of diluted conjugate for each slide (more if the slide contains several large sections). If the sections are small and the conjugate is precious the area to be covered by conjugate can be reduced or circumscribed with a chinagraph pencil, and a smaller volume applied. Examine the tube containing the conjugate each time it is used: a small precipitate in the bottom of the tube is often seen, but if this has markedly increased it may be a sign that the conjugate is deteriorating. Aggregation of the conjugate may also cause background problems (see 'Troubleshooting', *Table 6*).
12. Drain the slides of PBS + mouse serum, and carefully blot each in turn using paper tissues to remove moisture from all but the immediate area of the sections. This is to avoid dilution of the conjugate which is about to be added. Do *not* let the sections dry out — only blot two or three at a time before adding the conjugate by gentle pipetting on top of the sections. Make sure the conjugate is evenly distributed. Be careful not to wipe the sections off the slide when blotting dry — it is easy to forget which side of the slide is which. Incubate for 1 h at room temperature: place a lid on the staining tray, which should have plenty of liquid in the bottom to maintain a moist atmosphere.
13. Meanwhile, prepare the dilutions of the second antibody conjugate in PBS + 10% mouse serum, as above. If this is a peroxidase conjugate, all the diluting solutions must be free of azide, which inhibits peroxidase.
14. Drain the first antibody conjugate off the slides, and flood with PBS + BSA. Drain each slide of PBS + BSA and flood again, and repeat the cycle taking each slide in turn, three times.
15. Blot dry and add the second (peroxidase) conjugate, as above. Incubate for 1 h.
16. Drain the slides and flood with PBS; repeat the cycle four times. Collect the slides into a staining rack for immersion in a staining dish containing 200 ml of 0.15 M Tris pH 7.6: a similar dish will be used for the DAB substrate for the peroxidase stain.
17. Prepare the DAB substrate for the peroxidase staining. (Note: all operations involving DAB should be carried out in a carcinogen-designated fume hood. Wear gloves!). Weigh out approximately 100 mg of DAB. For each 100 mg of DAB place 200 ml of 0.15 M Tris pH 7.6 in a conical flask, add 400 μl of 100 volume hydrogen peroxide (store at 4°C and renew every few months). Immediately add the

DAB and stir gently to dissolve. Add NiCl$_2$ if required (see note on choice of substrates in text). Pour into staining dish and transfer the rack of slides from the dish containing Tris buffer. Stain for 5 min at room temperature, then transfer the rack back to the dish containing Tris buffer. Dispose of the DAB substrate solution and rinse the staining dish and other glassware thoroughly. Wash the slides through three successive changes of distilled water and return them to the original staining tray for the alkaline phosphatase stain.

18. Flood the slides with veronal acetate buffer pH 9.2 (VAB) (N.B. *not* PBS — the phosphate will inhibit the alkaline phosphatase reaction) and leave while preparing the alkaline phosphatase substrate solution. Weigh out 5 mg of Brentamine Fast Red TR (Sigma) into a small dry glass tube, and 5 mg of naphthol AS-BI sodium salt (Sigma) into a plastic universal container (20 ml size). Add 10 ml of VAB to the naphthol AS-BI, shake gently to dissolve (rapid). Immediately dissolve the Fast Red TR in three drops of dimethyl formamide (store protected from light) and add to the VAB/naphthol AS-BI. Rinse the remaining Fast Red TR from the glass tube with the VAB solution. Add 200 μl of levamisole (50 mM stock solution) to the 10 ml (final concentration of levamisole 1 mM) to inhibit endogenous alkaline phosphatase in all adult tissues except small intestine and stomach (33). Filter this substrate through Whatman No 1 paper in a glass funnel and use immediately.

19. Drain the slides of VAB and flood with substrate solution. Incubate for up to 1 h at room temperature. The intensity of staining can be increased by adding freshly prepared substrate at intervals of approximately 20 min.

20. Drain the substrate from the slides, wash several times in PBS or distilled water, counter-stain with haemalum (lightly if the red and brown colours are to be demonstrated using black and white photography) and mount in *aqueous* mountant.
    Note: the alkaline phosphatase reaction product is soluble in alcohol.

---

[a]For whole mounts of tissue use the same procedure but block endogenous peroxidase (step 8) for 30 min, and be very thorough with each washing step.
[b]See refs 15 and 30.

is required, decide on the strategy to distinguish the two monoclonals in the same section. This might be done by using monoclonals of differing subclass as first antibodies, and second antibodies specific for the IgG subclasses. Alternatively a hapten-sandwich system may be used [e.g. one monoclonal conjugated to biotin with a streptavidin conjugate as second step, and the other monoclonal conjugated with dinitrophenol (DNP) with an anti-DNP second antibody]. These systems are described in textbooks of immunohistochemistry. In practice, we have found that a third approach, direct conjugation of each monoclonal with an indicator molecule, either an enzyme or fluorochrome, has given the most reproducible results. Conjugated monoclonal antibodies are used in the double staining procedure given in *Table 5*.

(iv) Some monoclonal anti-H-2 antibody conjugates (e.g. to alkaline phosphatase, peroxidase, fluorochromes) are available commercially. These are expensive, but are probably the easiest answer if they are satisfactory. If no antibody conjugates are available, if those available do not stain adequately (some seem much better than others) or if you expect to use large quantities, you should consider making your own conjugates. Standard methods are given in textbooks of immunohistochemistry and in references 30 and 31. Note that to obtain sufficient antibody to work with, you will probably need to obtain the hybridoma cell line and produce either concentrated culture supernatant or ascitic fluid. Some technical hints on making enzyme−monoclonal antibody conjugates are given in *Table 4*.

(v)   Tissue preparation. The epitopes recognized by different anti-H-2 monoclonal antibodies will differ in their sensitivity to different fixatives, so the best regime must be found by trial and error. We and others have found that cryostat sections of unfixed tissue are necessary; the sections are then lightly fixed after cutting in dilute protein cross-linking fixatives (e.g. paraformaldehyde—lysine—periodate, *Table 5*, step 6) or in very cold ($-20°C$) acetone.

(vi)   The sequence of incubations with antibody conjugates and of staining must be determined by experiment. We obtain clearly better results if the alkaline phosphatase conjugate is added before the peroxidase conjugate, than if the two are added together or in the reverse order. The peroxidase staining reaction should be done before the alkaline phosphatase reaction.

(vii)   Choice of substrates to give contrasting double staining. Although a variety of substrates are available giving a range of colours for alkaline phosphatase and peroxidase, we have found that Brentamine Fast Red TR (for alkaline phosphatase) and diaminobenzidine (DAB) (for peroxidase: brown colour) give the best intensity of colour and are the most reliable. The red—brown contrast is not very good, and may be improved by adding heavy metal salts to the substrate solution for the peroxidase stain: for example, addition of 50 μl of 8% $NiCl_2$ per 10 ml of substrate gives a dark purplish peroxidase reaction product. β-galactosidase (turquoise blue product) (32) gives a good colour contrast with the red alkaline phosphatase stain, but in our hands the intensity is less than the brown of the peroxidase/DAB stain.

(viii)   Controls. As a counsel of perfection, flank the chimaeric tissue on every slide with sections of the corresponding tissue from normal animals of the strains pres-

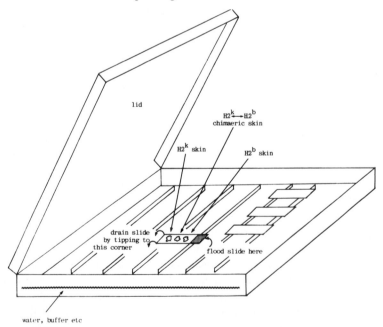

Figure 1. Staining tray for immunohistochemistry.

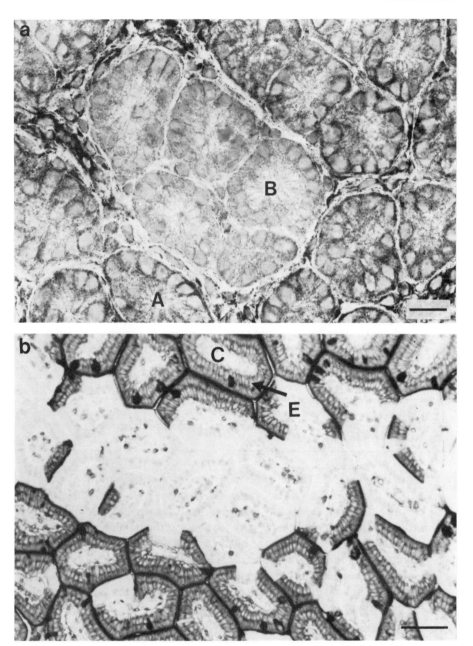

Figure 2. Visual markers of chimaerism. (**a**) Cryostat section of colonic epithelium from B10a↔B10ScSncc chimaera. The B10a crypts (A) are stained red with anti-H2^k alkaline phosphatase conjugate, and appear dark in this print. A group of B10ScSncc crypts (B) are stained yellow−brown with anti-H2^b peroxidase conjugate, and appears light. Bar = 40 μm. (**b**) DAB 4 μm paraffin section of the small intestinal epithelium of a DDK↔C57BL/6J (B6) chimaera, stained with DBA−peroxidase conjugate. The villi are cut across transversely, showing a core (C) surrounded by epithelium (E). The epithelium shows bands of cells of B6 origin (which bind DBA−peroxidase and appear black) and of cells of DDK origin (unstained). The cores of the villi contain some endothelial cells of DDK type, which also bind DBA−peroxidase and so appear black. Bar = 150 μm.

Table 6. Troubleshooting for H-2 antigen staining.

---

1. No stain or weak stain.
   One slide/group of slides.
       Error in dilution/addition of conjugate to slide.
   One antibody only.
       (a) Antibody conjugate deteriorated (this can happen quite suddenly). Check on thymus sections.
       (b) Substrate wrong/deteriorated. Check endogenous staining in 'unblocked' sections, exchange reagents with colleagues, or try another reliable antibody system if available.
       (c) Variation between mice. Some express H-2 antigens in a particular tissue more strongly than others. Check thymus which is usually reliably strong.
   Both antibodies
       (a) If thymus also weak there are probably errors in fixation or in dilution of conjugates.
       (b) If thymus is well stained the sections may be thinner than usual or there may be variation between mice (see above). Try sections of another tissue/mouse.
2. Non-specific staining.
       (a) If general, especially connective tissue this suggests 'dirty' conjugate. Try higher dilution: try spinning out aggregates (5 min in microcentrifuge); try gel filtration chromatography to purify fraction with less background (more likely to work if conjugate is new than if an old conjugate has deteriorated). If you think the conjugate is as good as you can get, you may have to try different fixatives. Otherwise, make new conjugate.
       (b) Specific sites, for example gut mucins. Different fixatives may solve a problem at a specific site. Otherwise, different 'blocking' agents can be tried empirically. The problem is often much worse with one conjugate than another, so a range of conjugates at different conjugation ratios, and purified by gel filtration, may provide the answer. If this fails, try changing the source of the reagents used (e.g. enzyme, glutaraldehyde).
3. 'Crystallization' of the alkaline phosphatase stain on storage of the sections.
   No remedy. Photograph your important results while they are fresh.
4. Poor resolution of the H-2 type of adjacent cells.
   At the light microscope level, this may be improved by using fluorescent rather than enzyme conjugates and viewing the different channels in rapid succession. Immunoelectron microscopy with H-2 antigens is possible (34).

---

ent in the chimaera. This provides both a positive control for the staining, and a control for specificity. Also, include at least two slides, each incubated with only one or the other antibody conjugate, but stained with *both* substrate solutions. This provides a further control for non-specific staining, and for an effect of one antibody conjugate on the binding of the other: this may be particularly useful when problems of non-specific staining suddenly arise as one of the conjugates deteriorates with time. Finally, always include a slide bearing thymus sections from each of the strains contained in the chimaera as a quality control for the staining run.

A method for double staining H-2^b and H-2^k antigens in sections of CBA↔C57 BL/6 chimaeric tissue is given in *Table 5* and an example of staining is shown in *Figure 2a*. Guidance on 'Troubleshooting' is given in *Table 6*.

3.2 Other antigens

(i)   Ia antigens can be demonstrated using monoclonal antibodies in the same way as the H-2 antigens. They are expressed in bone marrow and lymphoid tissues, endothelium and some epithelia. The best tissue processing and fixation conditions should be determined for each antibody.

(ii)   Thy-1 antigens also form a polymorphic system, AKR and other A strain mice expressing the Thy 1.1 allele, and other strains (the great majority) Thy 1.2. Monoclonal antibodies specific to these alleles are available. Thy-1 antigens are reported to be expressed on T cells, in the central nervous system, on fibroblasts and possibly on myoepthelial cells in the mouse: other species differ. The patterns of staining which are seen by immunohistochemistry are notoriously dependent on the fixative used, especially in brain. Some reports also suggest that Thy-1 antigens may be shed from some cells and picked up by others, so it is possible that Thy-1, in some tissues at least, is not a cell-autonomous marker.

3.3 **Carbohydrate polymorphisms revealed by lectins**

By analogy with polymorphisms such as the blood groups in man, one might expect carbohydrate polymorphisms among mouse strains which could be exploited as cellular markers. Only one, recognized by DBA, has so far been used in this way (17−19) (*Figure 2b*) and its application is described in detail below in Section 3.4. It may be worthwhile, faced with a specific requirement for a marker, to screen the appropriate tissues in the relevant mouse strains with a panel of lectins in case a new marker can be found. Although success is by no means guaranteed, the work involved is quite little. A general scheme for screening for a carbohydrate polymorphism using lectins is given below.

(i)   Prepare a panel of 12 (or more) lectins, conjugated with enzymic or fluorescent markers, chosen to cover as broad a range of carbohydrate specificities as possible. Kits of lectins for this type of screening are commercially available. If possible, titrate the lectins on some sections of tissues to which they are known to bind. A brief literature review will help. If in doubt mouse small intestine is a good bet for most lectins. Check that the binding is specific by including a control in which the lectin conjugate is diluted in a 0.2 M final concentration of the mono- or disaccharide for which it has binding specificity.

(ii)   Prepare cryostat sections of the relevant tissue from as many genetically different mouse strains as possible. Mount comparable sections from three or more different mouse strains on the same slide to save work in staining and to allow easier comparisons. You will need twice as many slides as you have lectins (to allow for specificity controls incorporating the competing sugar) plus a few spares. If possible, use the sections unfixed in the first screen: if this proves difficult, a mild cross-linking fixative (e.g. 10% formol saline for 1 min) probably carries the least risk of destroying a potentially useful binding difference.

(iii)   Incubate the lectin conjugates with the tissue sections for 1 h at room temperature, wash, and stain (if enzyme, rather than fluorochrome, conjugates). Look for unequivocal, all-or-none differences.

(iv)   If any differences are found, investigate further the stability of the staining to tissue fixation and embedding; and whether the polymorphism provides a truly cell-autonomous marker. This may be difficult to establish. Ideally, validation against a second, independent, marker is required. If the polymorphism is satisfactory on this score, investigate its stability under different physiological conditions, for example at different ages from embryo onwards, whether its distribution is uniform throughout the tissue, and its stability under pathological conditions, for example, neoplasia, if these are to form part of the experiment. Remember that cell surface carbohydrates might be expected to show changes under these conditions.

3.4 The use of *Dolichos* lectin (DBA) marker

This marker was discovered by a systematic screen, as described in Section 3.3, and validated against the H-2 marker. It is only applicable to some vascular endothelium and to intestinal epithelium (18). It can be used from about 12 days of embryonic development onwards and in adults (for details of anatomical distribution, see 18,19,35). The polymorphism has a curious reciprocal pattern. There are two alleles: *Dlb-1*^{a} and *Dlb-1*^{b} (*Table 7*). *Dlb-1*^{a} mice express DBA binding sites on endothelium but not on intestinal epithelium; in *Dlb-1*^{b} mice the distribution is reversed.

The DBA marker displays many advantages. It is extremely robust. The lectin binding site survives most fixation procedures (with the important exception of glutaraldehyde greater than 0.2% concentration), and embedding in paraffin wax or in resin provided it is not exposed to temperatures above about 65°C. The strong, clean staining is ideal for whole mount preparations. The staining procedure is also simple, and abolition of staining by incubation in 0.2 M *N*-acetylgalactosamine is an effective control for specificity.

The disadvantages are restricted tissue distribution, and the lack of a positive stain for the other component of the chimaera. The expression of DBA binding sites is also altered in dysplastic foci induced in intestinal epithelium by chemical carcinogenesis, and it is patchy in some capillary endothelium and in intestinal crypts adjacent to lymphoid follicles. This illustrates the caution which must be exercised in the evaluation of any new marker. The procedure for the use of *Dolichos* lectin as a cell marker is given below.

3.4.1 *Preparation of DBA conjugates*

Conjugates of DBA with alkaline phosphatase, peroxidase, fluorochromes, biotin and gold particles are available commercially or can be prepared by standard procedures outlined in references 30, 31 and in textbooks. Perform the conjugation in the presence

Table 7. Strain distribution of *Dlb-1* alleles.

Dlb-1^a	endothelium + intestinal epithelium −	DDK, GRS/A. LTS.Af, MAS, SM/Ja, STS, SWR, RIII-ro
Dlb-1^b	endothelium − intestinal epithelium +	AKR, A/Nimr, BALB/cHeA, C57L, C57BL/6J, C57BL/10ScSn, B.10A, C57BL/KS, C57BL/Nimr, CBA/Ca, CBA/HN, C3H/Bi, C3H/He, DBA/1, DBA/2, DBA/Af, GE, HTG, IF, LP/Nimr, LT, NZB, 020/A, SEA/J, SJL/J, 101/H, 129/RrJ

of 0.2 M *N*-acetylgalactosamine to protect the binding site of the lectin: the sugar can be dialysed out afterwards.

3.4.2 *Preparation of tissue for sections*

The best fixative is methacarn (methanol 60:inhibisol 30:glacial acetic acid 10). This fixative dissolves plastic, so glassware should be used. Fix the tissue overnight, and then transfer to 70% ethanol. Standard paraffin wax embedding can be used, but excessive heating of the wax should be avoided.

3.4.3 *Staining of sections*

Use freshly cut sections (3 μm paraffin or cryostat) if possible since staining intensity seems to diminish in stored sections. Note that fluorescent conjugates are not satisfactory on wax-embedded sections. De-wax (or air dry cryostat) sections and wash with phosphate-buffered saline (PBS).

(i)   For staining with DBA conjugated to peroxidase, first block endogenous peroxidase with 0.1% phenylhydrazine HCl in PBS, wash for 5 min with PBS, then incubate with PBS + 0.5% bovine serum albumin (PBS + BSA) for 10 min.

(ii)   Drain the slides of PBS + BSA, blot off excess moisture (see *Table 5*, step 12), and add DBA conjugate diluted in PBS + BSA.

(iii)   Incubate for 1 h at room temperature.

(iv)   Wash in several changes of PBS and stain as in *Table 5*, steps 16 and 17.

3.4.4 *Controls*

As a control for the specificity of staining include one section with DBA conjugate diluted in PBS + BSA with a final concentration of 0.2 M *N*-acetylgalactosamine. This will completely abolish specific staining unless the lectin conjugate is at very high concentration. As a quality control, include one section of a standard reference tissue (e.g. SWR mouse salivary gland in which capillaries stain) to assess quality of staining from run to run.

3.4.5 *Staining of whole mounts of tissue*

The procedure is the same as for sections, with the following modifications.

(i)   Fix the pinned out tissue (Section 6) in 10% formol saline (methacarn will dissolve the wax in the dissecting dish, and fixes and hardens the tissues rapidly, making removal of mucins difficult).

(ii)   Allow 30 min for blockage of endogenous peroxidase by phenylhydrazine HCl.

(iii)   Wash very thoroughly after removal of the conjugate (especially if the tissue is intestinal epithelium).

Whole mounts of chimaeric tissue are shown in *Figure 3*.

4. OTHER MARKERS GENERATING MOSAICISM

Some forms of tissue mosaicism generated naturally (X-inactivation), or artificially (extrinsic markers, induced somatic mutation or introduced DNA segments) will be men-

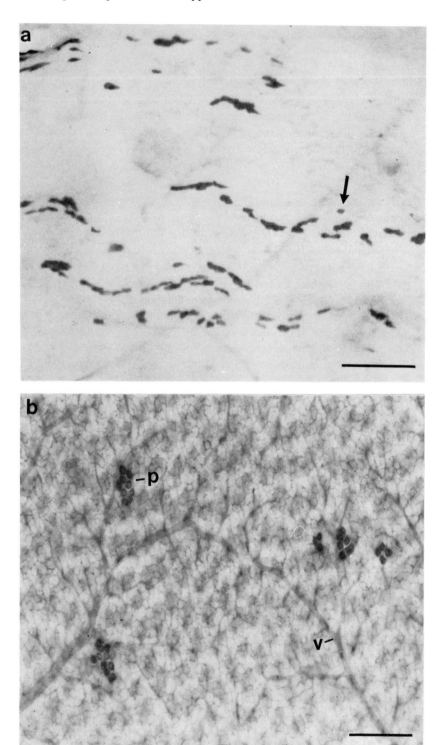

tioned only briefly here for completeness. Readers are referred to the literature cited for details of procedures and applications.

4.1 X-inactivation

Mosaicism for X-chromosome-linked genes in females arises from the random inactivation of one or other X chromosome in early development. Any polymorphic gene which is on the X chromosome, or can be made subject to X-inactivation as a result of X-autosomal translocations, is a potential candidate for use as a marker. The most widely used marker for studies of mouse development is the electrophoretic polymorphism for the enzyme phosphoglycerokinase, described in Chapter 7, Section 3. Recently, the use of DNA restriction fragment length polymorphisms combined with differences between the active and inactive X chromosome in methylation patterns have been used to provide clonal markers in studies of human cancers (36).

The advantages of X-chromosome mosaicism as a means of studying marked cell populations within an individual are:

(i)   a ready supply of mice without the need to make chimaeras;
(ii) minimal disturbance of normal tissue growth and function since the two marked populations differ only in the genes which are heterozygous and which are affected by X-inactivation.

The disadvantages of the X-chromosome mosaicism system are given below.

(i)   With the exception of pigmented tissues in 'flecked' mice with Cattenach's X-autosome translocation (see ref. 6) and of the enzyme ornithine carbamoxytransferase [which is useful only in liver (37)], the X-inactivation markers so far available cannot be readily visualized and so do not yield spatial information. It is possible that histochemical methods based on the 'null' allele for the X-linked enzyme glucose-6-phosphate dehydrogenase will shortly be developed for practical use in murine and human tissues.

(ii) X-inactivation mosaics are generally 'balanced', that is there are roughly equal populations of cells with the maternal X or the paternal X, chromosome active. Even with the use of the *Xce* genes, which result in preferential activation of one X, the mosaic cell populations are in the proportion of 70:30 (in contrast to the range of chimaerism that can be obtained: e.g. 95:5). This may complicate the interpretation of mosaic patches in terms of their clonal composition because adjacent clones of the same genotype will appear as single patches (see Section 5).

Figure 3. Whole mounts of chimaeric tissue to show clonal pattern. (**a**) Aortic endothelium of C57BL/6J (B6)◄►DDK chimaera. Stained with DBA − peroxidase. Endothelial cells of DDK origin bind DBA − peroxidase and appear black. The arrow indicates a single cell. The curved appearance of some of the groups of clones is due to contraction of the endothelium because of elastic fibres in the aortic wall. For detailed discussion of the patterns, see ref 46. Scale bar = 0.2 mm. (**b**) Small intestinal mucosa of C57BL/6J(B6)◄►RIII chimaera. Stained with DBA − peroxidase, viewed from the abluminal side after removal of the outer muscle layers (see Section 6). Five patches of B6 crypts (p) are seen. Because this chimaera was very 'unbalanced', with the B6 component in a minority, patch aggregation is at a minimum, and each patch is probably a discrete coherent clone. The coherent clones may, however, be all part of one descendent clone (see ref. 3 for discussion). The regular background pattern is due to villi on the other side of the sheet, and to RIII blood vessels (v) which bind DBA peroxidase. Bar = 0.5 mm.

4.2 Extrinsic single cell markers

Markers which can be visualized in tissue sections or whole tissues, such as horseradish peroxidase, latex microspheres, [^{3}H]thymidine and fluorochromes, have been used to label cells of the pre-implantation and early post-implantation embryos (see Chapters 2 and 3, and refs 1,11). These markers can be introduced either using cells which have been labelled *in vitro* and injected or aggregated into the unmarked embryos, or by direct injection of marker into single embryonic cells *in situ*. The markers introduced into the single cells are inherited by their descendents, although the labels do become diluted and so are suitable for short-term experiments only. The accuracy of the *in vivo* injections is not sufficient to specify precisely which cell or cells will be marked. However with this technique there is no constraint on the mouse strains which can be used.

4.3 Somatic mutation

In *Drosophila*, mitotic recombination induced by irradiation in flies heterozygous for a recessive marker mutation produces occasional homozygous mutant marked cells. These continue to divide and their progeny form a clone which can be recognized by the mutation. This technique has proved very valuable in the clonal analysis of development in the fly. Similar markers might, in principle, be generated in the mouse by mutation in animals heterozygous for appropriate marker genes. Mutations in coat colour genes have been used as the basis of the 'spot' test, which is used for assaying potential mutagens (38). The mutation beige (bg), which in the homozygous condition leads to the production of distinctively abnormal retinal melanocytes, may have a similar application (39). In principle, any of the other visual chimaeric markers which can be obtained in heterozygous form could be used in the same way. The likely problems would include:

(i)   the low frequency and unpredictable site of marked clones, which would require them to be readily assayable in large volumes of tissue;

(ii)   the damage to normal growth and development caused by the mutagen treatment; and

(iii)   the need for a large population of target cells to have a reasonable expectation of generating some marked cells. This latter requirement means that the method could not easily be used to mark cells in early embryonic development.

4.4 Transgenic mice

Transgenic mice are described in Chapter 11. In principle, it should be possible to introduce 'marker' genes whose product can be detected histochemically, immunologically or by *in situ* hybridization, at almost any stage of development from the oocyte or zygote to the adult, using direct microinjection, infection with defective retroviruses or introduction of genetically marked cells. The current problem for histochemical or immunological detection, which will certainly soon be solved, is to obtain gene constructs which will provide stable expression of the chosen gene in a wide spectrum of embryonic and adult tissues. Transmission of the marker gene through the germ line in some cases, will provide marked strains of mice; and integration of the gene in the X chromosome will provide an X-inactivation marker also.

5. INTERPRETATION OF RESULTS FROM MARKER STUDIES

5.1 Analysis of tissue extracts for markers of mosaicism

In this case, the marker provides no topographical information, but simply indicates the ratio of expression of the genotypes in the tissue sample. The potential problems are outlined below.

5.1.1 *Sensitivity*

How small a minority component in the tissue can be detected? Usually this seems to be of the order of $5-10\%$, although for glucose phosphate isomerase and phosphoglycerate kinase (Chapter 7, Section 3), greater sensitivity is claimed. Sensitivity may be assessed by reconstruction experiments, mixing tissue extracts from each genotype before assaying for the ratio of the two marker activities.

5.1.2 *Specificity*

There is no visual indication of which cell types are contributing which genotype to the tissue sample. Thus in the assessment of the composition of a distinct structure composed of several cell types (e.g. a tumour), monoclonal composition of one cell population may be obscured by admixture with other, polyclonal populations.

5.1.3 *Sampling problems*

These arise in two contexts.

(i)   The variance in composition of a tissue from a series of mice has been used to deduce the number of progenitor cells from which the tissue originates. The deduction requires the assumptions that the progenitor cells were sampled randomly from the original cell population, and that each has subsequently contributed to the same extent to the tissue which is being analysed (for discussion, see ref. 6). Probably neither of these assumptions will be true.

(ii)   The proportion of tissue samples of a given size which are of mixed composition has also been used, by similar reasoning, to deduce the clone size in the tissue. This argument requires that clones be of equal size and randomly arranged, which is clearly not the case in those tissues which have been studied by direct visualization of patches (3).

5.2 Analysis of visible mosaic patterns

Markers which provide a visual indication of the genotype of individual cells avoid many of the problems of non-visual markers. Even so, the interpretation of the mosaic patterns is not as straightforward as some accounts in the literature might suggest. A brief account is given here: readers are referred to the literature cited for further discussion.

5.2.1 *Definitions*

A 'clone' is a group of cells which share a common descent. A 'coherent clone' is a group of cells which have a common descent and which have remained contiguous; a 'descendent clone' is a group of cells which share a common descent but whose members have become separated so that their common origin may or may not be apparent from their spatial clustering. A 'patch' is a contiguous group of cells of like genotype in a mosaic tissue and is the unit of pattern which one actually sees: several adjacent coherent clones may aggregate to form a patch.

5.2.2 Analysis

(i) *Measurement of clone sizes.* There are two problems. The first is aggregation of clones into patches. Various mathematical methods have been proposed to allow calculation of clone sizes from the observed patch size, taking into account the proportions of each genotype in the tissue (40), but each method depends upon the clones being of regular shape, equal size and randomly distributed (41). The second problem is in the extrapolation of two-dimensional patch and clone sizes (e.g. in an epithelium) from the one-dimensional patterns seen in tissue sections. Once again this is only possible by reliance on unproven assumptions about the regularity of the patches, or by reconstruction from serial sections, which is likely to be impractically tedious. Even if true clone sizes can be deduced, their interpretation in developmental terms may not be very clear: this is discussed further in (iv) below.

(ii) *The spatial arrangement of clones.* More information than from sizes alone may be gained from the spatial distribution of clones. Clear patterns may be apparent on visual inspection [e.g. of the retinal pigment epithelium (42) or adrenal cortex (43) or melanocyte stripes in the coat]. In other tissues more subtle analysis may be required to define descendent clones. One approach is the Greig–Smith analysis of variance, widely used in plant ecology, and applied to chimaeric mouse tissues by Schmidt *et al.* (3). The frequency of occurrence of patches of one genotype is scored in each of the squares of a grid which is superimposed on the tissue: a random distribution of patches will give a Poisson distribution of numbers of patches per square of the grid; a non-random spatial distribution ('clustering' or 'over-dispersion') will be recognized by departure from the Poisson expectation. This approach has allowed the tentative recognition of descendent clones in intestinal epithelium and aortic endothelium (3).

(iii) *The use of whole mounts of tissue and 'unbalanced' chimaeras.* The interpretation of mosaic patterns is greatly simplified if whole mounts of tissue can be used (for preparation see Section 6). This avoids the problems of extrapolating from one dimension (an epithelium in section) to two (an epithelial sheet) and of the limited sampling imposed by tissue sections. It allows the spatial arrangement of patches to be examined over a wide area, which may reveal patterns unrecognizable on a smaller scale. The patterns will be even clearer if one component genotype is in a small minority, so that each patch represents a single coherent clone, and the relationship of coherent clones to one another in a descendent clone can be recognized without contamination by an adjacent clone. These conditions can be met either by using 'unbalanced' embryo chimaeras or by the introduction of single cell markers at later stages of development. If unbalanced embryo chimaeras are used, bear in mind that while the imbalance of the two component genotypes in the tissue may have arisen by chance, it may also reflect a growth advantage for one component over the other, which will complicate the interpretation of the patterns. If unbalanced chimaeras are frequent, and the imbalance is consistently in the same direction, this is very probably the case.

(iv) *What do clone sizes in mouse tissue mean?* Coherent clone sizes will reflect the effects of cell proliferation (and proximity of daughter cells) leading to larger clones, and of cell mingling, during growth. The relative contributions of each are not easily distinguished, so the interpretation of the final patterns in terms of clonal histories is likely to be uncertain. The subject is discussed by West (40) and by Schmidt *et al.* (3, 41). In the tissues which have been analysed, the frequency of clone sizes was not normally distributed, but highly skewed, with a preponderance of small clones and a few very large ones. The mean clone size thus provides a rather poor description of the tissues, which is better described by the median and upper and lower quartiles. More important, the skewed frequency distribution suggests that, probably by chance, some cells grow into larger clones than others. Thus, even if one ignores the possibility that one component of the tissue has a growth advantage over the other, one still cannot take a series of mosaic tissues, find that the minimum contribution of the minority genotype is, say, 1% and conclude that the whole tissue arises from a minimum of 100 progenitors. One progenitor cell in 100 will not necessarily give rise to 1% of the final tissue. Numbers of descendent clones also will not indicate the numbers of tissue progenitors, for two reasons. Firstly, the clones originate from a putative time of complete cellular mixing which is not defined in relation to commitment of progenitors to the tissue. Secondly, the total number of descendent clones in the tissue cannot reliably be inferred from the relative proportions of the minority and majority genotypes and the number of clones of the minority type, because of the possibility that the minority genotype is at a growth disadvantage.

6. PREPARATION OF WHOLE MOUNTS OF SOME EPITHELIAL TISSUES

6.1 **Equipment**

> Two pairs of fine forceps.
> Very fine scissors (e.g. Professor Kinmoth's scissors, Macarthy's Surgical Ltd, Dagenham, Essex, UK).
> Eye blade (Beaver KB-225-063, Downs Surgical Ltd, Mitcham, Surrey, UK)
> Entomological pins (Watkins and Doncaster, Hawkhurst, Kent, UK).
> Petri dishes with 5 mm wax base (Ralwax 2, Raymond A. Lamb, Sunbeam Road, London NW10, UK).
> Dissecting microscope, preferably with zoom lens and cold light source for illumination from above.

6.2 **Intestine**

(i)   Dissect out the entire intestine, trim off adherent mesentery and vessels.
(ii)   Cut at the ileo-caecal junction and at the ascending colon.
(iii)   Discard the caecum.
(iv)   Place in PBS in ice.

6.2.1 *Staining of the luminal surface for surface patterns on villi and colonic epithelium.*

(i)   Start at either end. Cut a length of intestine about 2/3 the length of the wax dish (to allow for stretching). Place into a wax-based dissecting dish well covered in cold PBS, pin each end to the wax under slight tension.

(ii)   Open the intestine with scissors along the mesenteric border (uppermost) to a length of 2−3 cm. Grasp the cut edges with fine forceps (or spear with the point of a pin) and pin to the wax with gentle stretching. Continue with pins every 8−10 mm until complete. Remove obvious faecal material by gentle pipetting (if you have a depth of 5 mm PBS over the gut you are less likely to damage the villi) as you go along.

(iii)   At the completion of each strip, fix briefly in 10% formol saline (30 sec−1 min) and rinse with PBS.

(iv)   When the dissection is complete, fix in formol saline for 45 min at room temperature.

(v)   Then incubate in 20% ethanol−80% 150 mM Tris pH 8.2 with 25 mM dithiothreitol for 45 min at room temperature. This helps to loosen mucins, which can be removed by gentle pipetting under the dissecting microscope. Avoid scraping the villi.

(vi)   Rinse and return to formol saline until ready for staining. Store in an air-tight box in formol saline.

6.2.2 *Staining the abluminal surface of the intestinal epithelium to demonstrate mosaic patterns formed by the bases of crypts (Figure 3b)*

Because in the adult mouse each crypt is monoclonal (44), each crypt represents a single 'cell' in the mosaic pattern (41).

(i)   Incubate the intestine in 2% (w/v) lignocaine hydrochloride in PBS at 4°C from 30 min (small intestine) to 2 h (colon) before dissection. This will loosen the circular and longitudinal muscle layers.

(ii)   Pin the intestine in the wax-based dissecting dish (a black wax may be easier). Starting at the duodenal end, use the dissecting microscope and eye blade to find a plane of cleavage between the muscle layers and the mucosa. The thin translucent layer of muscles can then be slowly stripped back with fine forceps like a stocking. All adventitious fat or mesentery must be carefully trimmed beforehand; the dissection plane may be lost at the sites of insertion of major vessels and around lymphoid follicles, and must be carefully re-established.

(iii)   When complete, open the mucosal tube longitudinally and pin out, villus (luminal) surface down.

(iv)   Fix in formol saline; no treatment is needed to remove mucins. One intestine will take 2−3 h, even after practice. See Schmidt *et al.* (45).

6.3 Aorta

(i)   Kill the mouse by ether anaesthesia (cervical dislocation may rupture the aorta).

(ii)   Dissect out the aorta from the arch to the diaphragm. Place in cold PBS in a dissecting dish preferably with black wax and incident illumination.

(iii)   Carefully trim off all fat and adherent tissue. Secure one end with a fine pin, carefully open the aorta longitudinally with fine scissors, and pin out. Gently pipette away adherent blood clot.

(iv)   Fix in formol saline or appropriate fixative.

(v)   Cell outlines can be stained with silver stains (0.2% silver nitrate, develop by exposure to bright light), or nuclei stained with haemalum as a check of the integrity of the endothelial layer. [See Schmidt *et al.* (46)].

6.4 Epidermis

(i) Remember to photograph the mouse before dissection if you want to correlate epidermal and coat colour patterns.

(ii) Kill the mouse, shave very gently with electric clippers, taking care not to abrade the skin surface, and dissect the required area of skin. For 'whole skin' preparations in which the midlines are to be examined, cut up one flank and around the pelvic and pectoral regions, and mark the midlines and one flank with a stitch before removing the skin—orientation subsequently may be difficult as the skin loses its shape.

(iii) Pin the shaved skin on a board and smear it with cosmetic depilatory cream to remove remaining hairs, leave for about 20 min and rinse off with PBS. Use the loose tip of a rubber glove on your finger to agitate a slurry of cream and PBS gently over resistant areas: this will remove remaining hair with minimum risk of rubbing a hole in the epidermis. Clear the cream and hair with a stream of PBS from a wash bottle until the skin surface is clean and smooth.

(iv) Unpin and incubate in Hanks buffered salt solution (HBSS) with 20 mM EDTA, adjusted to pH 7.2, at 37°C, for 20 h with gentle stirring. This will loosen the epidermis.

(v) Pin out in a black wax dish, epidermis uppermost, and cover in PBS. Under the dissecting microscope gently separate the epidermis by holding down the dermis with forceps while gently pushing back the epidermis with a second pair of closed forceps. An intact sheet of the entire body region can be dissected in 1−2 h.

(vi) Transfer the sheet (a paper or glass sheet support may help, or the sheet can simply be poured from one dish to the other) to a white wax dish, where it is pinned out, basal surface uppermost, for staining.

7. REFERENCES

1. Gardner,R.L. (1985) *Phil. Trans. R. Soc. Lond. B*, **312**, 163.
2. Le Douarin,N.M. (1985) *Phil. Trans. R. Soc. Lond. B.*, **312**, 153.
3. Schmidt,G.H., Wilkinson,M.M. and Ponder,B.A.J. (1985) *J. Embryol. Exp. Morphol.*, **88**, 219.
4. Fialkow,P.J. (1976) *Biochim. Biophys. Acta*, **458**, 283.
5. Mullen,R.J. (1977) *Nature*, **270**, 245.
6. McLaren,A. (1976) *Mammalian Chimaeras*. Cambridge University Press, Cambridge.
7. Le Douarin,N. and McLaren,A. (eds) (1984) *Chimaeras in Developmental Biology*. Academic Press, London.
8. Babinet,C. (1980) *Exp. Cell Res.*, **130**, 15.
9. Papaioannou,V.E. and Dieterlan-Lievre,F. (1984) In *Chimaeras in Developmental Biology*. Le Douarin,N. and McLaren,A. (eds), Academic Press, London, p. 3.
10. Green,M.C. (ed.) (1981) *Genetic Variants and Strains of the Laboratory Mouse*. Fischer Verlag, Stuttgart and New York.
11. West,J.D. (1984) In *Chimaeras in Developmental Biology*. Le Douarin,N. and McLaren,A. (eds), Academic Press, London, p. 39.
12. Feder,N. (1976) *Nature*, **263**, 67.
13. Dewey,M.J., Gervais,A.G. and Mintz,B. (1976) *Dev. Biol.*, **50**, 68.
14. Oster Granite,M.L. and Gearhart,J. (1981) *Dev. Biol.*, **85**, 199.

Cell marking techniques and their application

15. Ponder,B.A.J., Wilkinson,M.M. and Wood,M. (1983) *J. Embryol. Exp. Morphol.*, **76**, 83.
16. Enzine,S., Weissmann,I.L. and Rouse,R.V. (1984) *Nature*, **309**, 629.
17. Schmidt,G.H., Wilkinson,M.M. and Ponder,B.A.J. (1985) *Cell*, **40**, 425.
18. Uiterdijk,H.G., Ponder,B.A.J., Festing,M.F.W., Hilgers,J., Skow,L. and Van Nie,R. (1986) *Genet. Res. Camb.*, **45**, 125.
19. Ponder,B.A.J. and Wilkinson,M.M. (1983) *Dev. Biol.*, **96**, 535.
20. Kitamura,Y., Matsuda,H. and Hatanaka,K. (1979) *Nature*, **281**, 154.
21. Oliver,C. and Essner,E. (1973) *J. Histochem. Cytochem.*, **21**, 218.
22. Mintz,B. and Sanyal,S. (1970) *Genetics*, **64**, Suppl. 43.
23. Rossant,J. (1985) *Phil. Trans. R. Soc. Lond. B.*, **312**, 91.
24. Peterson,A.C. (1979) *Ann. N.Y. Acad. Sci.*, **317**, 630.
25. Mintz,B. and Baker,W.W. (1967) *Proc. Natl. Acad. Sci. USA*, **58**, 592.
26. Mintz,B. (1974) *Annu. Rev. Genet.*, **8**, 411.
27. Ford,C.E., Evans,E.P. and Gardner,R.L. (1975) *J. Embryol. Exp. Morphol.*, **33**, 447.
28. McLaren,A. (1975) *J. Embryol. Exp. Morphol.*, **33**, 205.
29. Lu,T.-Y. and Markert,C.L. (1980) *Proc. Natl. Acad. Sci. USA*, **77**, 6012.
30. Ponder,B.A.J., Wilkinson,M.M., Wood,M. and Westwood,J.H. (1983) *J. Histochem. Cytochem.*, **31**, 911.
31. Ponder,B.A.J. (1983) In *Immunocytochemistry*. Polak,J.M. and Van Noorden,S. (eds), Wright, PSG, Bristol, p. 129.
32. Bondi,A., Chieregatti,G., Gusebi,V., Fulcheri,E. and Bussolati,G. (1982) *Histochemistry*, **76**, 153.
33. Ponder,B.A.J. and Wilkinson,M.M. (1981) *J. Histochem. Cytochem.*, **29**, 981.
34. Van Ewijk,W., Rouse,R.V. and Weissman,I.L. (1980) *J. Histochem. Cytochem.*, **28**, 1089.
35. Ponder,B.A.J., Festing,M.F.W. and Wilkinson,M.M. (1985) *J. Embryol. Exp. Morphol.*, **87**, 229.
36. Vogelstein,B., Fearon,G.R., Hamilton,S.R. and Feinberg,A.P. (1985) *Science*, **227**, 642.
37. Wareham,K.A., Howell,S., Williams,D. and Williams,E.D. (1983) *Histochem J.*, **15**, 363.
38. Fahrig,R. and Neuhauser-Klaus,A. (1985) *J. Hered.*, **76**, 421.
39. Searle,A.G. and Stephenson,D.A. (1982) *Mutat. Res.*, **92**, 205.
40. West,J.D. (1975) *J. Theor. Biol.*, **50**, 153.
41. Schmidt,G.H., Garbutt,D.J., Wilkinson,M.M. and Ponder,B.A.J. (1985) *J. Embryol. Exp. Morphol.*, **85**, 121.
42. Schmidt,G.H., Wilkinson,M.M. and Ponder,B.A.J. (1986) *J. Embryol. Exp. Morphol.*, **91**, 197.
43. Weinberg,W.C., Howard,I.C. and Iannaccone,P.M. (1985) *Science*, **227**, 524.
44. Ponder,B.A.J., Schmidt,G.H., Wilkinson,M.M., Wood,M.J., Monk,M. and Reid,A. (1985) *Nature*, **313**, 689.
45. Schmidt,G.H., Wilkinson,M.M. and Ponder,B.A.J. (1984) *Anat. Rec.*, **210**, 407.
46. Schmidt,G.H., Wilkinson,M.M. and Ponder,B.A.J. (1986) *J. Embryol. Exp. Morphol.*, **93**, 267.

CHAPTER 7

Biochemical microassays for X-chromosome-linked enzymes HPRT and PGK

MARILYN MONK

1. INTRODUCTION

This chapter describes the measurement of enzyme activities in minute quantities of biological material, such as single mouse eggs, early cleavage embryos, small pieces of tissue representing precursor cell lineages and small numbers of stem cells or germ cells. The microassay techniques are designed to handle enzyme extraction and quantitation of activity in tiny volumes of supporting medium and at the utmost level of resolution and sensitivity. In my laboratory, we have concentrated on developing microassays for the X-linked enzymes hypoxanthine phosphoribosyltransferase (HPRT; EC 2.4.2.8) and phosphoglycerate kinase (PGK-1; EC 2.7.2.3). However, the general principles of the microassay techniques described here should be applicable and adaptable to the development of microassays for other enzymes.

The microassay described for HPRT is an activity assay that we have used extensively to monitor HPRT gene dosage in the course of studies of X chromosome activity during development of the female mouse embryo. The principle of the dosage studies is as follows. If two X chromosomes are active in females at a particular stage of development, or in a particular tissue, then the female sample will exhibit twice as much HPRT activity as the male sample ($X^+X^+:X^+Y = 2:1$). Where one X chromosome is inactivated in the female, then it will be equivalent to the male in HPRT avtivity ($X^+X^-:X^+Y = 1:1$).

A major advantage of the assay is that it can be used as a double microassay which simultaneously measures the activities of the X-linked HPRT and the autosome-linked adenine phosphoribosyltransferase (APRT; EC 2.4.2.7) in a single sample in a single reaction mixture. The HPRT and APRT enzymes are related in function, both of them being involved in the purine salvage pathway which re-incorporates purine bases into DNA. The double microassay enables the ratio of X-linked HPRT activity to autosome-linked APRT activity to be calculated. Thus an indication of gene dosage of the X-linked enzyme relative to an autosomal enzyme can be obtained in situations where the absolute quantity of tissue cannot be standardized by separate measurement (of, say, protein content) because of the paucity of tissue. A comparison of HPRT:APRT ratios eliminates sampling errors and allows comparison between samples. However, with standard practice and standard samples, such as pre-implantation embryos at a particular stage, or known numbers of cells, the microassay is sufficiently accurate

to give the activity of each individual enzyme (expressed per embryo or per cell).
Examples of the application of the HPRT:APRT double microassay are as follows.

(i) During early mouse development, up to the 8-cell stage, the increase in HPRT activity is attributable to translation of stored maternal mRNA for the enzyme (1,2). Early embryos derived from eggs of XX mothers show twice the HPRT activity compared with those from XO mothers (see also ref. 3). Studies using inhibitors of macromolecular syntheses support the argument for stored maternal mRNA coding for the early increase in HPRT activity (2).

(ii) By the morula stage the embryonic genes for HPRT enzyme synthesis become active. Female (XX) embryos have twice the HPRT activity as do male (XY) embryos, showing that both X-chromosome-linked HPRT genes are active in the female embryo at this stage (1,4).

(iii) Inactivation of one of the two X chromosomes occurs sequentially, first in the extra-embryonic trophectoderm, then in the primary endoderm and finally in the embryo precursor tissue (5).

(iv) Re-activation of the inactive X chromosome occurs in the female germ line just prior to meiosis (6).

(v) Recently, the double microassay for HPRT and APRT has been used to identify female mice which are X-chromosome-inactivation mosaics for HPRT-positive and HPRT-negative cells. The heterozygous females were sired by chimaeric males with germ cells derived from HPRT-negative embryonic stem cells. One half of the male progeny of the heterozygous females was shown to be HPRT-deficient (7). In humans, HPRT-deficiency is the basis of Lesch-Nyhan syndrome.

(vi) The microassay is sufficiently sensitive to diagnose HPRT-deficient male embryos by assay of a single blastomere of an 8-cell pre-implantation embryo (8). The rest of the embryo is transferred to a pseudopregnant foster mother to check the accuracy of the diagnosis. The sex of an embryo may also be accurately diagnosed by assay of HPRT dosage in a biopsied blastomere (9).

(vii) In addition, we have used the microassay to monitor the activity of cloned HPRT genes introduced into eggs by microinjection (Monk, Ao, Lovell-Badge and Melton, unpublished). The efficacy of different promoter functions can be assessed by using HPRT activity as a reporter function. In this case, the simultaneous measurement of APRT is a valuable control for normal development of the injected egg.

The microassay for PGK-1 described in this chapter has the advantage that it distinguishes the two X chromosomes as well as measuring activity levels. In the mouse the X-linked enzyme, PGK-1, has two distinct electrophoretic forms, PGK-1A (discovered in a wild mouse population by Nielsen and Chapman, 10) and PGK-1B. These PGK-1 isozymes can be separated by gel electrophoresis and used to quantitate the proportions of cells with either the maternal or the paternal X chromosome active in female embryos heterozygous for the *Pgk-1^a* and *Pgk-1^b* alleles. The quantitation of the two isozyme forms is achieved with considerable accuracy by repeated measurement of fluorescence as it develops in the staining reaction. This assay is highly sensitive and detects activity in single mouse eggs or embryos or in very few cells of different cell lineages dissected by microsurgery.

Examples of the application of the PGK isozyme quantitation to development are as follows.

(i) The paternally-inherited X chromosome is preferentially inactivated in the extra-embryonic lineages, the trophectoderm and the primary endoderm (11−13).

(ii) X-chromosome inactivation is random in the primordial germ cells of the mouse (14,15).

(iii) Mosaic studies show correlations in proportions of maternal and paternal X chromosomes active in the cells of the definitive germ layers and the germ cells, suggesting that these tissues derive from a common pool of X-inactivated cells (16).

(iv) The preferential expression of a translocated X chromosome is due to cell selection early in development (17).

(v) There is a preferential expression of the *Pgk-1^b* allele in oocytes of heterozygous females (18).

(vi) Single intestinal crypts in heterozygous females express one or the other isozyme form, suggesting that the crypts are monoclonal in origin (19).

(vii) In blood cell chimaeras, formed by injection of bone marrow into lethally ir-radiated mice, the PGK-1 isozymes have been used to monitor the re-population of bone marrow and peripheral blood (20).

(viii) PGK-1 isozymes have been used to investigate clonal interactions in tumours (21).

2. DOUBLE MICROASSAY OF HPRT AND APRT

HPRT is an X-coded enzyme (3,22) that catalyses the conversion of hypoxanthine and guanine to inosine monophosphate (IMP) and guanosine monophosphate (GMP). APRT is autosome-coded (23) and converts adenine to adenine monophosphate (AMP). Both enzymes use phosphoribosyl pyrophosphate (PRPP) as a source of the phosphoribosyl moiety. Gartler *et al.* (24) used a double-label assay to measure HPRT and APRT levels in hair follicles, separating the substrates from the products by thin-layer chromatography (t.l.c.). Other workers have employed high voltage electrophoresis to separate the substrates and products of the two reactions (25). The assay we have developed is based on the one described by Bakay *et al.* (26), and used also by McBurney and Adamson (27). It is a simple assay in which embryo extracts are added to a reaction mixture containing labelled substrates, [^{3}H]hypoxanthine (or guanine) and [^{14}C]adenine, and PRPP. The labelled nucleotide products, IMP (or GMP) and AMP, are precipitated with lanthanum chloride, collected on a glass fibre filter and counted in a scintillation counter (*Figure 1*).

Monk and Kathuria (28) developed the assay to sufficient sensitivity to measure activity in single pre-implantation embryos and over the years the sensitivity of the assay has been increased until now we can simultaneously measure less than 1 pmol/h of activity of both enzymes in a single reaction mixture. Activity of either enzyme, or both, can be measured in single eggs or even one blastomere at the 16-cell stage. Once the assay is set up and all the preliminary controls have been carried out, the method is accurate and reproducible. With practice and skill in microtechniques, up to 100 assays can be performed in a day.

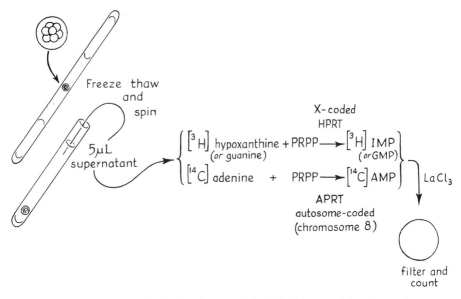

Figure 1. Diagram summarizing the double microassay for HPRT and APRT activities in an embryo extract. Details are given in the text. The diagram shows a single embryo being analysed. The embryo in PB1.PVP is transferred to the centre of a microcap which is then sealed at the ends and stored at −70°C. For assay the sample is freeze−thawed three times, centrifuged and the supernatant (~5 μl) added to the reaction mixture for assay.

2.1 Sample preparation

Eggs and pre-implantation embryos are isolated and manipulated as described in Chapter 2 (Sections 2 and 3). Inner cell mass and trophectoderm lineages are isolated from blastocysts as described in Chapter 2, Sections 3.4 and 3.5. Post-implantation embryos are isolated and dissected into component parts as described in Chapter 3, Section 4. Germ cells may be isolated as follows.

(i)   Isolate the gonads from post-implantation embryos. Wash and incubate them in PB1 (Chapter 13, Section 3.1.1) containing 0.4% polyvinyl pyrrolidone instead of albumin (PB1.PVP) and 0.2% EDTA, at room temperature for 30 min.

(ii)   Release the germ cells by rupture and squeezing of the gonads.

(iii)   Purify the germ cells by gently 'wafting' them away, in a stream of medium (using a mouth pipette), from gonadal somatic cells and red blood cells. Keep the germ cells always grouped together in the process.

(iv)   Estimate the number of germ cells in the pile, collect into microcaps (as below) and store at −70°C.

In all the cases the medium used for isolating embryos or washing cells or tissues is PB1.PVP. The PVP is included to increase the viscosity of the medium and thus reduce stickiness of small pieces of tissue, or of isolated cells, so that they are not lost due to sticking to the sides of the glass pipettes or microcaps. This is not so important for pre-implantation embryos which are encased in the non-sticky zona pellucida. Albumin is not used as it may show enzyme activity in the microassay. If stickiness continues to be a problem, it is helpful to keep the medium (and the watch glasses) cold and to keep the sample on the move till it is safely in the microcap.

Samples are finally collected into microcaps in 5 μl of PB1.PVP as follows.

(i) Take up plenty of PB1.PVP in a finely drawn Pasteur pipette controlled by a mouth pipette.

(ii) Take up the sample in the same Pasteur pipette. The sample may be a single blastomere, a single pre-implantation embryo, a single post-implantation embryo (or part thereof) or, say, 50 germ cells.

(iii) Insert the tip of the pipette into a Drummond 10 μl microcap and dispense the embryos into the microcap so as to half fill (5 μl) the microcap with fluid. Watch under the microscope that the embryo(s), tissue or cells enter the microcap.

(iv) Holding the microcap between finger and thumb, 'bump' the sample down into the middle of the microcap by gently knocking the heel of the hand on the bench.

(v) Immediately seal the ends of the microcap by melting the glass in a small flame.

(vi) Check under the microscope that the ends of the microcap are properly sealed.

(vii) Store the samples in small plastic Falcon tubes at $-70°C$.

(viii) Before assay, freeze−thaw the samples three times by transferring them from $-70°C$ to room temperature and back (avoid using liquid nitrogen which may cause the microcaps to break).

(ix) Centrifuge within the Falcon tube containers (2000 r.p.m. for 5 min at 4°C). (In cases where the samples may have stuck to the sides of the microcaps it is advisable to centrifuge the microcaps in the other direction also.)

(x) Remove the supernatant for assay as described below.

2.2 The reaction mixture

HPRT and APRT activities are measured simultaneously by incubating embryo extract at 37°C in 50 μl of reaction mixture consisting of sodium phosphate buffer (38.75 mM, pH 7.4), magnesium chloride (5 mM), PRPP (1 mM), [^{3}H]hypoxanthine (10 μM) and [^{14}C]adenine (10 μM). (In our earlier experiments [^{3}H]guanine was used instead of hypoxanthine as substrate for HPRT.) For maximum sensitivity the assay uses undiluted uniformly labelled [^{14}C]adenine, at the highest specific activity available. The other reagents are prepared and stored as sterile stock solutions in small volumes; this avoids contamination of reagents by repeated opening of the tube. Stock solutions are prepared as described below.

2.2.1 *Sodium phosphate buffer and [^{3}H]hypoxanthine*

Prepare sodium phosphate buffer (50 mM, pH 7.4) by mixing 1.75 ml of 0.2 M $NaH_2PO_4.2H_2O$, 8.25 ml of 0.2 M $Na_2HPO_4.12H_2O$ and 30 ml of water. Adjust the pH to 7.4 if necessary. Filter sterilize. Use this buffer component of the reaction mixture to dissolve the [^{3}H]hypoxanthine (which is supplied as a freeze-dried preparation, TRA.74, 1−10 Ci/mM, Amersham) as follows.

(i) Dissolve the freeze-dried preparation in the sodium phosphate buffer (50 mM, pH 7.4) to a concentration of 12.9 μM. For example dissolve 1 mCi of [^{3}H]-hypoxanthine at 6.2 Ci/mM in 12.5 ml of buffer.

(ii) Carefully take up the [^{3}H]hypoxanthine solution in a syringe and aliquot approximately 2 ml amounts through a disposable sterile filter (e.g. Millex, Millipore) into small Falcon tubes. Store at 4°C.

(iii)   If a reduction in specific activity is required, dilute the radioactive stock hypoxanthine with unlabelled 12.9 μM hypoxanthine (Sigma) dissolved in 50 mM phosphate buffer. The unlabelled stock hypoxanthine is sterilized by passage through a Millipore filter and stored as 10 ml amounts in Falcon tubes at 4°C.

(iv)   Use 38.75 μl of the [³H]hypoxanthine in buffer at the original specific activity, or lower specific activity as required, per 50 μl of reaction mix (final concentration of hypoxanthine is 10 μM and of buffer, 38.75 mM).

2.2.2 *Magnesium chloride*

Make up magnesium chloride (MgCl$_2$.6H$_2$O) to 50 mM and dispense 1 ml aliquots through a Millipore filter into sterile Falcon tubes. Store at 4°C. The magnesium chloride is made up separately and added at the time of preparation of the reaction mixture to avoid precipitation on storage. Use 5 μl per 50 μl of reaction mix (final concentration 5 mM).

2.2.3 *Phosphoribosyl pyrophosphate*

The PRPP (mol. wt 390.1, Sigma) is stored as a solid at -20°C and prepared fresh at 10 mM just before making the reaction mixture. Use 5 μl of 10 mM PRPP per 50 μl of reaction mixture (final concentration, 1 mM). Work out the total volume of PRPP solution required and dissolve at 0.39 mg/0.1 ml of sterile distilled water.

2.2.4 [¹⁴*C*]*Adenine*

The [U-¹⁴C]adenine (CFA.436; >220 mCi/mM, Amersham) is supplied as a liquid. Store it at 4°C. Uniformly labelled adenine is used because it is commercially available at higher specific activities than other labelled adenine preparations. An appropriate volume is added to 50 μl of the reaction mixture to give a final concentration of labelled adenine of 10 μM at the original specific activity (e.g. 2.35 μl of 235 mCi/mM adenine, at a concentration of 50 μCi/ml, to 50 μl of reaction mix).

2.3 **The reaction**

2.3.1 *Controls*

The microassay is extremely sensitive, measuring as little as 0.1 pmol/h of activity of HPRT. Consequently, all stock solutions must be sterile and all handling procedures scrupulously clean. Avoid contaminating tubes or microcaps with fingertips although it is impracticable to wear gloves.

To achieve maximum sensitivity, reactions may be incubated at 37°C for many hours, or overnight if desired, provided there is no risk of bacterial, or other, contamination, and provided the activity of each enzyme remains linear with time. Preliminary experiments should be done to monitor the linearity with time for the material to be assayed and also to determine the point at which the reaction mixture is exhausted (seen as a plateau of product counts in an activity versus time curve). These control experiments are essential, otherwise spurious HPRT:APRT ratios may be obtained.

The concentration of PRPP in the reaction mixture is important. Therefore, it is advisable to check various concentrations of PRPP for maximum recovery of product counts when setting up the reaction. The omission of PRPP from the reaction mixture should prevent the appearance of product counts in the lanthanum chloride precipitate.

In an experiment where a number of extracts are to be assayed, the samples are added to pre-prepared 50 μl aliquots of reaction mixture in turn, keeping note of the time of addition. It is essential to include blank controls of supporting medium (PB1.PVP) without extract at the beginning and end of the sequence of loading the samples. If the counts converted to product in the experimental reactions are not substantially increased over blank levels, then it is vitally important to include replicate blanks and to ensure that the counts in the replicate blanks are constant. This will depend on reproducible experimental techniques in setting up and stopping the reactions and in the washing procedures (see below).

It is also advisable to include a blank somewhere amongst the experimental samples to show that there is no carryover of counts from one sample to the next during the consecutive filtration and washing of the samples. Ideally, this blank control should be included after a sample that might be expected to have high product counts.

2.3.2 *Setting up an assay*

(i)   Work out a protocol, ordering the reactions, for example, blanks (PB1.PVP), 1−3; samples (say 20), 4−23; carryover control (50 μl reaction mixture only), 24; blanks (PB1.PVP), 25−27.

(ii)   Prepare the reaction mixture. In the example given in step (i), 27 aliquots of 50 μl of reaction mixture are required. Prepare the bulk reaction mixture as follows:

27 × 38.75 μl [^{3}H]hypoxanthine (12.9 μM) in phosphate buffer (50 mM, pH 7.4) . . . 1046.25 μl.

27 × 5 μl magnesium chloride (50 mM) . . . 135 μl.

27 × 5 μl PRPP (0.585 mg in 0.15 ml) . . . 135 μl.

27 × 2.35 μl [^{14}C]adenine (235 μCi/μM) . . . 63.45 μl.

Total volume = 1379.7 μl.

(iii)   Number the reaction tubes (e.g. disposable plastic test tubes, Falcon 2038, 10 × 75 mm) 1−27 and dispense 50 μl of reaction mix into the bottom of each tube.

(iv)   During the setting-up procedure, or some time before, freeze−thaw the samples in the microcaps three times, and centrifuge (in two directions if deemed necessary) as in Section 2.1.

(v)   Using a diamond knife, score the microcaps ready for breaking open.

(vi)   Start the clock.

(vii)   Break open each microcap in turn and, using a mouth pipette controlling a finely drawn Pasteur pipette, recover the supernatant (∼5 μl) from the microcap, add to the appropriate numbered reaction tube and place at 37°C. Do this in turn for each microcap until all samples are in the water bath, noting the start time for each reaction tube.

(viii) Leave at 37°C for the desired reaction time.

145

2.4 Stopping the reaction

The stop mix is an ice-cold solution of 0.1 M lanthanum chloride ($LaCl_3.7H_2O$, Sigma) containing 1 mM unlabelled hypoxanthine (Sigma) and 1 mM unlabelled adenine (Sigma), filtered and stored at 4°C.

(i)   At the appropriate times, remove each reaction tube in turn, add 1 ml of the stop mix and place the tube in an ice bath until all the samples are in the ice bath.

(ii)   Leave for 30 min after the last reaction has been stopped. The nucleotide monophosphate products (IMP and AMP) are precipitated by the heavy metal lanthanum.

2.5 Filtration and washing

The precipitated products are collected by filtration on glass fibre filters (2.4 cm, Whatman GF/C) using a Millipore filtration apparatus. The excess radioactive substrate is removed by extensive washing with distilled water.

(i)   Number the glass fibre filters with a biro, transfer to a glass Petri dish (filter number 1 on top) and cover with the lanthanum stop mix.

(ii)   Set up the Millipore filtration apparatus (ground glass support 3 cm diameter in a bored rubber bung set into a large side arm flask attached to a Millipore vacuum pump). The apparatus is set on a plastic tray (in case of spillage), together with two large beakers filled with distilled water for washing the funnel after each filtration, and a beaker of distilled water with a 10 ml syringe for washing the filters.

(iii)   Assemble the first filter on the glass support, clamp the funnel over the filter, and turn on the Millipore pump to create a vacuum.

(iv)   Vortex the first sample tube and pour the contents carefully down the funnel. Rinse the tube carefully several times with distilled water (a total of 10 ml from a 10 ml syringe) and add the washings to the funnel. Discard the tube.

(v)   Rinse around the funnel, washing through the filter, twice, with 10 ml of water each time.

(vi)   Remove the funnel and rinse by dipping it in the first, then the second, washing beaker filled with distilled water.

(vii)   Wash the exposed filter with another 10 ml of water and transfer the filter to an aluminium foil tray to dry.

(viii) Place the next filter in place, re-assemble the funnel and clamp, and repeat the filtration and washing procedure. Continue until all the samples are filtered and washed.

(ix)   Dry the filters in a hot air oven.

2.6 Counting and plotting the results

(i)   Transfer each filter to a plastic insert and add 5 ml of scintillation fluid (e.g. Liquifluor, New England). Insert the plastic cap and transfer the insert to a screw-capped glass scintillation vial.

(ii)   Count the ^3H and ^{14}C counts in the [^3H]IMP and [^{14}C]AMP products of the reaction, using the recommended dual-label counting settings of the scintillation counter.

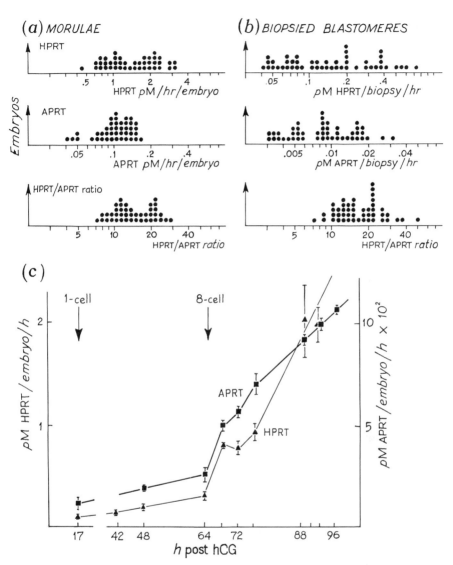

Figure 2. HPRT and APRT activities in pre-implantation embryos. (**a**) Distribution of HPRT and APRT and HPRT:APRT ratios in individual morulae. Two X chromosomes are active in female morulae and hence the HPRT activities and HPRT:APRT ratios show bimodal distributions of male and female embryos. In blastocysts, an X chromosome is inactivated in most cells of the female embryos and the HPRT and HPRT:APRT distributions become unimodal; data not shown. (**b**) HPRT, APRT and HPRT:APRT ratios in blastomeres biopsied from 8-cell embryos. (**c**) Increase in HPRT and APRT activities during pre-implantation development of the mouse embryo. Female mice were superovulated as described in Chapter 1 (Section 5.4) and embryos collected at various times following human chorionic gonadotrphin (hCG) injection. The first part of the increase of HPRT activity (up to the 8-cell stage) is coded by maternal mRNA, thereafter the increase is embryo-coded (2).

When the assay is first set up, and periodically thereafter, work out the spillover of ^3H counts appearing in the ^{14}C channel and vice versa. This must be done using reaction mixtures with ^3H alone (water instead of [^{14}C]adenine) and ^{14}C alone (sodium phosphate buffer instead of [^{3}H]hypoxanthine in buffer) and with experimental samples incubated, filtered and processed in the usual way. Do two blanks (supporting medium alone) with each of the 'single' reaction mixtures to subtract from the values obtained with the samples. To calculate the efficiency of counting (c.p.m./pM of substrate or product) for each isotope with the scintillation counter settings used, spot 5 μl of the ^{3}H reaction mixture and the ^{14}C reaction mixture to glass fibre filters. When dry these filters are counted directly, without lanthanum chloride precipitation and washing.

To calculate the activity of each enzyme in the double microassay, subtract the average blank values (from the complete reaction mixture plus supporting medium) from each channel, and subtract the fraction of counts attributable to spillover (calculated from samples incubated in the 'single' reaction mixtures). Convert the c.p.m. for either enzyme into activity measurements (pM/h/sample) using the efficiency of counting calculated above.

Typical results are shown in *Figure 2*. *Figure 2a* shows bimodal distributions (male and female embryos) of HPRT activities and HPRT:APRT ratios in individual morulae from a single litter. *Figure 2b* shows the enzyme activity and ratio distributions for blastomeres biopsied from 8-cell embryos of a single litter. The HPRT:APRT ratios in the biopsied samples clearly identify the male and female embryos (9). An analysis of HPRT and APRT activities throughout pre-implantation development is shown in *Figure 2c*.

2.7 Validation of the double microassay

The lanthanum chloride precipitation assay does not directly identify the products of the reaction and it could be argued that other enzymes might be utilizing or degrading the substrates or products. The validity of the assay is supported by a number of control experiments:

(i)   Omission of PRPP removes lanthanum-precipitable counts.

(ii)   An extract of HPRT-negative cells shows APRT activity but not HPRT activity.

(iii)   A 2-fold reduction of dosage is seen in HPRT activity in eggs from XO mothers compared with eggs from XX mothers (two X chromosomes are active during oogenesis) whereas the APRT activities are equivalent.

(iv)   Mary Harper (29) used t.l.c. to separate and quantify the purine substrates and nucleotide products in the assay, and compared recovery of products with that after lanthanum chloride precipitation. It is not the purpose of this chapter to provide methods for t.l.c. and only sufficient information will be given here to satisfy the reader that the substrate counts are indeed converted into product counts. Further details will be found in reference 29. The solvent system used [60 g $(NH_4)_2.SO_4$ dissolved in 100 ml of 0.1 M sodium phosphate buffer, pH 6.8, plus 2 ml of *n*-propanol] separates the substrates and products of the double microassay (see *Figure 3*). HPRT and APRT activities were measured separately in embryo extracts in single reaction mixes (see Section 2.6 above) with increasing concentrations of embryo extracts. In these experiments, guanine was used as the substrate for HPRT. The reactions were terminated by boiling, the substrate and product isotopes were separated by t.l.c., recovered from the paper and their radioactivity determined. The [^{3}H]guanine and [^{14}C]adenine substrates were con-

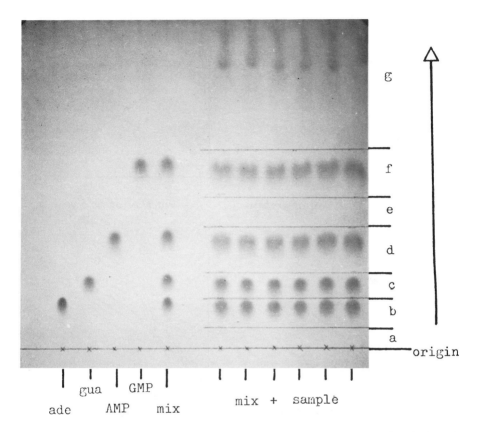

Figure 3. Separation of the substrates and products of the double microassay by t.l.c. Solvent is placed in a vertical chromatography tank and left to equilibrate for 1 h. Samples of 20 μl were spotted and dried onto the t.l.c. sheet. The t.l.c. was run at room temperature in the direction of the arrow until the solvent front reached the top of the sheet ($\sim$4 h). The sheet was removed and dried. The bases and nucleotides were visualized and their positions marked under short wavelength u.v. light (254 nm). The positions of adenine (ade), guanine (gua), AMP and GMP applied singly, or in a mixture of the four compounds (mix), are shown on the left of the figure. In an assay, the reaction mixture with the sample is dried and taken up in water containing the four compounds (mix plus sample). Sections of each track, labelled **a**−**g**, were counted for radioactivity. (Photograph kindly supplied by Mary Harper.)

verted to the products, [³H]GMP and [¹⁴C]AMP, respectively (see *Figure 4a*). For each reaction the combined counts in substrate and product account for over 90% of the total counts recovered from the chromatography sheet.

A comparison was made between the two methods of estimation of the products. The radioactivity of the t.l.c. [³H]GMP and [¹⁴C]AMP spots was compared with that in the GMP and AMP precipitated on the glass fibre filters. *Figure 4b* shows that the lanthanum chloride precipitation procedure provides a reliable measure of the products of the HPRT and APRT reactions.

3. MICROASSAY OF PGK

Beutler (30) has described the separation of electrophoretic variant forms of the X-linked glycolytic enzyme PGK-1 (EC 2.7.2.3) in human red cells by starch gel elec-

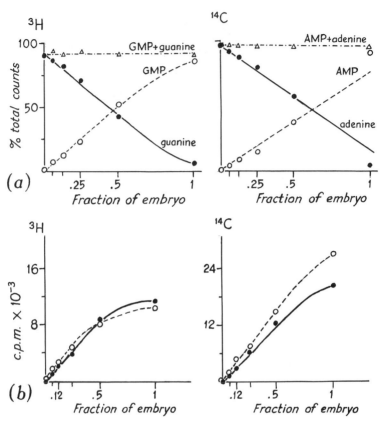

Figure 4. Conversion of substrates to products in the double microassay. (**a**)An extract of a post-implantation embryo was doubly diluted and 5 µl volumes of the diluted samples were added to 50 µl aliquots of reaction mixture. Enzyme reactions were stopped after 2 h by heating samples in a boiling water bath. 20 µl of distilled water containing standard solutions of guanine, adenine, GMP and AMP (each to 0.5 mM) were added to each tube and the whole volume spotted and dried (5 µl at a time) onto the t.l.c. sheet. Chromatography was performed as in the legend to *Figure 3*. After chromatography, the positions of the marker bases and nucleotides were marked and the t.l.c. cut into strips corresponding to the samples. The cellulose was scraped off according to positions of marker bands (b, c, d and f on *Figure 3*) as well as from the rest of the strip (a, e and g on *Figure 3*), and counted in scintillation fluid (toluene scintillant:Triton X-100:water, 4:2:1). The conversion of substrates (guanine and adenine) to products (GMP and AMP) is shown with increasing amounts of the embryo extract. The counts in each paired substrate and product are expressed as percentage total counts in the sample strip. (**b**) Several reaction tubes were set up with various amounts of embryo extract and duplicate samples were taken for separation by t.l.c. and by lanthanum precipitation. The radioactivity recovered in the GMP and AMP products separated by t.l.c. (○) are equivalent to those in GMP and AMP precipitated onto glass fibre filters with lanthanum chloride (●).

trophoresis. The enzyme bands were visualized using the backward reaction, where 3-phosphoglycerate and ATP are converted to 1,3-diphosphoglycerate and ADP (*Figure 5a*). The 1,3-diphosphoglycerate becomes the substrate for a second reaction catalysed by glyceraldehyde phosphate dehydrogenase (GAPDH, EC 1.2.1.12) in which fluorescent NADH is oxidized to non-fluorescent NAD. The position of PGK-1 activity, when viewed under long wavelength u.v. light (365 nm), appears as a dark band on a fluorescent background.

Meera Khan (31) described improved electrophoresis of several enzymes, including

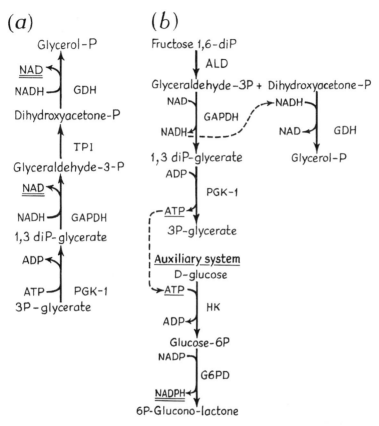

Figure 5. Backward (30) and forward (32) staining reactions for PGK activity are shown in (**a**) and (**b**), respectively. (**a**) The backward PGK-1 reaction is coupled to the conversion of fluorescent NADH to non-fluorescent NAD by glyceraldehyde phosphate dehydrogenase (GAPDH). The staining reaction can be made more efficient by driving the glyceraldehyde-3-phosphate through to α-glycerol phosphate with triose phosphate isomerase (TPI) and α-glycerophosphate dehydrogenase (GDH). (**b**) 1,3-diphosphoglycerate, the unstable substrate for PGK-1, is generated continuously from fructose-1,6-diphosphate by aldolase (ALD) and glyceraldehyde-3-phosphate dehydrogenase (GAPDH). ATP formed by the action of PGK-1 is coupled to an auxiliary enzyme system of hexokinase (HK) and glucose-6-phosphate dehydrogenase (G6PD) resulting in the conversion of non-fluorescent NADP to fluorescent NADPH. ADP is regenerated by HK and dihydroxy-acetone phosphate and NADH are eliminated by GDH.

PGK-1, on cellulose acetate gels, and Bücher *et al.* (32) have employed Cellogel electrophoresis and staining of the forward reaction (*Figure 5b*) to visualize PGK-1A and PGK-1B isozyme bands in tissue extracts from heterozygous mice. This method offers several advantages:

(i)   better separation of the bands;

(ii)   reduction of electrophoresis running time;

(iii) continuous generation of 1,3-diphosphoglycerate, the unstable substrate of PGK-1, from fructose-1,6-disphosphate.

The PGK-1 activity is demonstrated via an auxiliary enzyme system. Hexokinase uses the ATP formed from the reaction catalysed by PGK-1 to phosphorylate glucose

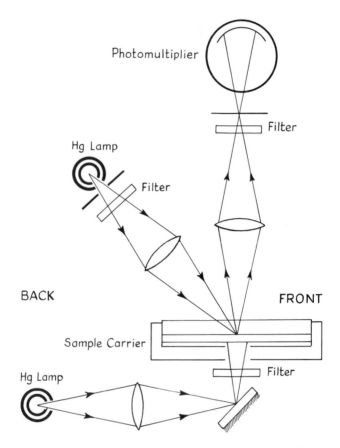

Figure 6. Measurement of fluorescence in the system described by Bücher *et al.* (32). A black polyvinyl chamber serves as a sample carrier and holds the Cellogel strip which lies face down on a cellulose acetate membrane soaked in stain. A glass plate seals the chamber. Light of 365 nm wavelength isolated by an interference filter from the beam of a mercury arc lamp (Hg) is directed through a slit onto the moving Cellogel strip from above. NADPH generated in the stain by the action of PGK-1 is excited to fluoresce by the u.v. light. The fluorescence is projected by an optical lens, through a cut-off filter to a photomultiplier which then sends its output to a computing integrator or a chart recorder. The Sigma FTR-20 also has a mercury lamp below the sample for use in the transmission mode. (Diagram reproduced by kind permission of Professor T.Bücher and Oriel Scientific.)

and thus regenerate ADP. The glucose-6-phosphate formed is then converted by glucose-6-phosphate dehydrogenase (G6PD, EC 1.1.49) to glucono-lactone with reduction of non-fluorescent NADP to fluorescent NADPH.

The fluorescence generated by the NADPH can be measured using a photomultiplier apparatus developed by Bücher *et al.* (32). The gel is placed on a moving conveyor (the chart recorder itself in the 'home-made' version of the scanner) and scanned with u.v. light of 365 nm wavelength focused through a slit onto the gel. The photomultiplier transmits the fluorescence of the NADPH to an amplifier connected to the chart recorder and a peak proportional to the fluorescence output from the band on the gel is recorded. A commercially available densitometer, the Sigma FTR-20 (Oriel Scientific), has been manufactured on these principles and is capable of scanning gels in modes of transmission and reflectance as well as fluorescence. The principles of the optics of the system are shown in *Figure 6*. Several measurements can be made of a single band so that the increasing peak areas provide a curve for activity as the stain develops on the gel with time. The areas under the peaks in the linear part of the activity/time curve provide a measure of the PGK-1 activity in the sample.

We have evaluated the Cellogel electrophoresis and quantitation of PGK-1 using tissue extracts of PGK-1A and PGK-1B mice (33). The enzyme activities in the extracts were previously assayed by spectrophotometric measurement and then mixed in different proportions and applied to the gel in varying amounts. The following parameters were established.

(i) *Calibration.* The gel system may be calibrated by applying samples of known volume and activity and quantitation of isozyme bands. This allows the total activity of PGK-1, as well as the ratio of the two isozymes, to be estimated directly from the gel.

(ii) *Sensitivity.* An applied PGK-1 activity down to 0.1 nmol/h can be detected.

(iii) *Resolution and accuracy.* One per cent of either isozyme in a mixture of the two can be resolved. In artificial mixtures of the two isozymes, there is good agreement between the proportions predicted by spectrophotometric measurement prior to mixing and the values recorded by fluorescence measurement following Cellogel electrophoresis.

The methods are described below (see also ref. 32 and 33).

3.1 Sample preparation

Repeated back-crossing to inbred mice has established the *Pgk-1^a* allele in C3H and CBA strains of mice (see ref. 34 for availability). Random-bred or inbred strains may be used as the source of the *Pgk-1^b* allele.

Eggs and embryos are isolated into microcaps as described in Section 2.1 except that the volume of supporting medium is kept at less than 1 μl in the microcap. We use PB1.PVP (Chapter 13, Section 3.1.1) or sample buffer (*Table 1*) as supporting medium.

(i)   Obtain enzyme extracts by freeze−thawing cell suspensions, homogenized tissue or microcap samples (cleavage stage embryos, eggs or different tissues dissected from later embryos).

(ii)   Centrifuge the extracts at 4°C (5 min, 1500 r.p.m.) to sediment the cell debris. Recover the supernatant for assay. Dilute samples of the supernatant if required.

(iii)   For tiny tissue samples (e.g. single oocytes) which will not be diluted, collect the sample initially in 1 μl of PB1.PVP in a 10 μl microcap and centrifuge in two directions as described in Section 2.1.

3.2 Electrophoresis

The constituents of the sample buffer, running buffer and stain are listed in *Table 1*.

The running buffer and stain solutions may be stored for up to 1 year at 4°C but the sample buffer is made up fresh each week. Electrophoresis is carried out on cellulose acetate strips 5.7 cm × 14 cm (Cellogel, Whatman). These are stored in 30% methanol prior to use. A Whatman electrophoresis tank is used (15 cm × 24 cm) with a gel support bridge cooled by water at 20°C circulated through the support from a tank with a pump and thermostat (*Figure 7*).

(i)   Take a Cellogel strip and carefully float it on the surface of 150 ml of running

Table 1. Preparation of solutions for PGK-1 Cellogel electrophoresis.

Running buffer pH 8.6		
5,5-diethylbarbituric acid	BDH	4.12 g/l
Sodium citrate	BDH Analar	2.94 g/l
$MgSO_4.7H_2O$	BDH Analar	1.23 g/l
EDTA	Sigma	0.74 g/l
1,4-dithioerythritol	Boehringer	25 mg added just before electrophoresis
5'AMP	Boehringer	45 mg added to cathodal reservoir before electrophoresis
Sample buffer		
Triethanolamine (50 mM), pH 7.6	Sigma	20 ml
1,4-dithioerythritol	Boehringer	6 mg
Bovine serum albumin	Miles	10 mg
Glycerol	Aldrich	20 ml
Stain solution A		
Triethanolamine (100 mM), pH 7.5	Sigma	14.92 g/l
$MgSO_4.7H_2O$	BDH Analar	4.93 g/l
Stain solution B		
Triethanolamine (50 mM), pH 7.6	Sigma	10 ml
$K_2HPO_4.3H_2O$	BDH	91 mg
NAD	Boehringer	9 mg
Fructose diphosphate	Boehringer	220 mg
Stain solution C		
Triethanolamine (50 mM),pH 7.6	Sigma	5 ml
Glucose	BDH Analar	135 mg
ADP	Boehringer	60 mg
NADP	Boehringer	165 mg
$MgSO_4.7H_2O$	BDH Analar	160 mg
Stain solution D		
Aldolase	Boehringer	20 μl
Glucose-6-phosphate dehydrogenase	Boehringer	20 μl
Glycerol phosphate dehydrogenase	Boehringer	10 μl
Glyceraldehyde-3-phosphate dehydrogenase	Sigma	10 μl
Hexokinase	Boehringer	10 μl

buffer (*Table 1*) in a plastic tank for 10 min. Floating the gel on the surface allows even hydration.

(ii)   Wash the gel for a further 10 min in 250 ml of running buffer containing 25 mg of dithioerythritol. As the dithioerythritol is unstable, it is added fresh. It is a reducing agent and serves to stabilize the enzyme. Gently shake the container to ensure total hydration of the gel.

(iii)   Transfer the running buffer from this second wash to the Whatman electrophoresis tank and tilt to evenly distribute the buffer in the two compartments, anodal and cathodal.

(iv)   Add 45 mg of AMP to the cathodal tank. (The AMP binds to adenylate kinase and causes it to run ahead of PGK-1 and off the gel. Adenylate kinase would otherwise react with the ADP and ATP in the stain.)

(v)   Place the gel on the cooling bridge in the tank, connect the electrodes to the power pack and pre-run at 20°C and 200 V for 10 min.

(vi)   Dilute the samples, if required, in sample buffer (*Table 1*). The sample buffer contains glycerol to reduce evaporation.

154

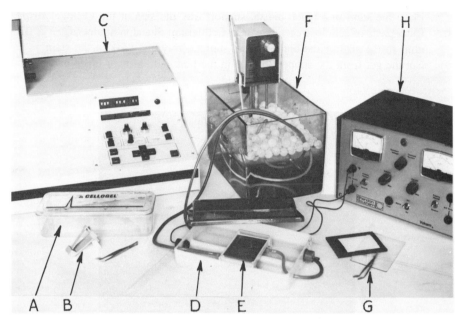

Figure 7. Equipment for PGK-1 Cellogel electrophoresis. **A**, Cellogel strips stored in methanol; **B**, semi-micro applicator; **C**, Sigma FTR-20 scanner-densitometer; **D**, Whatman electrophoresis tank; **E**, gel support bridge; **F**, tank circulating water at 20°C through gel support bridge; **G**, black plastic staining tray and glass cover; **H**, power pack. (Photograph kindly supplied by P.Maidens.)

(vii) Apply four samples of approximately 0.5 μl to a Cellogel semi-micro four-piece applicator (Whatman) and load onto the cathodal side of the Cellogel strip. For embryos in microcaps, and other microsamples, apply 0.5 μl or less of supernatant as a small drop directly onto the gel using a finely drawn Pasteur pipette. The volume of the applied sample may be estimated by pipetting the same volume into a 1 μl Drummond microcap (Scientific Supplies).

(ix) Electrophorese the gel for 1.5 h at 200 V at 20°C.

3.3 Staining

It must be stressed that cleanliness is essential for good results in the staining procedure described below. The Cellogel and Cellogram should not be touched by hand and the support tray and glass plate only handled at the edges. Most importantly, dust is highly fluorescent and must be cleared from the support and covering glass using non-fluorescent paper (e.g. Whatman filter paper). Keep one pair of forceps for handling the gel before staining and another pair for handling the staining Cellogram.

The reagents for the stain are listed in *Table 1*.

(i) Make up 660 μl of the stain 5 min before use by mixing:
     400 μl of stain solution A;
     200 μl of stain solution B;
      50 μl of stain solution C;
      10 μl of stain solution D.

Place the stain in a black plastic support tray the size of the Cellogel strip.

(ii)   Cut a piece of cellulose acetate paper (Cellogram; Shandon Southern, 57 × 127 mm) to the size of the black plastic staining tray and soak in the stain.

(iii)   Cut the gel from the support bridge in the tank with a razor blade and lay the

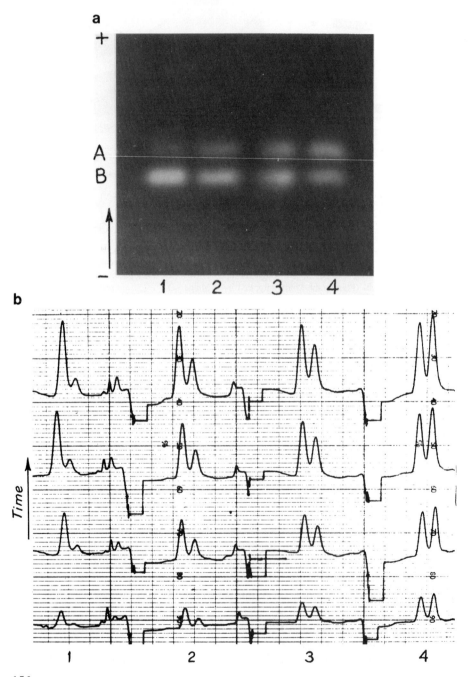

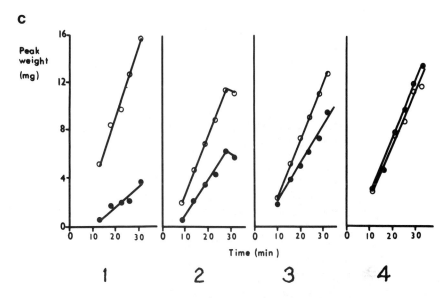

Figure 8. Cellogel electrophoresis and quantitation of artificial mixtures of PGK-1 isozymes. (a) Cellogel electrophoresis. The percentage PGK-1A activities as determined by spectrophotometric assay prior to mixing were: **lane 1**, 15%; **lane 2**, 30%; **lane 3**, 40%; **lane 4**, 50%. **A** and **B** show the migration positions of PGK-1A and PGK-1B. (b) Chart recordings of the development of fluorescence (increase in peak area with time of staining) of the tracks shown in **a**. (c) Activity versus time plot from the recordings in **b**. The percentage PGK-1A contributions from the fluorimetric gel quantitation are 16.3 ± 1.4, 31.8 ± 1.8, 42.0 ± 0.6 and 52.5 ± 0.9 for mixtures 1−4, respectively. The Cellogel quantitation thus agrees well with the composition of the artificial mixtures determined by spectrophotometric assay.

gel face down on the surface of the stain-soaked Cellogram, taking care to avoid any air bubbles. Place a clean glass cover over the support tray.

3.4 Quantitation of PGK-1 activity and proportions of the two isozymes

In setting up the gel system, it is wise to calibrate it initially for its sensitivity, its accuracy in response to activities of enzyme loaded, its power of resolution of a minority band in a mixture of known total activity and its reproducibility. To do this, take tissue extracts from PGK-1A and PGK-1B animals and assay the enzyme spectrophotometrically so that mixtures of known proportions and known activities can be tested on the gel system. A spectrophotometric assay for PGK-1 is given in Section 3.5 and sensitivity, resolution and accuracy are addressed in Sections 3.6−3.8. The present section describes the scanning and quantitation.

(i)   Incubate the gel for 8−10 min prior to recording to allow the fluorescent bands to start to develop. The u.v. source must also be allowed to warm up before recording.

(ii) Place the gel on the moving conveyor and scan the fluorescence along each track. Repeat the scan at various times up to about 30 min. A typical gel and timed readings for embryonic samples are shown in *Figure 8a* and *b*.

(iii) The isozyme contributions and PGK-1 activities in the samples are calculated by integration of the areas under the peaks. If a chart recorder is used, the peaks

157

may be carefully cut out and weighed; PGK-1 activity at a given time for a given isozyme band is proportional to peak weight. A graph of the peak weights against the time of staining of four artificial mixtures (see Section 3.8 below) is shown in *Figure 8c*. The generation of NADPH is linear with time, and 4−6 scans can be used to provide an accurate estimate of the proportions of isozymes present. Instead of cutting out peaks and weighing, the commercially available FTR-20 densitometer may be connected to an integrator (e.g. Shimadsu C-R3A Chromatopac) which will provide integrated areas under peaks by computation.

(iv)   The total PGK-1 activity per sample is obtained by comparison with known standards as described below.

3.5 Spectrophotometric assay of PGK-1 activity

The method of spectrophotometric assay is as described by Harris and Hopkinson (35).

(i)   For each assay the reaction mixture contains:
　　　　100 μl of 1.0 M Tris-HCl buffer, pH 8.0 (Sigma);
　　　　270 μl of distilled water;
　　　　400 μl of 20 mM ATP (Sigma);
　　　　100 μl of 0.1 M magnesium chloride (BDH);
　　　　 10 μl of GAPDH (Sigma);
　　　　100 μl of 2 mM NADH (Sigma).
　　　The substrate, phosphoglyceric acid, is prepared as a 0.1 M solution in distilled water.

(ii)   Add 10 μl of extract to 980 μl of reaction mixture lacking substrate in a clean quartz cuvette and equilibrate at 37°C for 3 min in a spectrophotometer (e.g. Cecil CE272) to give a straight base line recording on the chart.

(iii)   Add 10 μl of the substrate solution, mix thoroughly with a Pasteur pipette and measure the decrease in optical density (ΔOD) at 340 nM on the chart recorder (e.g. Servoscribe RE541) after 2 min equilibration. The reaction proceeds linearly for 20−30 min.

(iv)   The PGK-1 activity in μmol/min/ml is given by dividing the ΔOD/min/ml by 6.22 (molar extinction coefficient for NADH is 6.22×10^3 at 340 nM).

3.6 Sensitivity of Cellogel electrophoresis for PGK-1

The sensitivity of the gel electrophoresis and staining procedure is tested using various dilutions of tissue extract.

(i)   Measure the PGK-1 activity in the original sample in the spectrophotometer.

(ii)   Calculate the activities in diluted samples loaded onto the gel. For example, for an extract of mouse fibroblast cells (3×10^6 cells/ml) with a measured activity of 500 nmol/h per 10 μl, 0.5 μl of a 1 in 200 dilution showed detectable activity on the gel. The applied activity is therefore 0.125 nmol/h and represents activity extracted from about seven cells. From a series of such experiments it was found (33) that the lowest detectable PGK-1 activity is about 0.05 nmol/h.

3.7 Resolution of the minor band

Figure 9 shows artificial mixtures of thymocyte cell extracts from PGK-1A and PGK-1B

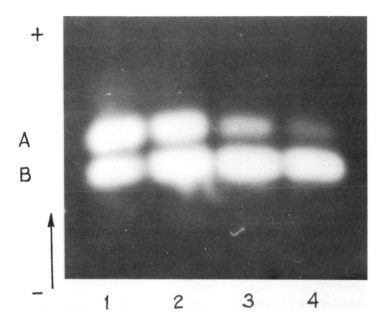

Figure 9. Resolution of PGK-1 isozymes by Cellogel electrophoresis. Four artificial mixtures of mouse thymocyte PGK-1A and PGK-1B were electrophoresed on Cellogel as described in the text. The percentage of PGK activity represented by PGK-1A was 50% **(lane 1)**, 20% **(lane 2)**, 5% **(lane 3)** and 1% **(lane 4)**. The minor PGK-1A isozyme band is clearly resolved when it is only 1% of the total activity.

mice. The extracts were assayed spectrophotometrically and the initial PGK-1 activity in each sample was found to be 900 nmol/h per 10 μl. Four artificial mixtures applied to the gel with the percentage of PGK-1A being 50%, 20%, 5% and 1%, are shown. The minor component, in this case the PGK-1A isozyme, is seen as a clearly defined band even when representing only 1% of the total activity. The reciprocal gel in which the PGK-1B activity is only 1% of the total activity should give similar resolution.

3.8 Determining the accuracy of quantitation of isozyme

(i)   Assay spectrophotometrically tissue extracts from PGK-1A and PGK-1B mice.
(ii)   Make a series of mixtures containing various proportions of the isozyme forms. In the example shown in *Figure 8* the percentage PGK-1A was 15%, 30%, 40% and 50%.
(iii) Following electrophoresis, stain the gels (*Figure 8a*) and measure the fluorescence by scanning at fixed time intervals (*Figure 8b*).
(iv)   Integrate the scan peaks (*Figure 8c*). A comparison of the estimated PGK-1A contribution in the mixtures measured by this method demonstrates good agreement with those predicted by spectrophotometric assay of the original samples (see legend to *Figure 8c*) and validates the use of the system on unknown samples.

3.9 Microassay for PGK activity and isozyme proportions in minute samples

(i)   Collect single eggs, embryos (or parts of them) or small numbers of cells, in

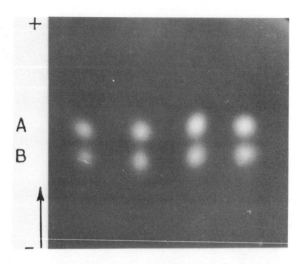

Figure 10. PGK-1 activity in single mouse oocytes from a female heterozygous for *Pgk-1ᵃ* and *Pgk-1ᵇ* alleles.

approximately 1 μl of phosphate-buffered saline (PBS) or PB1.PVP in microcaps (Section 3.1).

(ii)   Freeze − thaw and centrifuge the samples in the microcaps and apply a known volume (< 1 μl) of supernatant extract to the gel directly with a pulled Pasteur pipette.

(iii)   After electrophoresis, stain the gel in the usual way (Section 3.3) and quantitate the 'spots' of activity (Section 3.4). *Figure 10* shows PGK activity in single mouse oocytes.

(iv)   An accurate estimate of the absolute activity of PGK enzyme in these samples may be obtained by comparison with a set of standards applied to the same gel, that is reference spots of known volume (< 1 μl) from a range of samples whose activity is determined by prior spectrophotometric measurement.

4. ACKNOWLEDGEMENTS

I thank Hasu Kathuria and Mary Harper for their collaboration in the development and use of HPRT/APRT microassay, and Theodore Bücher, Ingrid Linke, Andy McMahon and Mandy Fosten for their collaboration in the development and use of the PGK-1 assay.

5. REFERENCES

1.   Monk,M. and Harper,M. (1978) *J. Embryol. Exp. Morphol.*, **46**, 53.
2.   Harper,M.I. and Monk,M. (1983) *J. Embryol. Exp. Morphol.*, **74**, 15.
3.   Epstein,C.J. (1972) *Science*, **175**, 1467.
4.   Kratzer,P.G. and Gartler,S.M. (1978) *Nature*, **274**, 503.
5.   Monk,M. and Harper,M. (1979) *Nature*, **281**, 311.
6.   Monk,M. and McLaren,A. (1981) *J. Embryol. Exp. Morphol.*, **63**, 75.
7.   Hooper,M., Hardy,K., Handyside,A., Hunter,S. and Monk,M. (1987) *Nature*, **326**, 292.
8.   Monk,M., Handyside,A., Hardy,K. and Whittingham,D. (1987) *Lancet*, in press.
9.   Monk,M. and Handyside,A. (1987) *J. Reprod. Fert.*, submitted.
10. Nielsen,J.T. and Chapman,V.M. (1977) *Genetics*, **87**, 319.
11. West,J.D., Frels,W.J., Chapman,V.M. and Papaioannou,V.E. (1977) *Cell*, **12**, 873.

12. Papaioannou,V.E., West,J.D., Bücher,T. and Linke,I.M. (1981) *Dev. Genet.*, **2**, 305.
13. Harper,M.I., Fosten,M. and Monk,M. (1981) *J. Embryol. Exp. Morphol.*, **67**, 127.
14. Johnston,P.G. (1981) *Genet. Res.*, **37**, 317.
15. McMahon,A., Fosten,M. and Monk,M. (1981) *J. Embryol. Exp. Morphol.*, **64**, 251.
16. McMahon,A., Fosten,M. and Monk,M. (1983) *J. Embryol. Exp. Morphol.*, **74**, 207.
17. McMahon,A. and Monk,M. (1983) *Genet. Res. Camb.*, **41**, 69.
18. McMahon,A. (1983) *Genet. Res. Camb.*, **42**, 77.
19. Ponder,B.A.J., Schmidt,G.H., Wilkinson,M.M., Wood,M.J., Monk,M. and Reid,A. (1985) *Nature*, **313**, 689.
20. Ansell,J.D. and Micklem,H.S. (1986) In *Handbook for Experimental Immunology*. Weis,D.M. (ed.), Blackwell Scientific Publications, Oxford, Chapter 56.
21. Woodruff,M.F.A., Ansell,J.D., Forbes,G.M., Gurdon,J.C., Burton,D.I. and Micklem,H.F. (1982) *Nature*, **229**, 822.
22. Chapman,V.M. and Shows,T.B. (1976) *Nature*, **259**, 665.
23. Kozak,C., Nichols,E. and Ruddle,F.H. (1975) *Somat. Cell Genet.*, **1**, 371.
24. Gartler,S.M., Scott,R.C., Goldstein,J.L., Campbell,B. and Sparkes,R. (1971) *Science*, **172**, 572.
25. Epstein,C.J. (1970) *J. Biol. Chem.*, **245**, 3289.
26. Bakay,B., Telfer,M.A. and Nyhan,W.L. (1969) *Biochem. Med.*, **3**, 230.
27. McBurney,M.W. and Adamson,E.D. (1976) *Cell*, **9**, 57.
28. Monk,M. and Kathuria,H. (1977) *Nature*, **270**, 599.
29. Harper,M.I. (1981) Patterns of enzyme activity reflecting genetic expression in mouse embryogenesis. PhD Thesis, University of London.
30. Beutler,E. (1969) *Biochem. Genet.*, **3**, 189.
31. Meera Khan,P. (1971) *Arch. Biochem. Biophys.*, **145**, 470.
32. Bücher,T., Bender,W., Fundele,R., Hofner,H. and Linke,I. (1980) *FEBS Lett.*, **115**, 319.
33. McMahon,A. (1981) Cell differentiation and X-chromosome activity in the definitive germ-layers and the germ-line of the mouse. PhD Thesis, University of London.
34. Festing,M.F.W. (1987) *International Index of Laboratory Animals*. 5th edition, Laboratory Animals Ltd, Newbury, Berkshire, UK.
35. Harris,H. and Hopkinson,D.A. (1976) *Handbook of Enzyme Electrophoresis in Human Genetics*. Elsevier/North Holland, Amsterdam.

CHAPTER 8

Qualitative analysis of protein changes in early mouse development

SARAH K.HOWLETT

1. INTRODUCTION

Fertilization transforms the relatively quiescent egg into a mitotically dividing cell capable of giving rise to a complete embryo. During the process of cleavage, growth and differentiation one might reasonably expect to be able to detect differential synthesis of particular protein species. Changes in the pattern of translated proteins within cells can result from changes in gene expression, mRNA availability, protein stability and protein modification (1). The contribution made by each of these mechanisms is open to investigation. Despite the problems imposed by the small quantity of protein within the early mouse embryo, techniques have been developed for detecting the presence and synthesis of proteins.

The electrophoretic separation of proteins in polyacrylamide gels (PAGE) is an ideal tool for the analysis of the many hundred gene products of any cell (see refs 2−4) and offers a convenient means of analysis of the changing patterns of proteins during development. Electrophoresis separates proteins on the basis of differential mobility related to size or to net charge. In the presence of sodium dodecyl sulphate (SDS), electrophoresis separates according to molecular weight. Isoelectric focusing (IEF) gives a separation based on the isoelectric point of polypeptides, which reflects the amino acid composition and modification, for example acetylation, glycosylation and phosphorylation. In combination, these two techniques will separate proteins into a two-dimensional array according to both properties. Considerable theoretical and technical details for running one- and two-dimensional gels are given in refs. 2 and 4 which are strongly recommended as sources of information.

It will depend upon the aim of the particular study as to which strategy of labelling and detection should be adopted and used in conjunction with PAGE. Different procedures are employed to investigate the presence of total, subpopulations or specific proteins, whilst others detect only those that are being synthesized at particular times.

A dynamic picture of the pattern of proteins being synthesized at any given time may be determined from the incorporation of radiolabelled amino acids (as mixtures or single amino acids, for example [^{35}S]methionine; Section 2.1.1) and visualization by fluorography (*Figures 1, 2* and *3*). In addition, the fate (e.g. stability) of the newly synthesized labelled species may be followed by washing the cells free from label and incubating them for several more hours (pulse-chase experiment; Section 2.1.3).

A static picture of global proteins may be assessed by silver staining unlabelled proteins in a gel (Section 4.2 and *Figure 4*), or alternatively by iodinating a lysed mixture of proteins (Section 2.2) and visualizing by autoradiography. In the case of the early embryo it is apparent that the pattern of proteins that are being synthesized is very different from the pattern of proteins that the embryo inherits, and that were presumably synthesized by, and stored in, the oocyte (see *Figure 4*).

In addition, it is possible to select for a particular subset of proteins, for example, by using ^{32}P for phosphoproteins (Section 2.3) and various ^{14}C- or ^{3}H-labelled sugar precursors (Section 2.4) or lectins to visualize glycoproteins. Where specific antibodies are available, individual species of proteins may be investigated. Thus, antibodies may be used to detect the presence of antigen from amongst electrophoretically separated proteins (immunoblotting; Section 4.3) or used to precipitate labelled antigen from amongst a mixture of metabolically labelled proteins (immunoprecipitation; Section 2.5).

In this chapter techniques are described for labelling different classes of proteins and for protein separation and identification. Although the techniques are universally applicable their specific use as micromethods for analysis of the developing embryo is emphasized.

2. LABELLING PROTEINS OF THE EARLY EMBRYO

Proteins can be labelled during synthesis (metabolic labelling) by the incorporation of radiolabelled amino acids. A major advantage of radiolabelling is that detection methods for PAGE-separated labelled proteins are far more sensitive than are methods for staining proteins (Section 4.2). The most convenient and probably the most commonly used labelled amino acid is [^{35}S]methionine, since methionine can be obtained at high specific activity. When labelling with a single amino acid such as methionine, it must be remembered that the number of methionine residues in each protein will affect labelling intensity; in particular, care must be taken with proteins that have a particularly low methionine content, such as histones. However, other labelled amino acids can be used, either singly or as mixtures. Methods of double labelling with two different amino acids each containing different radioisotopes (e.g. ^{14}C and ^{3}H) have been described (5−7). Double labelling may be used to detect differences in the types of proteins synthesized at different times or in different compartments of the early embryo (see also Section 4.1). Certain post-translational modifications can be detected by incorporation of, for example, radiolabelled sugars or phosphate. Use of specific antibodies will detect either the synthesis (by immunoprecipitation; Section 2.5) or the presence (by immunoblotting; Section 4.3) of particular proteins.

2.1 Metabolic labelling of total proteins

2.1.1 *Direct labelling*

Eggs or embryos are incubated in M16 culture medium containing bovine serum albumin (BSA) (see Chapter 2, Section 5.2) supplemented with labelled amino acids, for example [^{35}S]methionine, [^{3}H]leucine or [^{3}H]- or [^{14}C]amino acid mixtures. Embryos can be cultured in the radioactive medium for 1−5 h for one- or two-dimensional analysis with apparently no short term toxic effect. However, long term viability of embryos that have been subjected to as little as 50 μCi/ml [^{35}S]methionine for 30 min is impaired considerably (8). For [^{35}S]methionine with high specific activity ($\sim$1300 Ci/mmol) 1–3 μl added directly to 50 μl (120–360 μCi/ml) of culture medium is adequate (for examples of [^{35}S]methionine-labelled proteins see *Figures 1, 2* and *3*). Because of the lower specific activity of ^{3}H and ^{14}C, aliquots of label must be lyophilized and re-dissolved in culture medium in order to achieve a sufficiently high isotope concentration: for example, 50 μl of [^{3}H]leucine (specific activity $\sim$100 Ci/mmol) can be lyophilized and re-dissolved in 25–50 μl culture medium. It is important that labelling is performed in medium that is not supplemented with unlabelled amino acids since they will lower the specific activity of the labelled amino acids. Also, note that the presence of serum in the labelling medium may lower the isotope concentration. The low endogenous pools of methionine, leucine, phenylalanine and alanine in early embryonic cells make these the labelled precursors of choice (9).

2.1.2 *Double-isotope labelling*

It is possible to discriminate between two different populations of newly synthesized proteins by judicious use of two differently labelled amino acids (various methods are discussed in ref. 3). Thus, labelling one population of embryos with a mixture of [^{14}C]amino acids and another population with a [^{3}H]amino acid mixture should enable any differences in the two sets of newly synthesized proteins to be detected (5,6; see Section 4.1). The combination of [^{35}S]methionine and the γ-emitter, [^{75}Se]selenomethionine (7), offers a potentially rapid and powerful method for dual labelling of embryos since these two isotopes can be obtained at high specific activities. Two groups of embryos can be labelled with each of the amino acids and the radioactivity quantified (see Section 2.7) so that appropriate proportions of the mixtures can be loaded onto a gel. Obviously it is important to avoid artefactual differences and therefore, reciprocal combinations of both samples labelled with each isotope should be loaded.

2.1.3 *Pulse-chase labelling*

A large degree of the complexity of the pattern of synthesized proteins in the early embryo, and probably in most cells, results from post-translational modification to primary translation products. Pulse-chase experiments enable slow modifications to be assessed (modifications that occur almost immediately after release of a primary translation product from the ribosome will not be detected). The rationale behind this labelling procedure is to label proteins at a specific time and to follow their fate during a subsequent incubation period known as a 'chase'. It is important to lyophilize and re-dissolve the label at a reduced specific activity before use, otherwise embryonic development is often impaired and cleavage prevented during the 'chase' incubation period (8). An outline of a procedure that has been used (10) is given below.

(i)   Pulse label the embryos for 1 h in M16 + BSA culture medium (Chapter 2, Section 5.2) containing [^{35}S]methionine and 1–10 μM unlabelled methionine. The label should be at a specific activity of about 10 Ci/mmol and reduced further to 1 Ci/mmol for long chases.

(ii) Wash the embryos free from label through several drops of M16 + BSA medium supplemented with unlabelled methionine, phenylalanine and leucine, each at a concentration of 100 μM.

(iii)   Culture the embryos for specified periods of time in the same medium. The unlabelled amino acids exchange with the endogenous amino acids (9) during the chase period and so help to remove labelled methionine from the pools of free amino acids within the cells.

(iv)   Compare the samples of proteins collected after the initial pulse period with those at the end of the chase by electrophoresis. Changes in mobility of certain proteins indicate post-translational modification.

2.2 Non-metabolic labelling of total proteins

Non-metabolic labelling of existing proteins (as opposed to metabolic labelling of proteins being synthesized) offers a more sensitive alternative to staining of proteins in gels (see Section 4.2). Labelled material is separated by gel electrophoresis and visualized by autoradiography. In theory it should be possible to label the cell surface proteins alone, for example, by the use of the non-permeant protein label [^{14}C]isethionyl acetimidate (11) or by lactoperoxidase catalysed iodination (12). However, it is most important that cells are not damaged since this will invalidate the results. Therefore, such procedures should be done with a great deal of care and appropriate controls included.

To label total proteins under physiological conditions, [^{14}C]ethyl acetimidate may be used since it will penetrate cells without impairing membrane function (11). An alternative method uses the *N*-succinimidyl group of the Bolton and Hunter reagent (*N*-succinimidyl 3-(4-hydroxy 5-[^{125}I]iodophenyl)propionate) to condense with free amino groups of polypeptides (13). It is then informative to arrange the gel so as to compare the pattern of synthesized proteins (metabolically labelled with [^{35}S]methionine) on adjacent lanes with those present (non-metabolically labelled with ^{125}I) at given stages of embryogenesis. The method of non-metabolic labelling of proteins is given in *Table 1*. Note that appropriate precautions must be taken with the handling of ^{123}I and mouth pipetting of embryos should be avoided.

2.3 Phosphoproteins

Phosphoproteins may be identified by the incorporation of labelled phosphate. Embryos are cultured for 2 h in phosphate-free medium (Chapter 2, Section 5.2; M16 without

Table 1. Non-metabolic labelling of proteins.

1.	Wash embryos (or single embryo) free of BSA and place in 10 μl of 0.1 M borate buffer pH 8.5 in Eppendorf tubes. Samples can be stored at $-70°$C.
2.	Prepare the Bolton and Hunter reagent. The reagent is supplied by Amersham in benzene. Allow the benzene to evaporate to dryness and re-dissolve in borate buffer. Each ampoule contains enough reagent to iodinate $20-30$ samples.
3.	Add the Bolton and Hunter reagent to the embryo samples and leave for 20 min at room temperature[a]. Because of the very small quantities of protein in embryos the amount of reagent added per sample is not important, as it will always be in excess.
4.	Add 9 vol of ice-cold ethanol and leave the proteins to precipitate for 2 h at 4°C.
5.	Centrifuge the samples in an Eppendorf microcentrifuge for 10 min, remove the supernatant and allow the precipitate to dry. Dissolve the pellet in sample buffer (*Table 4*) and store at $-70°$C until electrophoresis (see Section 3.1).

[a]See ref. 13.

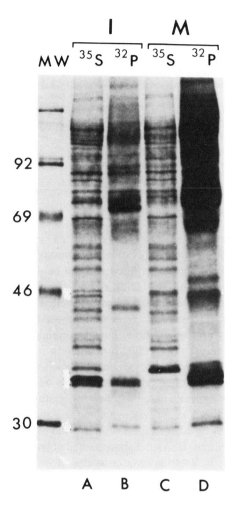

Figure 1. Phosphorylated proteins. Eggs were labelled in first S phase ('I', 10 h post-insemination; **A** and **B**) or as unfertilized eggs ('M', 14 h post-ovulation; **C** and **D**) with [^{35}S]methionine (**A** and **C**) or [^{32}P]phosphate (**B** and **D**) for 1 h or for 2 h, respectively.

KH$_2$PO$_4$) containing 1 mCi/ml of carrier-free [^{32}P]orthophosphate (Amersham International plc;). The ^{32}P is lyophilized and re-dissolved before use. Uptake of phosphate is poor, especially during interphase; consequently 100−150 embryos are required for one-dimensional analysis (see *Figure 1*). In future it may prove more convenient to microinject label directly into the cytoplasm of early stage embryos as is done for *Xenopus* eggs (14).

To verify that labelled proteins are indeed phosphorylated due to the action of protein kinases, phosphate groups can be removed *in vitro* by treatment with protein phosphatases (15) as described in *Table 2*. The shift in positions of [^{35}S]methionine-labelled bands can then be visualized by electrophoresis and fluorography. Such treatment will provide useful additional information to that provided from pulse-chase experiments (15).

Table 2. *In vitro* dephosphorylation of phosphoproteins.

1.	Label embryos for 1 h in [^{35}S]methionine-containing medium (as in Section 2.1.1), then wash them through several drops of M2 or M16 medium without BSA.
2.	Place a control group of labelled embryos directly into sample buffer (*Table 4*) ready for electrophoresis.
3.	Transfer three other groups of embryos to 5 μl of double-distilled water and leave to lyse at $-70°C$ for 30 min.
4.	Thaw one lysed sample and incubate at 37°C for 1 h with phosphatase (for example, alkaline phosphatase or potato acid phosphatase; Sigma) at a final concentration of 5 μg/ml (0.3 units/ml).
5.	Thaw two further control samples and incubate one with 5 μl of 2-mM phenylmethylsulphonyl fluoride (PMSF; protease inhibitor) and the other with 5 μl of double-distilled water alone.
6.	Add double-strength SDS sample buffer to each sample. Store samples at $-70°C$ until electrophoresis (see Section 3.1).

2.4 Glycoproteins

It is possible to precipitate proteins that are glycosylated by the use of specific lectins. However, for this method a large number of embryos are required for recovery of sufficient material for analysis and this method will not be described here (see ref. 16 for details). Glycoproteins can also be identified post-electrophoresis (see Section 4.3) or alternatively by metabolic labelling with radiolabelled sugars. Unfortunately the permeability of embryos to most sugars is poor until the blastocyst stage. It is possible to label early embryos with, for example, [^{3}H]glucosamine but large numbers must be pooled in order to analyse the labelled proteins (e.g. see ref. 17).

Glycoproteins synthesized by blastocysts may be labelled with fucose, mannose or glucosamine. Blastocysts should be washed in glucose-free medium (see Chapter 2, Section 5.2 for details of optimal culture medium for blastocysts) that is supplemented with alternative substrates such as pyruvate, glutamic acid and aspartic acid (18). Groups of 15−60 blastocysts may be labelled for 3−6 h in 50 μl of glucose-free medium containing bovine serum albumin (BSA) and about 25 μCi of lyophilized [^{3}H]glucosamine, [^{3}H]mannose or [^{14}C]fucose (with specific activities of 2−50 μCi/ml). The embryos are then washed through several drops of medium without label, several drops of medium without BSA and stored in sample buffer at $-70°C$ before being electrophoresed (see Section 3.1). It is important to remove serum, since large amounts of unlabelled protein will cause gel distortion.

2.5 Detection of specific synthesized proteins

Particular proteins, that are being synthesized or that have accumulated in the early embryo, may be detected by the use of combinations of specific antibodies and immunofluorescence, immunoprecipitation or immunoblotting. For indirect immunofluorescence individual embryos can be fixed and stained successfully, and these techniques are discussed in Chapter 2, Section 4. Unfortunately, for both immunoprecipitation and blotting, the very small quantity of protein in the early embryo necessitates the pooling of large numbers of embryos for analysis. Blotting is described in Section 4.3.

Immunoprecipitation is a means of identifying specific synthesized proteins by precipitation with antibodies (16). The most common means of precipitating the antibody/antigen complex involves binding it to *Staphylococcus aureus* protein A or Protein A − Sepharose. (In cases where the antibody is from a species which will not bind to Protein

Table 3. Immunoprecipitation.

1.	Label embryos with [³⁵S]methionine as in Section 2.1.1.
2.	Wash embryos free of BSA-containing medium and pool embryos in 200 μl of freshly prepared buffer[a]. A total of $1-2 \times 10^6$ c.p.m. per sample has been found to give good results. This may require, for example, $1-2000$ blastocysts that have been labelled for 3 h at the specific activities given in Section 2.1.1.
3.	Boil the sample for $1-2$ min and store at $-70°C$.

All subsequent procedures should be done on ice.

4.	Just before use, sonicate the sample to disrupt the cellular debris thoroughly. Retain a small volume (10 μl) that is stored at $-70°C$ as an untreated control. Split the remaining sample into two.
5.	Spin the embryo samples, the specific antiserum and the control serum at 15K for 15 min.
6.	Add an equal volume of specific or control serum to the two embryo samples. The optimal dilution of specific serum varies with each antibody and has to be established, but should be similar to that used for immunofluorescence.
7.	Preparation of Protein A–Sepharose beads: wash 500 mg of beads eight times in 5 ml of washing buffer (100 mM Tris, pH 8.3; 2 mM EDTA; 0.5% DOC: 0.5% NP-40; 0.1% SDS). Pellet each time with low speed spin and resuspend in approximately 5 ml. Finally, add an equal volume of buffer to the pellet (~ 2 ml) and the beads are ready for use.
8.	For each sample, spin down 100 μl of the prepared beads, remove the supernatant and add the protein/antibody mixture.
9.	Shake this mixture on ice (vigorously in order to keep the beads suspended) for 2 h. Then wash four times with 100 μl of the same washing buffer by repeated centrifugation (2 min at low speed) and resuspending (no need to vortex). Finally switch to 100 mM Tris, pH 8.3 for the last wash and spin.
10.	Recover the specific protein antigen from the beads by boiling with 50 μl of sample buffer (*Table 4*) for 3 min. Aspirate the solution from the beads using an Eppendorf micropipette with the tip at the bottom of each tube. Store at $-70°C$, before electrophoretic separation (see Section 3.1).

[a]100 mM Tris, pH 8.3; 2 mM EDTA; 0.5% sodium deoxycholate (DOC); 0.5% Nonidet P-40 (NP-40); 0.5% SDS containing freshly added 10 mM iodoacetamide, 1 mM PMSF, 1 μg/ml pepstatin, 1 μg/ml leupeptin and 10 μg/ml aprotinin. The percentage of SDS in the sample and washing buffers can be reduced to give a greater specific signal relative to background non-specific binding.

A, a second antibody must be added to induce precipitation.) A 'precipitation' with pre-immune or non-immune serum should be carried out in parallel in order to control for the extent of non-specific interaction. Antibodies produced against native protein are more likely to be useful for this technique than for blotting. A method developed for use with an auto-immune antiserum against lamin B (M.-N.Guilly and J.-C.Courvalin, personal communication) is described in *Table 3*. Other procedures are described in detail in ref. 16.

2.6 Preparation of samples

At the desired time for collection wash embryo samples through two drops of M2 or M16 culture medium + BSA (Chapter 2, Section 5.2), followed by several drops of M2 or M16 medium without BSA, and transfer in groups into Eppendorf tubes containing $10-30$ μl of sample buffer (*Table 4*; 19) for one-dimensional analysis, or of lysis buffer (*Table 7*; 20) for two-dimensional analysis. For large numbers of embryos larger volumes or double strength buffer may be necessary. It is important to carry over as little BSA as possible, since this unlabelled protein will cause gel distortion. Pierce the top of each tube with a needle and boil in a water bath for 2 min, and then store at $-70°C$. For labelled samples, it will depend upon the isotope used, the duration of the labelling period and the stage of embryo, as to the number of embryos that should be loaded per track of the gel. For eggs and embryos up to the 8-cell stage, following 1 h labelling in [^{35}S]methionine, groups of 10 for one-dimensional analysis, or 30−50 for two-dimensional analysis, give results that can be seen after about 4−7 days by fluorography. It is quite feasible to use single embryos for one-dimensional analysis, although the labelling period is best extended and the time for exposure will be longer.

2.7 Quantitation of incorporated label

It may be desired to quantify the amount of label that has been incorporated into protein, and/or the amount of labelled precursor that was taken up by the embryo. In this way it is possible to make a quantitative analysis of protein synthesis which may be essential if the rates of synthesis of individual proteins are to be assessed. To do this carry out the following procedure.

(i)   Wash embryos that have been incubated in medium containing labelled amino acids or sugars and collect as in Section 2.6.

(ii)   Take aliquots of 1−3 μl (from the total of 10−30 μl) of the sample or lysis buffer for quantitative analysis.

(iii)   For assessment of uptake of the labelled precursor, place this aliquot directly into scintillant.

(iv)   For assessment of the incorporation into protein, add the aliquot to a mixture of 500 μl of water plus 25 μl of 0.1% BSA (as a co-precipitant) and precipitate the proteins by addition of 500 μl of 10% trichloroacetic acid (TCA).

(v)   Leave to stand for 1−2 h at 4°C, then filter the samples through a Millipore filter, wash twice with 10% TCA and twice with 70% ethanol and allow to dry.

(vi)   Place filters individually into scintillant-containing vials and count in a liquid scintillation counter.

Values of 400−500 c.p.m. per embryo would be expected from labelling with [^{35}S]-methionine for about 1 h with the specific activities described in Section 2.1.1.

3. POLYACRYLAMIDE GEL ELECTROPHORESIS

Polyacrylamide gel electrophoresis (PAGE) is a powerful analytical tool for the separation and identification of proteins. The presence of SDS introduces negative charge to the proteins such that the charge per unit mass is approximately constant. Hence, for most proteins the electrophoretic mobility of the complex is dependent almost entirely upon molecular weight. Considerable technical detail for running one- and two-dimensional gels is given in Chapters 1 and 5 of a previous volume in this series, *Gel Electrophoresis of Proteins — A Practical Approach* (ref. 2). In addition, the application of gel electrophoresis to the proteins of the mammalian embryo has been discussed by Van Blerkom (4). Here I shall try to give sufficient detail for running the SDS dissociating system of vertical slab gels and include the modifications not covered by the above references.

It is becoming increasingly popular to run thinner (0.1−0.3 mm) and smaller (mini-) gels (for example see ref. 4). The same principles apply as described here for the conventional 0.8 mm thick gels except that volumes of solutions are smaller and running times are shorter. A recent development is a system that runs gels horizontally, rather than vertically (e.g. LKB apparatus), which is attractive since it cuts down running and processing times considerably. Both PAGE and IEF can be run on such flat-bed systems (21).

The source and purity of the ingredients used for electrophoresis will alter the degree of resolution dramatically. Where possible, ultrapure electrophoresis grades should be used and stock solutions should be made up every 1−2 months and stored as shown in *Tables 4* and *7*.

One-dimensional gels suffer from the disadvantage that they separate a limited number of protein species. Two-dimensional gels allow for the separation of an increased number of proteins since polypeptides are separated according to both their isoelectric point and their molecular weight. Thus, two-dimensional gels can identify up to about 1000−1200 individual species (see *Figure 3*). They do, however, suffer from the disadvantages of being more time consuming, of being more difficult to interpret and less reproducible.

3.1 One-dimensional gels

The distribution of proteins following gel electrophoresis depends upon the acrylamide concentration and unfortunately no single concentration can give a uniform distribution of proteins. Probably the best resolution is achieved using an exponential gradient gel.

Table 4. Stock solutions for one-dimensional gels.

Buffer	*Storage*
Sample buffer: 62.5 mM Tris-HCl, pH 6.8; 2% SDS; 5% 2-mercaptoethanol; 10% glycerol; 0.002% bromophenol blue	4°C
Acrylamide:bisacrylamide 30%:0.8% (A)	4°C in darkened bottle
0.5 M Tris-HCl, pH 6.8, 0.4% SDS (UGB)	4°C
1.5 M Tris-HCl, pH 8.8, 0.4% SDS (LGB)	4°C
10% Ammonium persulphate (AMPS)	4°C, can be kept for up to 1 month
Running buffer: 25 mM Tris-HCl, pH 8.3; 192 mM glycine; 0.1% SDS	Conveniently kept as 10× stock at room temperature diluted and adjusted with glycine to pH 8.3

Table 5. Recipe for the preparation of one-dimensional gels.

Stock solution (see *Table 4*)	*Resolving gel*	*Stacking*
A (ml)	8	1.5
UGB (ml)	−	2.5
LGB (ml)	8	−
Water (ml)	8	6
	De-gas	
AMPS (μl)	45	30
TEMED (μl)	15	20

Table 6. Assembly and running of one-dimensional gels.

1.	Rinse two clean plates with 70% alcohol and dry immediately prior to use.
2.	Place Teflon spacers (0.8 mm thick, and ~8−10 mm wide) between the plates at each side. To form the bottom edge use a narrow (0.5 mm bore, 0.25 mm wall) silicone rubber tubing (ESCO Ltd) as a seal that forms a U-shaped gasket running around the bottom of the plates and up the outside of the two side spacers. Assemble and clamp the plates. Ensure that the top of the gel will be horizontal.
3.	Prepare the lower, resolving gel mixture (see *Tables 4* and *5*) using acrylamide stock solution (A), lower gel buffer and double-distilled water, and de-gas. (Non-polymerized acrylamide is a neurotoxin and so should be treated with caution.) Add 10% ammonium persulphate followed by TEMED.
4.	Pour the gel mixture into the glass plate assembly to a level 15−25 mm below the top of the smaller of the two plates. Overlay with 1 ml of water-saturated 1-butanol. Insert the well-former 'comb' into the top of the assembly, but above the level of the overlay solution, so as to ensure that the plates are correctly spaced. Leave the gel to polymerize for 1 h. If the gel is not to be used directly replace the overlay solution with double-distilled water, or the lower gel buffer diluted 4-fold, cover with cling film to prevent evaporation and leave at 4°C (for up to 24 h).
5.	Fill the bottom well of the gel running tank with running buffer (*Table 4*).
6.	Remove the comb and overlay solution and rinse the top of the gel with water. Dry carefully (without disturbing the surface of the polymerized gel) with lint-free filter paper.
7.	Remove the tubing and load the plates into the tank by tilting the plates and lowering them gently from one side, this minimizes the number of air bubbles underneath the gel.
8.	Prepare the stacking gel mixture (see *Tables 4* and *5*) using acrylamide stock solution A, upper gel buffer and double-distilled water, and de-gas. Add 10% ammonium persulphate followed by TEMED. Layer this mixture on the top of the lower, resolving gel and insert the Teflon comb which serves as a slot former. Polymerization is complete after about 30 min.
9.	Remove the comb. Remove any non-polymerized mixture (conveniently done using a syringe fitted with a fine needle). Pour running buffer into the top reservoir and flood the wells.
10.	Ensure that air bubbles in the running buffer are removed from underneath the gel using a stream of running buffer ejected from a syringe fitted with a needle that has been bent upwards at the tip.
11.	Straighten the sides of the wells and mark and label them on the front plate with a marker pen. Load the samples using a Hamilton syringe. The syringe should be well rinsed between samples so as to avoid cross-contamination.
12.	Run labelled or unlabelled molecular weight markers on each gel. Load 10 μl of sample buffer in the empty lanes flanking the samples to avoid distortion of the sample tracks at the edges.
13.	A few drops of bromophenol blue can be added to the top reservoir as tracking dye, although the amount in the sample buffer provides sufficient evidence of the progress of the gel.
14.	Run the gel at a constant voltage of 150 V or at a constant current of 20 mA until the dye front is about 5mm from the bottom of the gel. This takes about 4 h.
15.	After completion, fix gels immediately to prevent diffusion of the proteins (unless the gel is to be transferred for blotting, see *Table 11*). This is done by shaking for 20−30 min in a 45% methanol, 10% acetic acid mixture or alternatively using 50% TCA.
16.	Use a system of cut corners to identify each gel when several are run at the same time. This also allows for correct orientation of the gel later. (If the gel is to be stored at this stage keep it in 7% acetic acid.) The gel can then be processed for fluorography, direct autoradiography (see Section 4.1) or for silver staining (see *Table 10*).
17.	Clean the gel plates with a non-scratch cream cleaner and small sponge, rinse well with water and then 70% alcohol and dry with lint-free tissue.

8−15% linear gradient gels are described in detail elsewhere (4) and will not be discussed here. A linear 10% acrylamide gel is simpler to run, gives a fairly uniform distribution of proteins and is consequently commonly used (see *Figures 1, 2* and *4*). Details are given in *Tables 4, 5* and *6* for running linear 10% polyacrylamide slab gels containing

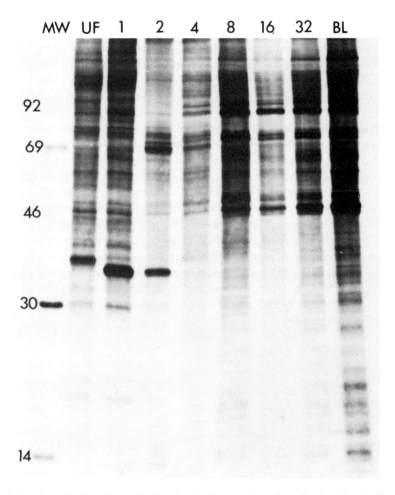

Figure 2. Protein synthesis during pre-implantation development. One-dimensional gel showing the pattern of [^{35}S]methionine-labelled proteins synthesized during 1-h pulses in the unfertilized egg (UF), 1-cell (1), 2-cell (2), 4-cell (4), 8-cell (8), 16-cell (16), early cavitating (32) and expanded blastocyst (BL). Each lane contains 10 embryos, except for (BL) which contains six embryos.

0.4% SDS and 0.5 M Tris-HCl (pH 8.8), with a stacking gel of 4.5% acrylamide containing 0.4% SDS and 0.125 M Tris-HCl (pH 6.8). (For details of other gel concentrations and other running systems see Chapter 1 of ref. 2.)

If the incorporation of label has been quantified, equal (or known) numbers of counts can be loaded onto adjacent tracks. This will allow quantitative relative changes of individual proteins amongst the population to be assessed more precisely. For semi-quantitative comparative analysis, the polypeptides from samples usually containing identical numbers of embryos, may be separated on adjacent tracks. An example is given in *Figure 2* showing the changing pattern of proteins synthesized during pre-implantation development. The gel shows equivalent numbers (10) of [^{35}S]methionine-labelled unfertilized eggs, embryos at the 1-cell, 2-cell, 4-cell, 8-cell, 16-cell and early cavitating stages and six embryos at the expanding blastocyst stage.

3.2 Isoelectric focusing and two-dimensional gels

High resolution two-dimensional gels allow the separation and resolution of up to about 1200 polypeptides. The technique was developed by O'Farrell (20), although various modifications have been made to improve resolution, reproducibility and convenience (22,23). The basic technique involves separation of proteins in the first dimension according to their isoelectric point in cylindrical, polyacrylamide gels (isoelectric focusing electrophoresis, IEF). Thus different proteins can be segregated into discrete, narrow zones (24). The new flat-bed system allows the IEF dimension of several samples to be run in a single slab gel, which should improve the reproducibility (21). Although IEF can be used alone it is more frequently coupled with electrophoresis in the second dimension, separating the bands into a series of spots, each of which usually corresponds to an individual polypeptide (see *Figure 3*). IEF is not the only first dimensional separation of a two-dimensional system; when dealing with basic proteins, for example, non-equilibrium pH gradient electrophoresis is useful (22). Despite the enormous potential of two-dimensional gels, reproducibility and comparison between gels remain serious problems. It is advisable to establish a set of reference polypeptides that can be used for localization. Without computer programs to standardize data from many gels (see refs 3,23), the complexity and lack of reproducibility of two-dimensional gels makes interpretation difficult.

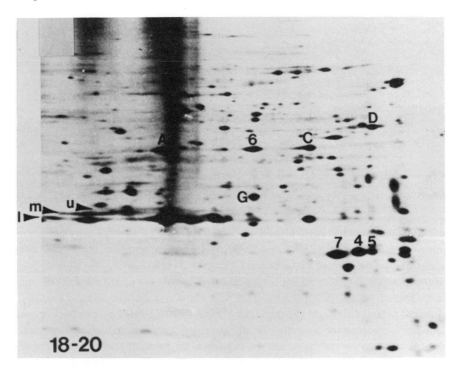

Figure 3. Two-dimensional separation of [^{35}S]methionine-labelled proteins. Two-dimensional gel showing the pattern of proteins being synthesized in a 2-h labelling period during first mitosis (18−20 h post-insemination). Polypeptide numbers and letters refer to reference polypeptides used for localization (annotation as for refs 15 and 39). Isoelectric focusing is from left (~pH 7.0) to right (~pH 4.5).

The overall resolution of two-dimensional gels depends largely on that obtained in the IEF dimension. The procedure for running IEF gels over the pH range $4.5-7.0$ is described in detail in *Tables 7, 8* and *9*. Note that mixtures of ampholines can provide a greater range of pH $2-10$, or can be used to extend a particular narrow pH range (in which case both lysis and sample buffers should contain the same pH range).

4. ANALYSIS OF PROTEINS SEPARATED BY GEL ELECTROPHORESIS

4.1 **Detection of radiolabelled proteins: autoradiography versus fluorography**

Detection of radiolabelled proteins separated by PAGE is far more sensitive than staining of unlabelled proteins. The type of isotope used will determine whether or not direct autoradiography of the dried down gel is sufficient or whether impregnation with a fluor (25) is necessary to enhance the efficiency of detection (see ref. 26 for detailed discussion). Direct autoradiography is suitable for the energetic emissions of ^{32}P and ^{125}I and an intensifying screen can be used to increase the sensitivity, although this

Table 7. Stock solutions for the preparation of isoelectric focusing gels.

Buffer	Storage
Lysis buffer: 9.5 M urea; 5% 2-mercaptoethanol; 2% NP-40; 1.6% ampholines, $5-7$; 0.4% ampholines $3.5-10$	Frozen
Acrylamide:bisacrylamide 28.83%:1.62% (B)	4°C in darkened bottle
10% Nonidet 40 (NP-40)	Room temperature
Ampholines,pH $3.5-10$ and $5-7$	4°C
10% Ammonium persulphate (AMPS)	4°C, for up to 1 month
10 mM H_3PO_4	1 M stock, room temperature
20 mM NaOH	Room temperature
Overlay solution: 8 M urea	Frozen
Sample overlay solution: 9 M urea; 0.8% ampholines, $5-7$; 0.2% ampholines, $3.5-10$	Frozen
SDS sample buffer: 62.5 mM Tris-HCl, pH 6.8; 2.3% SDS; 5% 2-mercaptoethanol; 10% glycerol	4°C

Table 8. Recipe for the preparation of isoelectric focusing gel.

Ingredient (see *Table 7*)	Amount[a]
Urea (g)	5.5
Solution B (ml)	1.33
NP-40 (ml)	2.0
Water (ml)	1.97
Ampholines $3.5-10$ (ml)	0.1
Ampholines $5-7$ (ml)	0.4
AMPS (μl)	10
De-gas	
TEMED (μl)	7

[a]Sufficient for 10 gels.

Table 9. Pouring and running isoelectric focusing and two-dimensional gels.

1.	Cast IEF gels in tubes sealed at one end with parafilm. It is critical that the tubes are clean, which can be ensured by soaking in chromic acid, extensive washing in double-distilled water, washing with absolute alcohol and air-drying.
2.	Make up the gel mixture (enough for 10 tubes with a pH range 4.5−7) in a 125-ml side-arm flask with the urea, acrylamide stock solution B (*Table 7*), NP-40, double-distilled water and the mixture of ampholines (LKB or Servolyte) as described in *Table 8*. Swirl the flask (it can be warmed slightly) until the urea is completely dissolved. Add the ammonium persulphate. De-gas the mixture under vacuum for about 1 min. Add the TEMED and immediately load the mixture into the tubes using a long-needled syringe. Take care to eliminate all air bubbles by loading the mixture from the bottom of the tubes.
3.	Overlay with 20 μl of urea overlay solution (*Table 7*). After 1.5 h replace overlay with 20 μl of lysis buffer (*Table 7*) followed by 10 μl of water. To avoid urea crystallizing out of solution, place the tubes in front of an infrared lamp.
4.	After a further 1.5 h fill the lower reservoir of the gel tank with 10 mM phosphoric acid. Remove the parafilm from each tube and place the polymerized gels into the gel tank.
5.	Remove the solution from the top of each gel and replace with 20 μl of fresh lysis buffer.
6.	Fill to the top of each tube with de-gassed 20 mM NaOH. Fill the top reservoir with the same solution. Pre-run gels for 15 min at 200 V, 30 min at 300 V and 30 min at 400 V.
7.	Turn off the power and discard the lysis and top reservoir buffers. The gels are now ready to be loaded.
8.	Ensure that the samples are saturated with urea and freeze−thaw three times before loading with a Hamilton syringe or an Eppendorf micropipette.
9.	Overlay samples with 10 μl of sample overlay solution (*Table 7*), top up each tube carefully with 20 mM NaOH and refill the top chamber. Run gels for a total of 6000 V h, the last hour of which is at 800 V (for example, run at 325 V for 16 h plus 800 V for 1 h).
10.	Extrude the gels from the tubes using a water-filled syringe attached to a short piece of tubing that fits over the end of the tube. Push the gel out slowly by gentle pressure applied to the syringe[a].
11.	Where the proteins are simply to be analysed by IEF dry the gels directly.
12.	For separation in the second dimension, equilibrate the gels with 5 ml of SDS sample buffer (which is changed after 15 min) for 2 h. Store the gels in SDS sample buffer at −70°C or electrophorese directly in the second dimension.
13.	For second dimensional separation place each tube gel along the top of a 10% acrylamide slab gel (prepared exactly as described in *Tables 5* and *6*) that has been cast in a glass plate assembly that is modified to accommodate the tube gel (see refs 4 and 20). Prepare the stacking gel as in *Table 6* except that the well-former is replaced by a piece of Teflon that will provide a flat surface onto which the tube gel can be laid. Remove the Teflon former after polymerization of the stacking gel. Remove any non-polymerized material. Pour about 1 ml of molten 1% agarose onto the top of the stacking gel (to hold the tube gel in place) and carefully place the tube gel onto the agarose (being careful not to trap air bubbles under the gel). After about 2 min fill the top reservoir with running buffer (*Table 4*). Run the gel in the second dimension to separate polypeptides according to molecular weight as described in *Table 6*.

[a]If it is required to estimate the pH gradient, an unloaded IEF gel may be sliced, the slices eluted with 1 ml of de-gassed distilled water for several hours and the pH measured for each.

does decrease the resolution. Compared with direct autoradiography, fluorography increases the efficiency of detection of ^{3}H 1000-fold and of ^{35}S and ^{14}C about 15-fold, with only a slight loss in resolution.

The hazardous and laborious method of impregnation with 2,5 diphenyloxazole (PPO) in dimethyl sulphoxide (DMSO) (outlined in ref. 2 and discussed in ref. 26) has largely been superseded by the use of commercially available products such as Enhance,

Autofluor, Enlightening and Amplify. We have had most experience using Amplify (Amersham) which has proved to be reliable and as sensitive as PPO impregnation. With Amplify, gels are transferred directly from the fixing solution into Amplify (sufficient to cover the gel, approximately 15 ml per gel) and shaken for $15-30$ min before being dried down under constant vacuum at $60-80°C$ onto filter paper. For direct autoradiography gels are dried down directly after fixation. It is most important that the vacuum is not released until the gel has dried completely, as this will cause the gel to crack. Use of a radioactive marking and/or number system for the gels avoids confusion over the contents of each gel. This can be done simply by using a fibre-tipped pen dipped into ink to which has been added a small amount of ^{14}C- or ^{35}S-containing liquid (such 'hot' pens are available commercially, e.g. from Amersham) or by using an adjustable stamp with a 'hot' ink pad. Exposure of the gel to pre-flashed medical film, such as Fuji RX or Kodak X-Omat AR X-ray film and storage at $-70°C$ increases the sensitivity of the film to light emissions from small amounts of radioactivity (27). To hypersensitize the film, it is pre-exposed to a flash of white light from a flash gun of duration about 1 msec; this will increase absorbance to $0.1-0.2$ A_{540} units above that of unexposed, developed film. The optimal distance between the film and flash gun and the duration of the flash will have to be determined (for method see ref. 2).

The image produced on the film after developing can be photographed and/or scanned densitometrically which enables a certain degree of quantitation. There are a number of different types of scanners available commercially that can be set to scan one- and two-dimensional gels directly (after staining or by scanning at 280 nm) or the autoradiographic images (discussed in ref. 3). When used for two-dimensional gels, computer analysis of the output from such a scanner is really necessary. There are a number of commercial computer programs available (for example, apparatus and programs by Joyce-Loebl) that enable two-dimensional gels to be analysed. For one-dimensional gels, an alternative to scanning involves slicing of the wet gel and determining the radioactive protein content of each slice by scintillation counting (see ref. 2). In theory, this approach can be used in dual-isotope labelling experiments; however, resolution is limited by the slice sizes. Higher resolution is achieved by careful choice of isotopes that can be distinguished in the intact gel.

As mentioned in section 2.1.2, a method is available for discriminating between ^{14}C- and ^{3}H-labelled proteins (5,6). The gel is fluorographed and exposed to pre-flashed film at $-70°C$ to detect both isotopes. This is followed by direct autoradiography on untreated film at room temperature to detect only the ^{14}C. This method, however, suffers from the disadvantage that long exposure times are necessary due to the low specific activity available for these two isotopes (discussed in ref. 3). The combination of [^{35}S]methionine and [^{75}Se]selenomethionine provides a more rapid means of discriminating between two populations of newly synthesized proteins that have incorporated different relative amounts of each of the two isotopes. The gel is exposed to two pieces of pre-flashed X-ray film, separated by a piece of exposed, developed (blackened) film. Both isotopes will be detected on the first film, whereas only the more energetic γ^{75}Se emissions have sufficient energy to pass through to the second film (and can be enhanced by the use of an intensifying screen). A band that is only seen amongst the ^{35}S-labelled

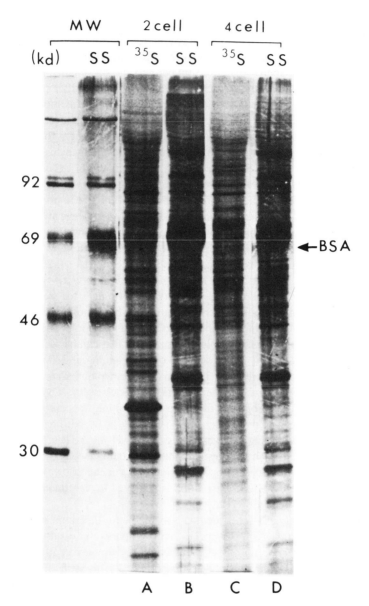

Figure 4 . Proteins accumulated versus those being synthesized in the early embryo. One-dimensional gel comparing the [³⁵S]methionine-labelled (**lanes A** and **C**) and silver stained (**lanes B** and **D**) patterns of proteins in the 2-cell and 4-cell embryo.

proteins will be present only on the first film and a negative of the first film when compared with the second film will detect such proteins (7).

4.2 Staining of total proteins

Coomassie Brilliant Blue R-250 is probably the most well known stain and can detect

Table 10. Silver staining of gels.

1.	Following electrophoresis and fixing (in 45% methanol:10% acetic acid), wash the gel in 50% methanol for at least 30 min followed by double-distilled water (3 × 30 min). This ensures complete removal of all interfering substances such as glycine, glycerol and Triton.
2.	Stain each gel for 15 min with freshly prepared ammoniacal silver nitrate solution (0.8 g of silver nitrate in 4 ml of double-distilled water plus 21 ml of 0.36% NaOH and 1.4 ml of ammonia solution, made up to 100 ml with double-distilled water).
3.	Wash the gel thoroughly with double-distilled water (3 × 2 min).
4.	Develop for 8−15 min in 500 ml of a solution containing 0.05 mg/ml citric acid and 0.18 mg/ml formaldehyde solution, before transferring to 50% methanol (30 min), followed by 20% methanol or double-distilled water (2 × 45 min).
5.	Dry the gel and photograph.

Table 11. Western blot method for identification of specific proteins.

For identification of specific proteins, large numbers of eggs or early embryos (for example, samples of 800−1000 unfertilized eggs per lane of a one-dimensional gel) are required.

1.	Pre-soak a gel-sized piece of Biodyne paper or nitrocellulose, plus two larger pieces of Whatman's filter paper in a buffer containing 2 mM sodium acetate, 5 mM MOPS and 20% ethanol (if Biodyne paper is used) or in a buffer containing 25 mM Tris, pH 8.3, 192 mM glycine and 20% methanol (if nitrocellulose is used).
2.	Immediately after electrophoresis place the gel against the paper to which the proteins are to be transferred (blot) and sandwich this between the two pieces of filter paper. Place the whole pile into the transfer apparatus (e.g. Bio-rad Trans-Blot Cell) ensuring that the blot is towards the positive electrode. Fill up the reservoir with the same buffer that was used to soak the papers. Proteins are then transferred at 50/60 V for 2 h or 30 V overnight.
3.	Following transfer dismantle the apparatus and remove the blot (the gel can be kept to check for complete transfer by silver staining, see *Table 10*).
4.	Block non-specific protein binding sites by shaking overnight with 10% fetal calf serum, 3% BSA or a 10% non-fat milk powder solution in phosphate-buffered saline (PBS: 0.14 M NaCl, 2.7 mM KCl, 1.5 mM KH_2PO_4, 8.1 mM Na_2HPO_4, with added 0.02−0.1% sodium azide).
5.	Rinse the blot for 10 min in PBS containing 0.1% Triton X-100 or 0.1% Tween. This is used for each wash step.
6.	Dilute the first layer antibody[a] as required (dilutions will depend upon the specific antibody; for example, monoclonal antibodies often need to be used undiluted) in 3% BSA in PBS and incubate with the blot for 45−60 min in a sealed bag or purpose-built small volume (0.5−2 ml) boxes or troughs. Wash the blot thoroughly (4 × 10 min) before incubating with the second layer antibody.
7.	For the second layer, use 5 µl of the appropriate biotinylated anti-immunoglobulin diluted in 100 µl of 3% BSA in PBS and made up to 0.5 ml. Incubate for 45−60 min.
8.	Wash the blot thoroughly (4 × 10 min) before incubating with 5 µl of [35S]streptavidin diluted in 50 µl PBS plus 3% BSA for 60 min in a sealed bag.
9.	Wash the blot thoroughly (4 × 10 min), air dry and expose to pre-flashed film at −70°C for 3−20 days.

[a]Each diluted antibody should be spun at 15K for 2−5 min prior to use.

0.2−0.5 µg of protein in a sharp band (see ref. 2); unfortunately this is not sensitive enough to be of much use for detecting the very small quantity of protein contained within early embryos. Improved techniques of silver staining have been developed (e.g. 28,29) which claim to be 100−200 times more sensitive, and so are applicable to the early embryo. There is some dispute as to the use of silver staining for quantitative analysis (see ref. 3). However, it appears that provided sufficient care is exercised so as not to under-estimate over-stained bands and providing reference proteins are run so that the densities of other spots can be normalized, quantitative integral analyses can be achieved by densitometric scanning.

Material from about 60 pre-implantation embryos is required on each track of a one-dimensional gel, and 250 embryos for two-dimensional gels, for detection of protein by silver staining (see *Figure 4*). A method (developed from ref. 29) is given in *Table 10*.

4.3 Detection of specific proteins

Staining methods have been developed for the detection of specific classes of poly-peptides (reviewed in Appendix I of ref. 2). For example, labelled lectins provide an approach towards the identification and partial characterization of glycoproteins in gels. The lectins can be visualized by a variety of methods, either directly, if they are congu-gated to a fluorescent moiety (30) or to horseradish peroxidase (31), or indirectly using anti-lectin antibodies (32). Such methods have not as yet been applied extensively to early embryos.

Antibodies can be used to detect specific proteins after electrophoresis; however, the rate of diffusion of antibodies into gel matrices is very poor. Hence, methods have been developed for transferring proteins (usually electrophoretically, see ref. 33) from polyacrylamide gels to nitrocellulose. Once the pattern of proteins has been blotted onto nitrocellulose antibodies can be applied directly. Immunoblotting works with polyclonal antisera and some monoclonal antibodies that recognize the appropriate an-tigen despite denaturation (34). Antibody—antigen interaction is then detected either by virtue of labelled (usually ^{125}I) antibody or by amplification with labelled second or third layer antibodies. Several amplification systems have been developed recently:

(i)   a second layer antibody coupled to horseradish peroxidase is visualized by incu-bation with 3,3′-diaminobenzidine and hydrogen peroxide to produce a visible stained band (35);

(ii)   a radiolabelled second layer provides more sensitivity, either directly as a labelled second layer, or as the ^{125}I-labelled *S. aureus* Protein A which binds to the first layer immunoglobulin (36);

(iii)   a biotinylated second layer is visualized by a third layer of [^{35}S]streptavidin.

The latter system (available from Amersham) has proved to be sensitive and success-ful and is included in the procedure for electrophoretic transfer of proteins and sub-sequent antibody amplification described in *Table 11*.

5. CONCLUDING REMARKS

Electrophoretic techniques have been applied to the study of proteins in the early embryo. It is likely that future developments will be directed towards techniques that will aid the identification, characterization and isolation of particular proteins. The improve-ment in high resolution two-dimensional gel analysis, possibly by the use of flat-bed systems, will be invaluable. The application of allied techniques, such as the electro-elution of particular protein bands after electrophoresis (37,38), opens up the possibili-ty of analysing individual species by peptide mapping and amino acid sequencing (discussed in ref. 3). In addition, the isolation of individual proteins will aid the production of antibodies that can be directed towards the embryo for immunofluorescence, immunoblotting or immunoprecipitation. Subfractionation and/or selective solubilization will assist localization and possibly identification of specific proteins at particular stages of development.

6. ACKNOWLEDGEMENTS

I should like to thank the following people for their help during the writing of this chapter: Azim Surani, Evelyn Houliston, Bernard Maro, Bob Moor and Martin Johnson.

7. REFERENCES

1. Johnson,M.H., McConnell,J. and Van Blerkom,J. (1984) *J. Embryol. Exp. Morphol.*, **83**, suppl. 197.
2. Hames,D. and Rickwood,D., (eds) (1981) *Gel Electrophoresis of Proteins − A Practical Approach.* IRL Press, Oxford.
3. Dunn,M.J. and Burghes,A.H.M. (1983) *Electrophoresis*, **4**, 173.
4. Van Blerkom,J. (1978) In *Methods in Mammalian Reproduction.* Daniel,J.C., Jr (ed.), Academic Press, New York, p. 68.
5. McConkey,E.H. (1979) *Anal. Biochem.*, **96**, 39.
6. McConkey,E.H. (1982) *Proc. Natl. Acad. Sci. USA*, **79**, 3236.
7. Lecocq,R.E., Hepburn,A. and Lamy,F. (1982) *Anal. Biochem.*, **127**, 293.
8. MacQueen,H.A. (1979) *J. Embryol. Exp. Morphol.*, **52**, 203.
9. Schultz,G.A., Kaye,P.L., McKay,D.J. and Johnson,M.H. (1981) *J. Reprod. Fertil.*, **61**, 387.
10. Bolton,V.N., Oades,P.J. and Johnson,M.H. (1984) *J. Embryol. Exp. Morphol.*, **79**, 139.
11. Whiteley,N.M. and Berg,H.C. (1974) *J. Mol. Biol.*, **87**, 541.
12. Johnson,L.V. and Calarco,P.G. (1980) *Dev. Biol.*, **77**, 224.
13. Bolton,A.E. and Hunter,W.M. (1973) *Biochem. J.*, **133**, 529.
14. Taylor,M.A., Robinson,K.R. and Smith,L.D. (1985) *J. Embryol. Exp. Morphol.*, **89**, 35.
15. Howlett,S.K. (1986) *Cell*, **45**, 387.
16. Johnstone,A. and Thorpe,R. (1982) *Immunochemistry in Practice.* Blackwell Scientific Publications, Oxford.
17. Van Blerkom,J. (1981) *Proc. Natl. Acad. Sci. USA*, **78**, 7629.
18. Surani,M.A.H. (1979) *Cell*, **18**, 217.
19. Laemmli,U.K. (1970) *Nature*, **227**, 680.
20. O'Farrell,P.H. (1975) *J. Biol. Chem.*, **250**, 4007.
21. Burghes,A.H.M., Dunn,M.J. and Dubowitz,V. (1982) *Electrophoresis*, **3**, 354.
22. O'Farrell,P.Z., Goodman,H.M. and O'Farrell,P.H. (1977) *Cell*, **12**, 1193.
23. Garrells,J.I. (1983) In *Methods in Enzymology.* Wu,R., Grossmann,L. and Moldave,K. (eds.), Academic Press, New York, Vol. 100, p. 411.
24. Drysdale,J.W. (1975) In *Methods of Protein Separation.* Catsimpoolas,N. (ed.), Plenum Press, New York, p. 93.
25. Bonner,W.M. and Laskey,R.A. (1974) *Eur. J. Biochem.*, **46**, 83.
26. Laskey,R.A. (1984) *Radioisotope Detection by Fluorography and Intensifying Screens.* (Review 23), Radiochemical Centre, Amersham, UK.
27. Laskey,R.A. and Mills,A.D. (1975) *Eur. J. Biochem.*, **56**, 335.
28. Switzer,R.C., Merril,C.R. and Shifrin,S. (1979) *Anal. Biochem.*, **98**, 231.
29. Wray,W., Boulikas,T., Wray,V.P. and Hancock,R. (1981) *Anal. Biochem.*, **118**, 197.
30. Furlan,M., Perret,B.A. and Beck,E.A. (1979) *Anal. Biochem.*, **96**, 208.
31. Wood,C.M. and Sarinana,F.O. (1975) *Anal. Biochem.*, **69**, 320.
32. Glass,W.F., Briggs,R.C. and Hnilica,L.S. (1981) *Anal. Biochem.*, **115**, 219.
33. Burnette,W.N. (1981) *Anal. Biochem.*, **112**, 195.
34. Anderson,N.L., Giometti,C.S., Gemmell,M.A., Nance,S.L. and Anderson,N.G. (1982) *Clin. Chem.*, **28**, 1084.
35. Olden,K. and Yamada,K.M. (1977) *Anal. Biochem.*, **78**, 483.
36. Burridge,K. (1978) In *Methods in Enzymology.* Ginsburg,V. (ed.), Academic Press, New York, Vol. 50, p. 54.
37. Wallis,M.H., Kramer,G. and Hardesty,B. (1980) *Biochemistry*, **19**, 789.
38. Niemann,M.S., Volanakis,J.E. and Mole,J.E. (1980) *Biochemistry*, **19**, 1576.
39. Howlett,S.K. and Bolton,V.N. (1985) *J. Embryol. Exp. Morphol.*, **87**, 175.

CHAPTER 9

Construction of cDNA libraries for pre-implantation mouse embryos

CHRISTINE J.WATSON and JOSIE McCONNELL

1. INTRODUCTION

The long cell cycles and the resilience to manipulation of the mouse embryo have allowed a fairly detailed description of the cell biology of early embryogenesis to be made (1,2, Chapters 2 and 3). However, the pre-implantation mouse embryo is not an auspicious system for the analysis of the molecular events underlying the morphological changes that have been observed. It is only recently, with improvements in the sensitivity of recombinant DNA techniques, that it has become possible to begin to analyse embryonic development in molecular terms.

The principal difficulty encountered in any biochemical or molecular analysis of development arises from the small size of the embryo and the limited number of embryos available: it is estimated that each mouse oocyte contains approximately 0.35 ng of total RNA (3) and even with the aid of superovulation (Chapter 1, Section 5.4) only around 30 oocytes can be obtained from a single mouse.

Despite these limitations, a considerable body of biochemical data about the early stages of development has accumulated (4). It has been established that maternally-inherited mRNA is functional during the first two cell cycles; after this stage most if not all of this mRNA is destroyed to be replaced by mRNA transcribed from the embryonic genome (5−8). A closer analysis of the proteins encoded by the maternally-inherited transcripts has shown that a subset of these are modified in a cell cycle-dependent fashion (9).

One approach to understanding the flow of molecular information from genes to proteins is to analyse mRNA populations at various stages of development. Selective destruction or activation may become apparent through such analyses and the presence of a particular mRNA at one stage may imply a role for this mRNA in certain developmental decisions. In order to facilitate a molecular analysis it is necessary to construct complementary DNA (cDNA) libraries from mRNA extracted at various stages of embryonic development. However, as mentioned above, such an approach is seriously hampered by the limiting quantities of material available. It is not feasible to use a model system such as the embryonal carcinoma (EC) cell type (10), which is frequently used in studies of differentiation, since EC cell lines are not zygote derived and cannot be expected to mimic the earliest embryonic events. Given that the use of RNA from oocytes and later stages of development is required, it becomes clear that if a molecular analysis of development is to succeed, all techniques used must be optimized to compensate for the very limiting quantities available.

Using a bacteriophage vector system, very high efficiencies of cloning can be achieved, of the order of 2×10^7 plaque forming units (p.f.u.) per μg of double-stranded cDNA (11), thereby requiring very small amounts of starting tissue. The number of individual cloning events which would yield a 99% probability of cloning a particular mRNA within a population may be calculated according to the following formula:

$$N = \ln(1 - P)/\ln(1 - 1/n)$$

where N is the number of clones required, P is the probability of obtaining a given sequence and $1/n$ is the fractional proportion of the total mRNA population that a single type of mRNA represents (12). For example, in an SV40-transformed human fibroblast cell line, the number of different sequences in the low abundance class (~ 14 copies per cell) was estimated to be 10 670 representing 29% of the mRNA (13). Therefore, in this case $n = 10\ 670/0.29$ which is 36 790. So in order to have a 99% probability of cloning a low abundance mRNA, approximately 169 000 clones are required. Assuming that the complexity of the RNA population of an embryo is similar to the above example, around 200 000 clones should constitute an essentially complete library. However, it should be noted that different relative efficiencies of cDNA synthesis can occur (14) making this number a minimum estimate. *Table 1* lists the predicted number of mouse embryos from different developmental stages that would yield the minimal amount of RNA required for the construction of a complete library by state-of-the-art techniques. This number may be reduced with further improvements in the efficiency of each reaction step, but is unlikely to be decreased markedly.

Embryo libraries should prove invaluable in several respects.

(i)   Using the technique of differential screening, sequences which are expressed at different relative abundance levels at different developmental stages can readily be isolated (15).

(ii)   It may be possible to study the patterns of expression of particular cloned sequences throughout early development in an approximately quantitative manner by screening representative cDNA libraries, a more sensitive method than transfer hybridization analysis.

(iii)   Clones isolated from cDNA libraries should also be useful for studying multigene

Table 1. RNA content of pre-implantation embryos.

Developmental stage	Total RNA per embryo ng[a]	Estimated poly(A)+ RNA per embryo pg[b]	Number of embryos for complete library[c]
Oocyte	0.35	7.0	14 286
1-cell	0.42	8.3	12 048
Late 2-cell	0.24	2.6	38 461
8−16-cell	0.69	4.4	22 727
32-cell	1.47	14.2	7 042

[a]Based on data of Piko and Clegg (3).
[b]Based on data of Piko and Clegg (3) assuming that poly(A)+ RNA is 10 times the poly(A)+ content of the embryo.
[c]Complete library is estimated to require 2×10^5 events and approximately 0.1 μg of polyadenylated RNA (this will depend on the efficiency of each step). See text for details.

families. Sequence analysis would unequivocally identify the individual members of a family which are expressed at a particular time.

The availability of cDNA libraries from various pre-implantation stage embryos should therefore be a powerful tool for the study of the molecular events associated with early development.

2. EVALUATION OF CARRIER RNA

In order to overcome the technical difficulties associated with manipulating very small amounts of material, a carrier system must be employed to minimize losses of the mRNA and cDNA which would otherwise be unacceptably high. A suitable carrier for use in the selection of polyadenylated RNA and subsequent synthesis of cDNA must satisfy certain criteria. These include the following.

(i)   It should be polyadenylated thereby participating in all the reactions in a manner analogous both quantitatively and qualitatively to the mRNA being cloned.
(ii) It should contain a defined number of species, preferably as few as possible.
(iii) These sequences should not be held in common with the RNA of interest.
(iv) The entire sequence of the carrier RNA(s) should be known.
(v)   The carrier must not clone preferentially.

These conditions are met by the genome of cowpea mosaic virus (CPMV). This is a type member of the comovirus group and infects only plants. Its genome is composed of two single-stranded RNA molecules of positive (messenger) polarity which are separately encapsidated. The RNAs are of different sizes, the larger being 5889 bp long while the shorter RNA is 3481 bp in length. Unusually for plant viruses, both RNAs are polyadenylated (16) and their entire sequence has been determined (17,18). Thus pre-conditions (i), (ii) and (iv) are satisfied by this carrier. A trial cloning experiment, in which CPMV RNA (kindly supplied by Drs J.Davies and G.Lomonossof, John Innes Institute, Colney Lane, Norwich, UK) was mixed with mouse liver RNA in a weight ratio of 9:1, revealed that approximately nine times more plaques hybridized to a CPMV radiolabelled RNA probe than to a mouse liver RNA probe and that no plaques were observed to hybridize to both probes (19). This satisfies the remaining two criteria (iii) and (v). CPMV RNA therefore fulfills all the requirements of a carrier and since it is easily purified in large amounts, it should be emminently suitable for use in cDNA cloning experiments.

3. COLLECTION OF OOCYTES AND EMBRYOS AND ISOLATION OF RNA

Superovulation of female mice is described in Chapter 1, Section 5.4, and the collection of oocytes and pre-implantation embryos described in Chapter 2, Section 2, of this volume. We will therefore not describe these procedures in detail but focus on points of particular importance to the isolation of RNA.

Note that collection of oocytes and embryos should be performed under sterile conditions and care taken to avoid contamination with ribonuclease.

3.1 Collection of oocytes

Around 16 000 oocytes (from ~600 superovulated, but unmated, female mice) will be required to enable the isolation of sufficient RNA for the construction of a complete oocyte library (see *Table 1*). In order to obtain a good library, it is essential that the mRNA is undegraded. Therefore, oocytes should be collected in batches and as quickly as possible to minimize degradation. Any oocytes which have an abnormal appearance should be discarded. It is particularly important to separate the oocytes from cumulus cells since these are of maternal origin and would therefore generate libraries which are not representative of the oocyte. When embryos have been collected free of cumulus cells, wash these once in M16 medium without bovine serum albumin (BSA) (Chapter 2, Section 5) and then deposit them on the side of a 1.5 ml Eppendorf tube in the minimum amount of medium possible (2 μl). In our experience, we find that it is convenient to collect 8000 oocytes in one tube which should then be immediately snap frozen in liquid nitrogen and stored at $-70\,^\circ$C. It is extremely important that the oocytes are not allowed to thaw at any point between collection and RNA extraction since this would lead to lysis of the cells with the concomitant release of endogenous ribonucleases. This may cause degradation of the RNA.

3.2 Collection of pre-implantation embryos

These should also be collected exactly as described in Chapter 2, Section 2. Discard any retarded or morphologically abnormal embryos. Since it is more time consuming to collect these samples, batches of around 1000 embryos should be collected at a time.

Table 2. Isolation of total RNA.

---

1. Thaw samples on ice and immediately add 100 μl of extraction buffer (EB)[a]. Mix by pipetting up and down twice.
2. Add 10 μl of EB containing 0.1% SDS (w/v) followed by 100 μl of redistilled buffered phenol[b].
3. Add a further 0.5 ml each of EB and phenol. Vortex briefly then separate the phases by centifugation in a microcentrifuge for 5 min.
4. Recover the aqueous phase (note that there is only a small interface) and re-extract the phenol phase with 100 μl of EB. Pool the aqueous phases.
5. Remove remaining traces of phenol by shaking with three changes of an equal volume of water-saturated ether or until the aqueous phase is clear.
6. DNA may be removed at this stage by treating with ribonuclease (RNase)-free deoxyribonuclease (DNase)[c]. Add 350 μl of 3 $\times$ RQ1[d] buffer followed by 10 units of DNase. Incubate at 37°C for 10 min.
7. Add an equal volume of phenol and repeat extraction as in steps 4 and 5.
8. Precipitate the RNA by transferring the aqueous phase to a sterile siliconized Corex tube and adding 2.5 volumes of cold absolute ethanol plus 0.1 vol of 3 M sodium acetate.
9. Keep at $-20\,^\circ$C overnight.
10. Collect the precipitated RNA by centrifugation at 10 000 r.p.m. in a Sorvall SS34 rotor for 45 min at 4°C. Wash the pellet with cold 70% (v/v) ethanol, dry briefly under vacuum and resuspend in 0.5 ml of ice-cold sterile ddH$_2$O[e].
11. Determine the quantity of total RNA recovered by spectrophotometric measurement using a sterile autoclavable cuvette. Also, measure the A_{260}:A_{280} ratio. A value of $1.8-2.0$ indicates a protein-free nucleic acid preparation, although ratios of $1.7-1.8$ are acceptable. From the A_{260} measurement, estimate the yield of total RNA (1 OD = 40 μg/ml for RNA).

---

[a]EB is 0.1 M NaCl, 0.001 M EDTA, 0.02 M Tris-HCl pH 7.4.
[b]See reference 26.
[c]DNase (product RQ1[TM]) is purchased from Promega Biotech. 1 unit of enzyme is that amount required to completely degrade 1 μg of DNA in 10 min at 37°C.
[d]3 $\times$ RQ1 buffer is 120 mM Tris-HCl pH 7.9, 30 mM MgCl$_2$.
[e]ddH$_2$O is glass double-distilled water.

Wash each batch in M16 medium without BSA and collect the material on the side of an Eppendorf tube in as small a volume as possible (as for oocyte samples). Snap freeze immediately in liquid nitrogen and store at $-70°C$ until sufficient material has been obtained for the extraction of RNA.

3.3 Isolation of RNA

Total nucleic acid is isolated from oocytes and embryos by a modification of the method of Braude and Pelham (20), as described in *Table 2*. It is important that carrier tRNA is not added to maximize recovery since this contains small amounts of double-stranded DNA which would subsequently be cloned.

3.4 Selection of poly(A)$^+$RNA and addition of carrier

From the estimated yield of total RNA (*Table 2*), calculate the amount of poly(A)$^+$ RNA (see *Table 1*). Add sufficient CPMV carrier RNA to give a total of 1 μg of polyadenylated RNA.

Polyadenylated RNA may then be isolated either by chromatography on oligo(dT)−cellulose (Type 3, Collaborative Research) or binding to messenger affinity paper (Hybond™-mAP, Amersham). The latter method is recommended for small amounts of RNA. The manufacturer's protocol should be adhered to, with the inclusion of an additional wash in 0.1 M NaCl for 5 min before the 70% (v/v) ethanol rinse.

4. CONSTRUCTION OF LIBRARIES

Methodologies for cDNA library construction have been presented in detail in a previous volume of this series (21) and so will not be reiterated here. This section will focus on aspects that are particularly important when handling small amounts of material.

4.1 Choice of vector

The use of plasmid vectors in the construction of cDNA libraries has now been largely supplanted by lambda bacteriophage vectors for three main reasons.

(i)   A very high efficiency of cloning is achieved with phage due to the superiority of *in vitro* packaging as compared with transformation of plasmids. Approximately 1% of phage genomes can be recovered (22) whereas transformation is about 10 times less efficient. Excellent packaging extracts are now available commercially and these are continually being improved. Since the cDNA cloning efficiency depends on the packaging efficiency, the best obtainable extracts should be used when only small amounts of cDNA are available.

(ii)   Screening of phage plaques on nitrocellulose filters with nucleic acid probes is more sensitive and reproducible, with a higher signal-to-noise ratio, than bacterial colony hybridization. This is especially important when carrying out a differential hybridization analysis.

(iii)   When immunity insertion vectors such as λgt10 or NM607 are utilized, non-recombinant molecules (i.e. reconstituted vector molecules without a cDNA insert) can be eliminated from the library by plating on an appropriate strain of *Escherichia coli*.

The vector of choice for constructing cDNA libraries is λgt10 (23). This is a vigorously growing phage which produces large uniform plaques, a considerable aid in differential screening. All cDNA fragments up to 7.6 kb in size can be successfully cloned with this vector. The majority of eukaryotic mRNAs fall within this size class. Where full-length copies of very large RNAs, such as laminin B2 (8 kb), are required a different phage vector such as NM1149 (24), which is capable of accepting larger inserts, should be employed. If the only available method of screening utilizes antibody probes which recognize the protein encoded by a cDNA, the cDNA molecules should be cloned in a phage expression vector such as λgt11 (23).

4.2 Synthesis and cloning of double-stranded cDNA

A large number of protocols have been devised for the synthesis of double-stranded cDNA (e.g. 11,23,25,26) and any one of these may be used. The method that we recommend is that of Watson and Jackson (11) which incorporates the novel approach introduced by Okayama and Berg (27) for synthesizing the second strand of cDNA. In this, the mRNA/cDNA hybrid product of the first strand reaction is treated with RNase H to produce short RNA fragments which act as primers for *E. coli* DNA polymerase I which will synthesize a second cDNA strand and replace the RNA with DNA. Nicks are sealed with *E. coli* DNA ligase. This approach favours the synthesis of full-length cDNA and reduces losses and artefacts (such as cDNAs which are not co-linear with the mRNA) around the 5' end sequences. These problems frequently occur as a result of digestion with the single strand-specific nuclease S1 (28) used in alternative approaches.

A schematic representation of the cDNA synthesis and cloning protocol appears in *Figure 1* and detailed experimental procedures are provided in Chapter 3 of reference 21. Two modifications should be made when applying this protocol to starting samples of less than 1 μg of polyadenylated RNA.

(i)   As described in Section 3.4, CPMV carrier should be added to the sample RNA before selection of poly(A)$^+$ RNA.

(ii)   A 1 μg aliquot of CPMV RNA should be added to the linker-ligated double-stranded cDNA before the Ultragel column chromatography step (which removes excess linkers) to prevent losses of cDNA caused by non-specific binding to the column matrix.

It is essential, in all the manipulations involving cDNA, that siliconized microcentrifuge tubes be used. Also, for the column chromatography step (to remove the *Eco*RI linkers, *Figure 1*), both the column and the glass wool plug should be siliconized. This may be carried out by rinsing the tubes and glassware several times in a 2% solution of dichlorodimethylsilane in 1,1,1-trichloroethane (supplied by Hopkin and Williams) followed by several rinses in sterile distilled water and then autoclaving. Note that this solution is toxic and should be used in a fume hood.

In vitro packaging of the hybrid phage DNA should follow the directions of the supplier (e.g. Amersham, Gigapack) and phage particles plated out as described in reference 23. For valuable samples it is essential that the packaging extracts are assayed before use.

To determine the complexity of the library with respect to the RNA of interest, a number of controls should be performed during the packaging reaction. A background

CLONING cDNA INTO λgt 10

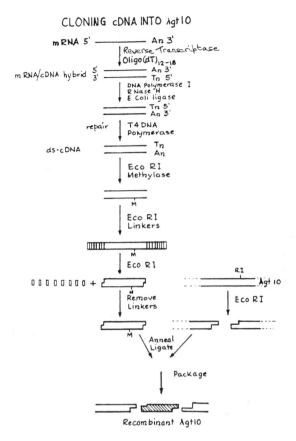

Figure 1. Procedure for synthesis and cloning of double-stranded cDNA in the bacteriophage vector λgt10. Detailed experimental procedures are presented in reference 11.

of clear non-recombinant plaques is unavoidable and the frequency of these can be determined by packaging cut and re-ligated λgt10 vector alone. The ratio of CPMV carrier cDNA clones to embryo clones depends on the ratio in which they were mixed. The number of CPMV clones can be determined easily by screening a portion of the library with a radiolabelled probe made from CPMV RNA using reverse transcriptase and calf thymus random primers (see *Table 3D*). Since only two species of CPMV RNA are present, in approximately equal numbers, the signals obtained will be of the same intensity thereby allowing exact numbers to be determined. Thus the number of embryo-specific recombinants can be estimated by subtracting background and CPMV plaques from the total number of clear plaques obtained. A typical filter, prepared from a mouse oocyte library and hybridized to a CPMV probe, is shown in *Figure 2*. However it should be noted that since the mean size of the two CPMV RNA species is 4.7 kb, more than twice the size of the average eukaryotic mRNA, the number of CPMV molecules per gram will be less than that for embryo RNA and so the number of embryonic clones will be underestimated.

The following section deals with the various procedures that can be used to isolate clones of interest from a cDNA library.

189

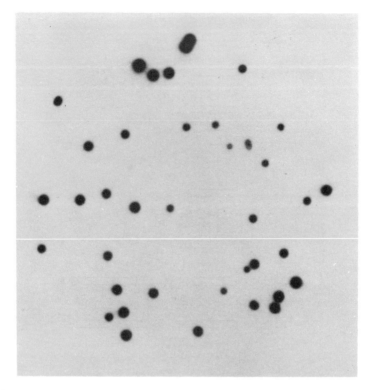

Figure 2. Hybridization pattern obtained when a nitrocellulose filter, carrying 65 plaques from an oocyte/CPMV (1:19) cDNA library, was hybridized to a radiolabelled CPMV probe. The number of hybridizing plaques is 40, in close agreement with the predicted number of 46 (65 less 16 background non-recombinant plaques and 3 oocyte-specific clones, see reference 19).

5. SCREENING OF THE LIBRARY

Generally, cDNA libraries are screened by hybridizing radiolabelled nucleic acid probes to nitrocellulose or nylon membrane filters carrying phage DNA. The preparation and hybridization of filter replicas of library plaques is described in detail in reference 26. The following sections will discuss the variety of probes and labelling techniques available. In addition, alternative methods for isolating clones of interest will be described.

5.1 Test for representative nature of the library

It is essential to confirm that embryo-specific sequences are represented in the library. This could be achieved by screening a small portion of the library with high specific activity single-stranded cDNA prepared from the mRNA used for the library construction. Since this would utilize very valuable embryo RNA it is preferable to carry out preliminary screens with a cloned probe encoding an RNA known to be expressed in most cell types or at the particular stage of embryogenesis from which the library material was obtained. Probes for abundantly expressed mRNAs such as β-actin are recommended.

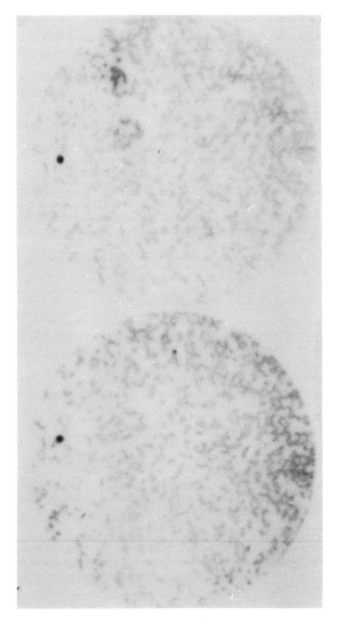

Figure 3. Test for representative nature of the library. One pair of replica filters of approximately 3000 plaques, from an oocyte/CPMV cDNA library, were hybridized to the insert fragment from pPE386 which was radiolabelled by the random primed polymerization method (*Table 3A*). This cDNA clone contains 1.1 kb of laminin B1 sequences (29). One positively hybridizing clone is detected. The high background results from the hybridization of poly(A) tails in the library clones to poly(A) tail sequences in the probe.

Table 3. Preparation of radiolabelled nucleic acid probes.

A.   Random primed polymerization

This method can be used to label DNA, plasmids containing the sequence of interest or specific DNA fragments excised from low gelling temperature (LGT)[a] agarose gels (without prior purification of the DNA). For cDNA clones, the insert fragment should be excised prior to radiolabelling.

1.   Following restriction enzyme digestion and electrophoresis of the cloned DNA[b], cut the required band out of the gel in as small a piece as possible, weigh the gel segment and place it in an Eppendorf tube.

2.   Dissolve the gel and denature the DNA by placing the tube in a boiling water bath for 10 min.

3.   Incubate at 37°C for 10−60 min. Set up the labelling reaction as follows, adding the components in the stated order:

      10.0 μl of oligolabelling buffer[c]

      2.0 μl of bovine serum albumin (10 mg/ml)

      35.5 μl of DNA/agarose solution (not greater than 30 μl from step 2 + ddH$_2$O).

      2.0 μl of [^{32}P]dCTP (20 μCi) (Amersham, sp. act. 400 Ci/mmol)

      0.5 μl of Klenow fragment (2 units) (nbl enzymes Ltd)

4.   Incubate at room temperature for 5−16 h then add 200 μl of stop solution [20 mM NaCl, 20 mM Tris-HCl pH 7.5, 2 mM EDTA, 0.25% (w/v) SDS, 1 μM dCTP].

5.   Phenol extract once, ether extract the aqueous phase three times then add 10 μg of sonicated salmon sperm DNA and ethanol precipitate by adding 3 vol of 96% (v/v) ethanol and 0.1 vol of 2 M sodium acetate.

6.   Centrifuge for 15 min in an Eppendorf microcentrifuge, remove the supernatant, dry the pellet and resuspend it in 200 μl of TE[d].

7.   Boil for 5 min before hybridization.

B.   Kinase 5′ end-labelling of synthetic oligonucleotides

The 5′ termini of oligonucleotides can be labelled using T4 polynucleotide kinase and [γ-^{32}P]ATP. This enzyme catalyses the transfer of the γ-phosphate from the ATP to the 5′ hydroxy terminus of the oligonucleotide. Set up a 50 μl labelling reaction as follows.

1.   Mix:

      1 μg of oligonucleotide

      5 μl of 10 × kinase buffer[e]

      50 pmol of [γ-^{32}P]ATP (150 μCi) (Amersham, sp. act. 3000 Ci/mmol)

      10 units of T4 polynucleotide kinase[f]

      to 50 μl of ddH$_2$O

2.   Incubate at 37°C for 30 min.

3.   Add 2 μl of 0.5 M EDTA.

4.   Extract once with phenol:chloroform (1:1).

5.   Remove unincorporated [γ-^{32}P]ATP by chromatography on a column of Sephadex G50 run in 10 mM Tris-HCl pH 7.5, 100 mM NaCl, 1 mM EDTA. Pool the fractions containing the first peak which contains the labelled oligonucleotide.

C.   Filling-in labelling of oligonucleotides

Alternatively, oligonucleotides may be labelled by synthesizing the complementary oligomer, using a short (usually 8-base) oligonucleotide as a primer. The radiolabelled nucleotide should be chosen carefully to maximize the number of labelled bases incorporated. This will vary considerably depending on the sequence of the oligo. Set up the reaction as follows.

1.   Anneal

      1 μl of template oligonucleotide (30 ng)

      1 μl of primer oligonucleotide (10 ng)

      10 μl of TM buffer[g]

2. Incubate at 65°C for 5 min.
3. Cool at room temperature for 5 min then centrifuge briefly to collect droplets of condensation.
4. Add 1 μl each of the three cold dNTPs (at 1 mM) and 4 μl (40 μCi) (Amersham, sp. act. 3000 Ci/mmol) of the remaining base as [^{32}P]dNTP.
5. Add 4 units of Klenow fragment of DNA polymerase I.
6. Incubate on ice for 2 h then chase the 'hot' base by adding 1 μl of the same 'cold' base (at 1 mM) and incubating on ice for 45 min.
7. Add 5 μl of de-ionized formamide, heat at 65°C for 5 min then run on a 10% acrylamide sequencing gel for approximately 3 h or until the bromophenol blue dye (loaded in a parallel lane) is about 10 cm from the bottom.
8. Wrap in cling film, place X-ray film on top (under red safe lights) and expose for 2 min.
9. Develop the film, align it with the gel and cut out the labelled band as cleanly as possible.
10. Re-expose the gel to film to check that the band has been correctly removed. It is important to remove only the region of the gel containing the labelled band in order to avoid contamination with the unlabelled template oligonucleotide as this would reduce the hybridization efficiency.
11. Add the gel slice to 500 μl of 1 mM EDTA containing 20 μg of sonicated salmon sperm DNA. Homogenize briefly and elute overnight then add directly to hybridization solution.

D. Single-stranded cDNA from specific RNA populations

This procedure is a modification of the first strand synthesis reaction for cDNA cloning and utilizes random primers, prepared from sonicated, strand-separated calf thymus DNA (Pharmacia).

1. Vacuum dry 50 μCi (sp. act. 3000 Ci/mmol) each of dCTP and dTTP plus 1 μl of actinomycin D[h].
2. Add:
     5.0 μl of 5 × reverse transcriptase buffer[i]
     2.5 μl of random primers (2 mg/ml)
     0.5 μl of dithiothreitol (1.0 M)
     1.0 μl of dTTP (0.1 mM)
     1.0 μl of dCTP (0.1 mM)
     2.5 μl of dGTP (5.0 mM)
     2.5 μl of dATP (5.0 mM)
     1.0 μl of RNasin (30 units)[j]
     1.0 μg of the specific poly(A)$^+$ RNA
     1.0 μl of reverse transcriptase (22 units)[k]
   to 25.0 μl of ddH$_2$O.
3. Incubate at 37−42°C for 2 h.
4. Remove unincorporated nucleotides by chromatography on Sephadex G50. Boil the probe for 5 min before adding to the hybridization solution.

---

[a]Purchased from SeaKem or Sigma.
[b]See ref. 26.
[c]Prepare oligolabelling buffer as follows: solution O, 1.25 M Tris-HCl pH 8.0, 0.125 M MgCl$_2$; solution A: 1 ml of solution O + 18 μl of 2-mercaptoethanol + 5 μl of dATP, 5 μl of dTTP, 5 μl of dGTP (each 0.1 M in ddH$_2$O); solution B: 2 M Hepes (Sigma) pH 6.6; solution C: hexadeoxyribonucleotides (Pharmacia); suspended in TE (see footnote d) at 90 OD units/ml. Mix solutions A:B:C in the ratio 100:250:150.
[d]TE is 10 mM Tris-HCl, pH 7.5, 1 mM EDTA.
[e]10 × kinase buffer is 0.5 M Tris-HCl pH 7.6, 0.1 M MgCl$_2$, 50 mM dithiothreitol, 1 mM spermidine, 1 mM EDTA.
[f]Supplied by BRL. One unit of T4 polynucleotide kinase transfers 1 nmol of the terminal phosphate from ATP to the 5′ hydroxy termini of micrococcal nuclease-treated DNA in 30 min at 37°C.
[g]TM buffer is 0.1 M Tris-HCl pH 8.0, 0.1 M MgCl$_2$.
[h]Supplied by Calbiochem. Stock solution is 1 mg/ml in 80% (v/v) ethanol.
[i]5 × RT buffer is: 250 mM Tris-HCl pH 8.3 (at 42°C), 40 mM MgCl$_2$ 250 mM KCl.
[j]Supplied by Biotech. 1 unit of RNasin will inhibit by 50% the activity of 5 ng of ribonuclease A.
[k]Supplied by nbl enzymes limited. 1 unit of reverse transcriptase will incorporate 1 nmol of dTMP into an acid-insoluble product in 10 min at 37°C.

Screening with clones for rare RNAs is also informative, indicating the complexity of the library. *Figure 3* shows the hybridization signals obtained when one pair of filters from a mouse oocyte library (plated at high density) were probed with a cDNA clone containing 1.1 kb of laminin B1 3' end sequences (29). The B1 polypeptide constitutes approximately 0.06% of oocyte protein (30). Although the abundance of the corresponding RNA is not known, the number of B1 hybridizing oocyte clones was approximately 10-fold less than the number of actin hybridizing clones, a result which accords well with expectations based on protein levels.

5.2 Screening of library for specific sequences

There are five main approaches to screening cDNA libraries, three of which utilize nucleic acid probes. These are outlined below and protocols for preparing probes are presented in *Table 3*.

5.2.1 *Cloned probes*

Probes consisting of previously cloned sequences are frequently used for isolating that particular sequence element from a cDNA library of a different species or from a different developmental stage. These DNA probes may be conveniently labelled either by nick translation (31) or by random primed polymerization (32). The latter procedure is preferred as it yields higher specific activity probes. If a cross-species hybridization is being performed, the stringency of washing after hybridization should be reduced accordingly.

5.2.2 *Oligonucleotide probes*

These are useful for isolating individual members of multigene families or where only the protein sequence (either partial or complete) is known. Generally a 20-base oligomer, homologous to the sequence of interest, is sufficient. This may be labelled either at the 5' end, using polynucleotide kinase (*Table 3B*), or by 'filling-in' using an 8-base oligomer as primer for complementary strand synthesis (*Table 3C*).

5.2.3 *Single stranded cDNA probes*

When it is desired to isolate sequences which are expressed at elevated levels at one developmental stage compared with another, the technique of differential screening should be adopted. For this, high specific activity cDNA probes (prepared from the two RNA populations being compared) are required and phage should be plated at low density (~ 400 plaques per 90 mm dish). Single-stranded cDNA probes are synthesized as described in *Table 3D* and specific activities in excess of 10^8 c.p.m./μg should be achieved routinely. A typical hybridization pattern is shown in *Figure 4*. This pair of replica filters, from an EC cell cDNA library, were hybridized to probes prepared from either differentiated (differentiation induced by retinoic acid treatment) or undifferentiated EC cell RNA. Many differences are apparent between the two filters, some clones encoding RNA species expressed at much higher levels in undifferentiated cells and *vice versa*. Exactly the same procedure should be applicable to the differential screening of embryo libraries provided sufficient material is available. Note that random primed cDNA probes are preferable to oligo(dT)-primed cDNAs for two reasons. Firstly, a

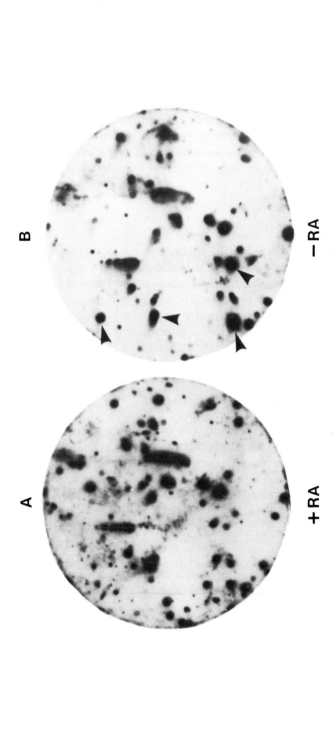

Figure 4. A typical pair of replica filters, from an EC cell cDNA library, carrying approximately 400 plaques were hybridized to single-stranded cDNA probes (*Table 3D*) transcribed from either undifferentiated (−RA) or differentiated (+RA) EC cell RNA. Plaques hybridizing more strongly to the undifferentiated cell probe are arrowed. RA = retinoic acid.

higher specific activity can be obtained (~ 10-fold) and secondly, all homologous clones will be detected since there is no bias in favour of the 3' end sequences [as is the case with oligo(dT)-primed probes].

5.2.4 *Hybrid arrest*

When it is desired to isolate a clone encoding a particular protein which has been identified by *in vitro* translation, the procedure of hybrid arrest of translation may be used (33). Such an analysis depends on the ability of a particular mRNA to anneal with its complementary cDNA clone. Recent improvements in hybrid arrest (see ref. 34 for details) make it a much more likely analytical technique for systems where material is limiting than hybrid selection and translation of mRNA, which is both expensive on RNA and troublesome to perform (and scarcely feasible in a system such as the mouse embryo).

5.2.5 *In situ hybridization*

It may also be possible to identify tissue- or cell-specific expression of the mRNA encoded by a particular cDNA clone by *in situ* hybridization (see Chapter 10). This technique is also useful for studying the temporal pattern of expression of a cloned sequence of interest and has already been successfully performed on certain early stages of mouse development (35). The distinct advantage of this method over other screening procedures is that it uses only small numbers of mouse embryos. *In situ* hybridization has proven to be an invaluable tool for studying gene expression in *Drosophila* embryos. It is to be hoped that future technical improvements will enable *in situ* hybridization to mouse embryo sections to be used as a screening procedure for cDNA clones.

6. ACKNOWLEDGEMENTS

We wish to thank Jim Jackson for his collaboration and major contribution to the original cDNA cloning protocol. We are grateful to J.Davies and G.Lomonossof for their kind gift of CPMV RNA and to colleagues in Cambridge for their invaluable assistance in collecting mouse oocytes and blastocysts. We also thank Martin Johnson, Peter Rigby and Keith Willison for helpful discussions. The work of C.J.W. and J.M. is supported by the Cancer Research Campaign and the Medical Research Council.

7. REFERENCES

1. Johnson,M.H., McConnell,J. and Van Blerkom,J. (1984) *J. Embryol. Exp. Morphol.*, **83** (suppl.), 197.
2. Howlett,S.K. and Bolton,V.N. (1985) *J. Embryol. Exp. Morphol.*, **87**, 175.
3. Piko,L. and Clegg,K.B. (1982) *Dev. Biol.*, **89**, 362.
4. Johnson,M.H. (1981) *Biol. Rev.*, **56**, 463.
5. Flach,G., Johnson,M.H., Braude,P.R., Taylor,R.A.S. and Bolton,V.N. (1982) *EMBO J.*, **1**, 681.
6. Clegg,K.B. and Piko,L. (1983) *J. Embryol. Exp. Morphol.*, **74**, 169.
7. Bolton,V.N., Oades,P.J. and Johnson,M.H. (1984) *J. Embryol. Exp. Morphol.*, **79**, 139.
8. Giebelhaus,D.H., Heikkila,J.J. and Schultz,G.A. (1983) *Dev. Biol.*, **98**, 148.
9. Howlett,S.K. (1986) *Cell*, **45**, 387.
10. Martin,G.R. (1980) *Science*, **209**, 768.
11. Watson,C.J. and Jackson,J.F. (1985) In *DNA cloning − A Practical Approach*. Glover,D.M. (ed.), IRL Press, Oxford, Vol. I, p. 79.
12. Clarke,L. and Carbon,J. (1976) *Cell*, **9**, 91.
13. Williams,J.G. (1977) *Cell*, **17**, 903.

14. Rougeon,F. and Mach,B. (1977) *J. Biol. Chem.*, **252**, 2209.
15. Brulet,P., Condamine,H. and Jacob,F. (1985) *Proc. Natl. Acad. Sci. USA*, **82**, 2054.
16. El Manna,M.M. and Bruening,G. (1973) *Virology*, **56**, 198.
17. Lomonossof,G.P. and Shanks,M. (1983) *EMBO J.*, **2**, 2253.
18. Van Weezenbeek,P., Verver,J., Harmsen,J., Vos,P. and Van Kammen,A. (1983) *EMBO J.*, **2**, 941.
19. McConnell,J. and Watson,C.J. (1986) *FEBS Lett.*, **195**, 199.
20. Braude,P.R. and Pelham,H.R.B. (1979) *J. Reprod. Fertil.*, **56**, 153.
21. Glover,D.M. (ed.) (1985) *DNA Cloning — A Practical Approach*. Volume I, IRL Press, Oxford.
22. Enquist,L. and Sternberg,N. (1979) In *Methods in Enzymology*. Wu,R. (ed.). Academic Press, New York, Vol. 68, p. 281.
23. Huynh,T.V., Young,R.A. and Davis,R.W. (1985) In *DNA Cloning — A Practical Approach*. Glover,D.M. (ed.), IRL Press, Oxford and Washington, DC, Vol. I, p.49.
24. Murray,N.E. (1983) In *Lambda II*. Cold Spring Harbor Laboratory Press, New York.
25. Gubler,U. and Hoffman,B.J. (1983) *Gene*, **25**, 263.
26. Maniatis,T., Fritsch,E.F. and Sambrook,J. (1982) *Molecular Cloning — A Laboratory Manual*. Cold Spring Harbor Laboratory Press, New York.
27. Okayama,H. and Berg,P. (1982) *Mol. Cell. Biol.*, **2**, 161.
28. Land,H., Grez,M., Hauser,H., Lindenmaier,W. and Schutz,G. (1981) *Nucleic Acids Res.*, **9**, 2251.
29. Barlow,D.P., Green,N.M., Kurkinen,M. and Hogan,B.L.M. (1984) *EMBO J.*, **3**, 2355.
30. Cooper,A.R. and McQueen,H.A. (1983) *Dev. Biol.*, **96**, 467.
31. Rigby,P.W.J., Dieckman,M., Rhodes,C. and Berg,P. (1977) *J. Mol. Biol.*, **113**, 237.
32. Feinberg,A.P. and Vogelstein,B. (1984) *Anal. Biochem.*, **137**, 266.
33. Paterson,B.M., Roberts,B.E. and Kuff,E.L. (1977) *Proc. Natl. Acad. Sci. USA*, **74**, 4370.
34. Minshull,J. and Hunt,T.H. (1986) *Nucleic Acids Res.*, **14**, 6433.
35. Vasseur,M., Condamine,H. and Duprey,P. (1985) *EMBO J.*, **4**, 1749.

CHAPTER 10

In situ hybridization to messenger RNA in tissue sections

REBECCA HAFFNER and KEITH WILLISON

1. INTRODUCTION

The possibility of hybridization between nucleic acid probes and their complementary DNA or RNA sequences within a cytological preparation has led to the development of *in situ* hybridization. The technique was originally described by Gall and Pardue (1) for use in the localization of cloned DNA sequences to specific chromosomes. *In situ* hybridization involves the incubation of a labelled DNA or RNA probe together with a tissue section or cell preparation, followed by the removal of unhybridized and non-specifically bound probe by washing. The position of the hybridized message is located either by autoradiography, for radioactively labelled probes, or by immuno-histochemical techniques for biotin-labelled probes.

In situ hybridization is now routinely used in the mapping of cloned DNA sequences to particular chromosomal sites on metaphase chromosomes (2, and see Chapter 5, Section 6). The technique has been extended to allow the detection of nucleic acids in tissue sections. Clonal analysis of cell lineages in interspecific mouse chimaeras may be carried out in tissues of embryonic and adult chimaeric animals, by using probes specific to each of the original parent cells (3−5). Hybridization to RNA in a tissue section allows the localization of mRNA sequences to single cells, so that the particular cell types transcribing a specific message can be identified. *In situ* hybridization has become a powerful tool in the study of regulatory genes during development, particularly in the localization of homeotic and segmentation gene expression during early development in *Drosophila melanogaster* (6), and more recently in the visualization of expression of homeobox-containing genes in newborn and adult mice (7).

This chapter outlines the various procedures employed in *in situ* hybridization to cellular RNAs in tissue and their localization by autoradiography. Protocols for *in situ* hybridization to metaphase chromosomes are not described here but detailed procedures for this technique are given in reference 8, and in Chapter 5, Section 6, in this book.

Under optimum conditions *in situ* hybridization enables the detection of RNA species present at cellular concentrations in the range of $0.1−0.01\%$ of the poly(A)$^+$ RNA population in a particular cell (6). The sensitivity of localization of a particular RNA in a cytological preparation depends on the following parameters:

(i)   the abundance and retention of RNA sequences within the tissue;
(ii)   the accessibility of the cellular sequences to the hybridization probe;

(iii)   the specificity of binding of the probe to the complementary mRNA species;
(iv)   the type of probe used — single- or double-stranded DNA, or RNA;
(v)   the form of label, whether radioactive label (where specific activity and mean path length of emission must be taken into account) or immunofluorescence or peroxidase labels;
(vi)   the efficiency and sensitivity of the method used for the detection of the signal.

Optimization of the above variables will be discussed under the relevant headings in this chapter. However, it is important to stress at the outset that the appearance of silver grains following hybridization and autoradiographic exposure does not necessarily indicate the presence of the RNA sequence of interest within the tissue, and therefore a very strict regimen of controls (described in Section 11) must be employed to avoid the possibility of erroneous conclusions.

2. PROBES

2.1 Choice of probe

Most early studies employing *in situ* hybridization used radioactive cDNA probes synthesized from purified mRNA templates, or nick-translated recombinant DNA probes complementary to the target mRNA. However, the use of single-stranded probes complementary to the mRNA has allowed a dramatic increase in hybridization efficiency, as much as eight times that achieved with double-stranded DNA probes (9). Single-stranded DNA probes can be synthesized from single-stranded M13 templates. This procedure involves a Klenow polymerization reaction, a subsequent restriction enzyme digestion and a gel purification procedure to separate the labelled probe from the template (10). Single-stranded RNA probes are synthesized from sequences cloned into vectors containing specific RNA polymerase promoters such as the SP6 promoter from the *Salmonella typhimurium* bacteriophage or the promoters of the *Escherichia coli* bacteriophages T7 and T3. The single-stranded RNA probes can be labelled to very high specific activities (11), they are as easy to prepare as nick-translated probes, and they provide more stable hybrids with cellular RNA than DNA probes. A disadvantage of single-stranded RNA probes is that the desired sequence must be cloned into the appropriate vector.

The most commonly available vectors for RNA probe synthesis are the SP6 vectors pSP64 and pSP65. These are pUC-derived plasmids containing the SP6 promoter and a polylinker with unique restriction sites for 11 restriction enzymes. pSP65 differs from pSP64 in the orientation of the polylinker which allows for transcription of both strands of the insert if both plasmids are used (11). The transcription of the RNA molecule is carried out with the enzyme SP6 RNA polymerase.

A more convenient plasmid for RNA probe synthesis, now available, is the Bluescript vector which contains both the T7 and the T3 promoters inserted in opposite orientations on either side of the multiple-cloning site. This allows transcription of either strand of the insert from the same plasmid by simply using the appropriate phage RNA polymerase with the correctly linearized plasmid. Furthermore, Bluescript contains the *lacZ* gene so that recombinant plasmids can be identified by X-gal colour screening. Bluescript cloning kits are available from Stratagene, San Diego, CA and NBL Enzymes

Ltd, Northumberland. Amersham International also markets Amprobe Kits which contain a range of oncogenes and peptide hormone genes cloned into the SP6 plasmids for use in the synthesis of these probes for *in situ* hybridization. Our laboratory has available the following probes cloned into the Bluescript vector: mouse actin (C-10), a mouse MHC sequence from an H-2 sub-clone spanning exons 3, 4 and 5 from pH-2IIa, a 300-bp fragment of the 18S ribosomal gene and a sequence from the mouse t-complex TCP-1 gene.

Recent advances in the production of synthetic oligonucleotides have allowed the construction of specific oligonucleotides for use as probes in *in situ* hybridization. Probes as short as 25 bases have been shown to give specific hybridization (12,13), although longer probes show increased hybridization signals. A potential advantage of using short oligonucleotide probes is that it should be possible to detect polymorphisms and single base mutations in RNA molecules. For example they could be used to determine the expression of individual members of a multigene family such as the immunoglobulins, or the Class I major histocompatibility antigens. It should also be possible to identify the presence of point mutations in activated oncogenes or other transcripts within a tissue.

2.2 Choice of label

Tritium is most commonly used to label probes for *in situ* hybridization because of the extremely low energy of the beta particle that is emitted. The particles travel less than 1 μm through the autoradiographic emulsion so that the position of the silver grain on the autoradiograph is an accurate indication of the location of the radioactive molecule. However, because of its low specific activity and its low energy of emission, long exposures of several weeks or even months are required to achieve an adequate signal.

This problem of low specific activity can be avoided by using ^{35}S label in place of tritium. High specific-activity probes can be obtained which give good cellular resolution. Although the path length of ^{35}S emissions ($\sim$20 μm) is longer than tritium, autoradiographic resolution can be increased to 4 μm by using thin emulsions. Clear results can be obtained in a matter of days rather than weeks. Employment of ^{35}S-labelled probes does increase the background level of labelling. However, the RNA probe system allows the use of RNase to remove unhybridized probe, so that this problem can be overcome.

In summary it may be necessary to use tritium-labelled probes to determine the precise location of RNAs within a cell, but for determination of which cells within a population contain the relevant molecules, ^{35}S probes are recommended. A useful technique employed by some investigators, is to include a trace amount of [^{32}P]UTP in the ^{3}H or ^{35}S probe synthesis reaction. Sections can be coarsely examined for hybridization using X-ray film and those sections showing positive hybridization can then be treated with liquid emulsion for fine resolution.

Non-autoradiographic techniques for *in situ* hybridization localization have also been developed. These involve the use of biotin-labelled probes which can be detected by biotin-specific antibodies conjugated to fluorescent or enzymatic reagents, whose activity is detectable in sections (14). It has been reported that biotinylated probes provide superior sensitivity to radiolabelled probes (5), although this has been disputed (8). Obvious advantages of biotin-labelled probes are the speed with which the signal can

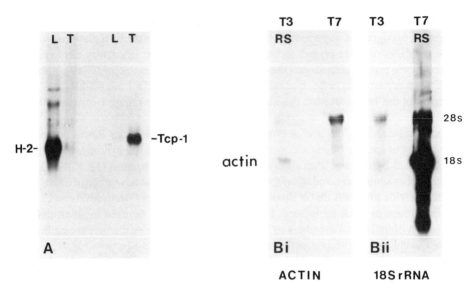

Figure 1. Northern analysis with ^{32}P-labelled single-stranded RNA probes. (**A**) Hybridization to total liver (L) or testis (T) RNA: ^{32}P-labelled H-2 (class 1) and ^{32}P-labelled Tcp-1 probes, synthesized from pSP6 vectors (11), with pSP6 RNA polymerase. Hybridizations were carried out in 50% formamide, 6 × SSC at 42°C and the filters were washed in 0.1 × SSC at 65°C. (**B**) Hybridizations to total testis RNA: ^{32}P-labelled actin probes (i) and ^{32}P-labelled 18S rRNA probe (ii), synthesized from Bluescribe vectors with T3 or T7 RNA polymerase. The hybridizing probe (RS, reverse strand) for actin is the T3 synthesized transcript, and for 18S rRNA is the T7 synthesized transcript. Hybridizations were carried out in 50% formamide, 6 × SSC at 42°C, and the four filters were washed in 0.1 × SSC at 60°C. Note the non-specific hybridization of both strand probes to 18S and 28S rRNA. This was first noted as a potential artefact by Zinn *et al.* (16) when performing hybridizations in formamide buffers, and might also apply to *in situ* hybridizations under similar conditions.

be visualized (results can be obtained on the same day that the washings are carried out) and the reduced background generated by these probes. Nevertheless, all *in situ* hybridizations to RNA in *Drosophila* embryos published so far have used auto-radiographic techniques. Detailed protocols for *in situ* hybridization with biotin-labelled probes are given in reference 15.

2.3 Preparation of radiolabelled probes

To synthesize single-stranded RNA probes using Bluescript or SP6 vectors, the desired plasmid must be linearized with an enzyme which cleaves downstream of the insert in the multiple cloning site. If full length probes are not required, a truncated probe may be synthesized by cutting with a restriction enzyme which cuts within the insert (see Section 2.5). The digest must be phenol-extracted, and the DNA precipitated with ethanol before use as a template for transcription.

It is vital to ascertain the orientation of the insert so that the strand synthesized is complementary to the relevant mRNA molecule. Characterization of the synthesized molecules is done by Northern blot analysis (*Figure 1 B*). The appropriate vector, in the case of SP6 plasmids, or the correct RNA polymerase, in the case of Bluescript vectors, may then be selected.

Table 1. Probe synthesis.

1. Thaw all reagents to room temperature. Add the components in the given order

5 × transcription buffer[a]	4 µl
0.2 M dithiothreitol (freshly prepared)	1 µl
ribonuclease inhibitor	20 units
nucleotide mix[b]	2 µl
linearized plasmid (1 µg/ml)	1 µl
labelled UTP[c]	
water	
Final volume	20 µl
including RNA polymerase	

2. Mix at room temperature, not on ice, because the spermidine can cause the DNA to precipitate at 0°C.
3. Remove 0.5 µl of the reaction mix and spot onto a Whatman 540 filter (hardened, ashless), to determine the background level of counts before the addition of the polymerase. Wash and count filters as described in Section 2.4.
4. Add 5 units of the appropriate RNA polymerase and mix.
5. Incubate the reaction mixture at 40°C for SP6 polymerase, and at 37°C for T7 and T3 polymerases, for 1 h.
6. Following incubation remove 0.5 µl of the reaction and spot onto a second Whatman filter to determine the degree of nucleotide incorporation.
7. Add 10 µg of carrier tRNA.
8. Phenol extract and ethanol precipitate the reaction mix. Resuspend the probe pellet in 20 µl of double-distilled water. Store at −20°C.
9. Use 1 µl of the reaction per slide treatment for *in situ* hybridizations (see Section 6).

NOTE. If desired, the DNA template may be removed prior to phenol extraction by incubation at 37°C for 10 min with 2 units of RNase-free DNase. However, we have found no necessity for this procedure and prefer to exclude it because of the possibility of RNase-contaminated DNase.

[a]5 × transcription buffer is 200 mM Tris pH 8.0, 40 mM $MgCl_2$, 10 mM spermidine, 250 mM NaCl, for T7 or T3 polymerase. For SP6 polymerase the buffer is as above but does not contain NaCl.
[b]The nucleotide mix is composed of equal volumes of 20 mM solutions of rATP, rCTP and rGTP.
[c]50 µCi of [^{3}H]UTP, 40.9 Ci/mmol; 60 µCi of [^{35}S]UTP, 410 Ci/mmol; 50 µCi of [^{32}P]UTP, 800 Ci/mmol.

The transcription reaction and all subsequent reactions must be free of ribonuclease. Therefore, it is essential to bake or autoclave all tubes and pipette tips. The double-distilled water used for all buffers should be treated with 0.1% diethylpyrocarbonate (DEPC). Gloves must be worn at all times to avoid contamination with ribonucleases.

The procedure for synthesis of radiolabelled single-stranded RNA probe is given in *Table 1*.

2.4 Calculation of the specific activity of the probe

The weight of the probe synthesized and its specific activity can be calculated from the percentage of nucleotide incorporation. Incorporation is determined by trichloroacetic acid (TCA) precipitation. Wash the filters containing a sample of the probe reaction (*Table 1*, steps 3 and 6) consecutively in 5% TCA, in 5% TCA with 0.5% tetrasodiumpyrophosphate, and in 90% ethanol, for 1 min each. Dry and count the filters in a scintillation counter. Incorporation is usually between 30 and 70%. The mass of the probe produced can be calculated by determining the incorporation of nucleotides in nmol, and using the formula that 1 nmol of nucleic acid weighs approximately

0.33 μg. For example, if a reaction used 60 μCi of [^{35}S]UTP at 410 Ci/mmol this would equal 0.15 nmol (410 Ci/mmol = 410 μCi/nmol, therefore 60 μC = 0.15 nmol). If incorporation of the nucleotide were 50%, then 0.075 nmol were incorporated. Assuming all UTPs were labelled, the total probe would equal 0.3 nmol (since there are four nucleotides). 1 nmol weighs 0.33 μg, therefore the weight of the probe synthesized would be 0.099 μg, or approximately 100 ng. Say the total counts of the probe, determined by counting the filters, were 6×10^7 c.p.m., the specific activity of the probe would be 6×10^7 c.p.m./100 ng probe, or 6×10^8 c.p.m. per μg of probe. A good reaction will yield 70−100 ng of probe. Specific activity ranges between 6.0 and 8.0×10^8 c.p.m./μg of probe synthesized.

2.5 Optimum length of probe

It is of major importance that the probe employed is able to reach the locations of mRNA within the section. The efficiency of penetration of the probe will depend upon its length. The optimum length of probes for *in situ* hybridization is 75−200 bases long. The length of the probes can be reduced by alkaline hydrolysis (9).

(i)   Hydrolyse the samples in 40 mM NaHCO$_3$, 60 mM Na$_2$CO$_3$ pH 10.2 at 60°C for an appropriate length of time (see below).

(ii)   Neutralize the samples by addition of sodium acetate pH 6.0 to 0.1 mM and glacial acetic acid to 0.5% v/v.

The hydrolysis time required is calculated by the following formula:

$$t = \frac{Lo-Lf}{kLo.Lf}$$

where t = time in min, Lo = initial fragment length in kb, Lf = final fragment length in kb, and k = rate constant for hydrolysis, which is approximately 0.11 kb/min.

The probe lengths can be directly measured by electrophoresis on a 4% agarose gel (Nusieve, Miles Scientific, UK) containing 0.1% SDS, or a 10% polyacrylamide gel in 8 M urea at 60°C.

Truncated probes can also be synthesized by digesting the plasmid with an enzyme which cuts the insert at a site the required number of bases downstream from the initiation site. It should be noted that shorter probes incorporate radiolabelled nucleotides less efficiently.

3. SLIDE PREPARATION

A major problem encountered during *in situ* hybridization to mammalian tissues is the loss of the tissue section from the slide at some stage during the procedure, usually when washing away the unhybridized probe. Different tissues vary in their ability to adhere to glass slides. In our experience mouse testis sections will remain in position throughout most treatments. Sections of mouse liver or embryos on the other hand are less well retained. Although for many treatments subbing with a gelatin−chrome alum solution is adequate, poly-lysine-coated slides provide a more powerful surface for tissue retention.

Thoroughly clean glass microscope slides in chromic acid or alcohol and then coat with a layer of poly-L-lysine at 100 μg/ml. The most efficient method for coating employs a simple air-brush (Badger Air Brush Co., IL) available from any good art store, which sprays the poly-lysine over the surface of the slide in a fine and even layer. Alternatively the slides can be dipped in a Coplin jar, containing the solution, for 3 sec, and then left to dry in a vertical slide rack. Gelatin-subbed slides are prepared by dipping the slides in a solution of 0.5 g gelatin, 0.5 g chrome alum dissolved (at 60°C) in 200 ml of distilled water.

4. TISSUE FIXATION AND SECTIONING

Paraffin-embedded sections retain a high degree of cellular morphology but they tend to be less sensitive to *in situ* hybridization probes. Frozen sections will generally give a stronger signal; however, the degree of cellular resolution provided by these sections is sometimes insufficient for conclusive results. For example frozen cryostat sections provide inadequate resolution to distinguish the intermediate cell types of spermatogenesis. Therefore, in each case a balance must be decided between clear morphology and hybridization sensitivity. The priorities will vary for different experiments.

4.1 **Paraffin-embedded sections**

(i)   Immediately after removal from the body, place the tissue to be sectioned in 10% formaldehyde, 150 mM sodium chloride and fix for at least 5 h prior to sectioning.

(ii)   Carry out the tissue pre-treatment (see *Table 2*) under an alternating pressure vacuum system which improves the efficiency of the wax impregnation.

(iii)   Following embedding of the tissue, cut $2-3$ μm sections on a microtome and float them off in warm water onto subbed slides.

(iv)   Store slides in a dust-free slide box. They can be kept for at least 6 months before use.

4.2 **Frozen sections**

(i)   Freeze the tissue in liquid nitrogen and embed in a drop of Tissue-Tek OCT (Lab-Tek Products, Illinois).

(ii)   Raise the tissue temperature to -30°C and leave for 1 h in a frozen cryostat.

(iii)   Cut serial 6 μm sections on a blade kept at -15 to -30°C, with the knife at

Table 2. Preparation of tissue for paraffin-embedded sections.

Number of treatments[a]	Solution	Time (min)	Temperature (°C)	Function
7	Isopropanol	10	40	Dehydration
1	Xylene	14	40	Linking agent
1	Xylene	20	40	
1	Wax	14	60	Impregnation
3	Wax	20	60	

[a]Transfers to a fresh solution.

Table 3. Pre-teatment of tissue sections prior to hybridization.

Care should be taken that all solutions used in this procedure are RNase-free. Unless specified all steps in this table are carried out at room temperature.

1. Slides embedded in paraffin wax must be first deparaffinized. To do this transfer them through two washes of xylene for 10 min each.
2. If previously dehydrated, rehydrate the slides by passing them through an alcohol series of decreasing ethanol concentrations (100, 90, 80, 60, 30%) for 3 min each.
3. Rinse the slides in distilled water for 1 min.
4. Place the slides in a staining rack in a dish containing 0.2 M HCl for 20 min. This removes basic proteins from within the section.
5. Transfer to 2 × SSC[a] for 10 min.
6. Incubate in 0.1 × SSC, pre-warmed, for 30 min at 50°C.
7. Drain the slides and allow them to dry briefly.
8. Place the slides horizontally in a container and cover the sections with 200 μl of Proteinase K (1 μg/ml)[b]. Incubate in a covered container for 30 min at 37°C.
9. Rinse in PBS for 30 sec.
10. Treat one batch of control slides (placed horizontally in a container as in step 8) with RNase A (100 μg/ml) in RNase buffer[c]. Incubate for 30 min at 37°C. Continue to step 12 for other slides.
11. Incubate the RNase-treated slides in RNase buffer for 30 min at 37°C.
12. Fix the slides in a solution of 4% paraformaldehyde in PBS for 20 min.
13. Rinse twice in PBS for 30 sec.
14. Place the slides in 100 mM triethanolamine, 25 mM acetic anhydride freshly made up, for 10 min. This helps to reduce background.
15. Rinse twice in PBS for 30 sec.
16. Dehydrate by passing along an alcohol series of increasing concentration.

[a]The composition of 20 × SSC is 3 M NaCl, 0.3 M trisodium citrate, pH 7.
[b]The buffer for Proteinase K is 0.1 M Tris-HCl pH 8.0, 50 mM EDTA.
[c]The buffer for RNase A is 0.5 M NaCl, 10 mM Tris-HCl pH 8.0.

an angle of 14° (the optimum angle may vary for different cryostats).

(iv) Pick up sections directly onto subbed slides and air dry for 10 min.

(v) The most satisfactory fixation for frozen sections is a fixative of 4% paraform-aldehyde in phosphate-buffered saline (PBS; 130 mM NaCl, 7 mM Na_2HPO_4, 3 mM NaH_2PO_4) applied for 20 min following sectioning.

(vi) Dehydrate the slides through an alcohol series of increasing ethanol concentrations, and store at $-70°C$ for up to 14 days before use. Alternatively use slides immediately after sectioning in which case dehydration is not required.

It is advisable to collect two or more sections per slide as this allows an estimation of variation between sections undergoing identical treatments. Slides should be labelled with a diamond pen and the outline of the sections scored on the underside of the slide.

5. PREPARATION OF TISSUE SECTIONS FOR HYBRIDIZATION

The retention of cellular RNA within the sections and the accessibility of the target RNAs to the hybridization probe are two very important parameters in the pre-treatment of tissue sections for hybridization. The exact protocols will vary for different tissues as their response, particularly to the protein digestion and heat pre-treatments, will be different. The procedure outlined in *Table 3* has been adapted for sections of mouse testis. Variations for other tissues should be determined empirically.

Table 4. Hybridization of probe to sections.

1.	Prepare hybridization buffer. Final concentrations are: 50% formamide, 0.3 M NaCl, 10 mM Tris-HCl pH 8.0, 1 mM EDTA, single strength Denhardts solution (0.02% bovine serum albumin, 0.02% Ficoll, 0.02% polyvinylpyrolidone), 500 μg/ml yeast tRNA. 10% dextran sulphate can be included to increase effective probe concentration, although some workers find considerable variation in background on the autoradiographs caused by different batches of dextran sulphate.
2.	Aliquot sufficient buffer to cover each slide into Eppendorf tubes. 30 μl of hybridization mix will cover an area of 700−1000 mm². Add the probe at the desired concentration.
3.	To the pre-treated slide, which should be completely dry, add 30−40 μl of the mix, placing it over the centre of the section.
4.	Place a coverslip, of minimum size required to cover the section, over the drop. Place one edge of the coverslip behind the drop and gently lower the coverslip over it with a pair of watchmakers forceps so that any bubbles are forced outward.
5.	Immediately surround the edges of the coverslip with a line of mineral oil, or rubber cement, to seal in the probe and prevent evaporation during hybridization (Sigma supply mineral oil in convenient 6 ml plastic bottles).
6.	Label the slides with a diamond knife.
7.	Hybridize in a moist chamber at between 40 and 50°C[a], overnight[b]. A moist chamber is most easily constructed from a plastic lunchbox by taping a folded paper towel to the lid of the box and soaking the towel with water. The towel should be moist but should not drip onto the slides below. The cover of the box is sealed with electrical tape.

[a]Cox's (9) thermal denaturation studies have calculated that the Tm for RNA−RNA duplexes *in situ* is about 5°C lower than the Tm in solution for probes 150−200 bases long. Maximum *in situ* hybridization rate occurs at 25°C below the normal Tm. The Tm reduction for RNA−RNA duplexes is 0.235°C/% formamide, as opposed to 0.65°C for DNA−DNA duplexes. RNA−RNA hybrids are also about 10°C less stable than DNA−DNA duplexes in the absence of formamide. Together this data indicates that the optimum temperatures for RNA−RNA hybridization are between 40 and 50°C for hybridizations in 50% formamide.
[b]For most hybridizations the reaction appears to terminate after about 5 h. Hybridizations are normally carried out overnight for convenience.

The following notes will clarify some of the steps in *Table 3*.

(i) The heat treatment in step 6 is critical. Different tissues and different subbing methods will show different responses to heat treatment. In general, mammalian tissue will float off the slides if exposed to temperatures higher than 55°C; some tissues will float off at temperatures lower than this.

(ii) The Proteinase K digestion in step 8 should remove cellular proteins with minimum disruption of morphology. For mouse testis and liver tissues the optimum treatment is digestion with 1 μg/ml of proteinase K for 30 min. If pronase is used the concentration is less critical, with 0.25 mg/ml being satisfactory in most cases. Dissolve pronase in water to give a final concentration of 40 mg/ml, and incubate at 37°C for 4 h to inactivate contaminating nucleases. Place 200 μl of pronase at a final concentration of 0.25 mg/ml (in 50 mM Tris-HCl pH 7.5, 5 mM EDTA) over the section and incubate for 10 min at room temperature. Following pronase digestion blot the slides and immerse them in a solution of 2 mg/ml glycine for 30 sec to block further proteolytic activity.

(iii) The second fixation treatment in step 12, following protein digestion, cross-links RNA molecules exposed in the previous stages.

(iv) Acetylation treatment of the slides in step 14 reduces the background of silver

Table 5. Washing procedure.

1.	Remove the slides from the moist chamber following hybridization. Remove the mineral oil by washing the slides in two changes of chloroform for 10 min each. Surface tension prevents the coverslips from floating off during these washes even if held vertically. If rubber cement has been used, peel off the cement.
2.	Immerse the slides in 4 × SSC. After 1 min the coverslips can be removed by gently lifting away with a pair of fine forceps.
3.	Wash the slides in three changes of 4 × SSC for 5 min each.
4.	Remove unbound probe by incubating in 20 μg/ml RNase A for 30 min at 37°C, as in *Table 3*, step 10.
5.	Incubate in RNase buffer for 30 min at 37°C.
6.	Incubate in 2 × SSC for 30 min at 45°C.
7.	Incubate in 0.1 × SSC for 45 min at 45°C[a].
8.	Dehdyrate by passing through a series of alcohol solutions of increasing concentrations, for 3 min each.

[a]The stringency of washings for short oligonucleotide probes is critical, and will be at a much lower stringency than recommended above. See note on thermal denaturation studies, *Table 4*, footnote a and refs 12 and 13.

grains in the developed autoradiograph by blocking basic groups of insoluble proteins in the section which otherwise are capable of binding labelled RNA (17). Non-specific binding of probe may also be reduced by rinsing the slides in a solution of 50 μg/ml heparin for 5 min just prior to adding the probe and by including heparin in the hybridization buffer at a concentration of 50 μg/ml (18). Heparin shows stoichiometric binding affinity for DNA binding proteins.

6. HYBRIDIZATION

The hybridization mixture containing the labelled probe is applied to the sections on the slide, as described in *Table 4*. The optimum probe concentration which gives the best signal-to-noise ratio, is the concentration just sufficient to saturate the target RNAs. For most reactions 0.5 μg/ml of probe will saturate cellular RNAs. Cox *et al.* (9) have calculated that RNA saturation occurs at 0.2−0.3 μg/ml of probe per kb of probe length.

7. POST-HYBRIDIZATION WASHES

The washing procedure is given in *Table 5*.

8. AUTORADIOGRAPHY

Following the removal of the unhybridized probe the slides are coated in a layer of liquid emulsion for autoradiography, as described in *Table 6*. The type of emulsion used will depend on the isotope employed. ^{32}P is unsuitable for this type of autoradiography because of its high energy and path length. For ^{3}H and ^{35}S the recommended emulsions are Ilford K-2, or Kodak NTB-2, which have a grain size of 0.2 μm and 0.26 μm, respectively. These emulsions do not accumulate background as rapidly as other emulsions (19), nevertheless they should not be stored for more than 3 months, and storage should be at 4°C. Kodak AR-10 stripping film can also be used instead of a liquid emulsion. Although harder to apply, stripping film reduces variation in sensitivity due to the constant thickness of the emulsion layer over the section (19).

The length of exposure will vary with the label used. A signal should be visible with

Table 6. Coating slides for autoradiography.

1.	Make up a solution of 0.2% glycerol. Filter at least 10 ml into a glass measuring cylinder. Mark the point which is double the volume of the glycerol solution with a thick line which will be visible under a safety light.
2.	In a dark-room fitted with a safelight filter (Ilford 902) illuminated with a 15 W bulb, dilute the emulsion 1:1[a], by adding the emulsion pellets to the measuring cylinder till the fluid volume reaches the marked line.
3.	Melt the emulsion at 42°C. Stir gently but do not allow bubbles to form.
4.	Pour the emulsion slowly into a dipping chamber avoiding bubbles. The dipping chamber can be a special container sold by Electron Microscopy Sciences for this purpose, or we have found a slide mailer, or the containers in which pH indicator paper is supplied, to be adequate. A solution of 18 ml will fill the chamber and is sufficient to coat 15−20 slides.
5.	Dip the slides by immersing them slowly twice, holding the slides for about 2 sec in the emulsion.
6.	Remove the slide smoothly from the emulsion and hold vertically, blotting the bottom edge on a paper tissue to remove excess emulsion. Wipe the back of the slides clean of emulsion.
7.	Allow the slides to dry vertically for at least 10 min in total darkness.
8.	Transfer the slides to a light-proof plastic slide box, containing a dessicant and seal with electrical tape.
9.	Allow to expose at 4°C.
10.	It is useful to dip some blank slides spotted with 1 μl of the isotope employed for determining developing times (see Section 9.1). Dip these slides last!

[a]Over some sections a thinner emulsion will provide a more even coating. In these cases the emulsion should be diluted 5:7 or 5:9 in water.

tritium after 5 days. ^{35}S gives more rapid results, with a signal apparent after 3 days or even 2 days where the target message is particularly abundant. It is advisable when setting up the hybridizations to make replicate copies of each treatment so that identical slides can be allowed to expose for different lengths of time.

9. SLIDE DEVELOPING AND STAINING

9.1 Developing

After a suitable period of exposure the slides are developed in Kodak D-19 developer, made up according to the manufacturer's instructions. (The developer is kept in a dark bottle and the air above the solution is gassed with nitrogen each time the bottle is opened.) The optimum developing time can be determined by developing test slides (*Table 6*, step 10) over a range of time periods, between 2 and 3.5 min. Determining the optimum time is important since the strength of the signal increases to a point where background rises rapidly, and obscures the signal (19).

The developing procedure is carried out in a dark-room under a safety light, as follows.

(i) Dip the slides in developer for a predetermined length of time, agitating occasionally by gently rocking the slide holder or Coplin jar so that the developer does not become depleted around areas of high grain density.

(ii) Transfer the slides to stop solution (1% glacial acetic acid) for 30 sec. Agitate.

(iii) Fix the slides in fixative (Amfix, May and Baker, Dagenham, UK) for 1 min, agitating occasionally.

(iv) Wash the slides under running water for 5 min. This step can be carried out in the light.

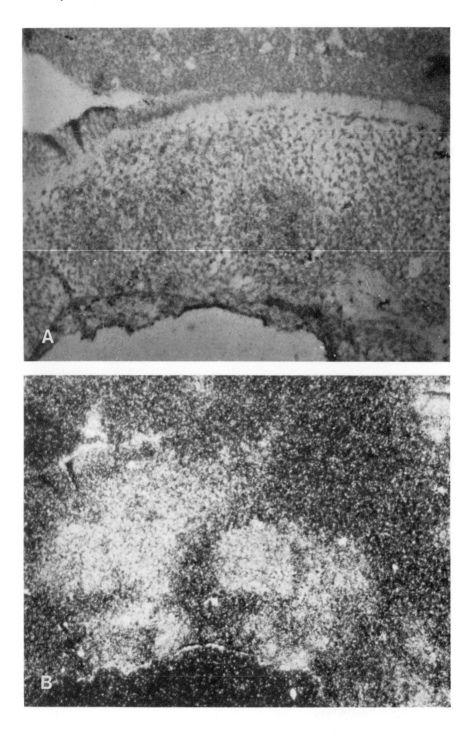

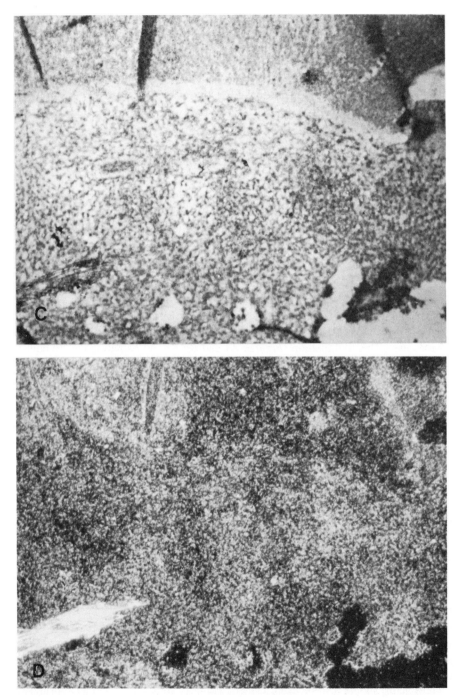

Figure 2. Hybridization to spinal cord area of 11-day mouse embryo of Tcp-1 probe shown together with negative control, Tcp-1 reverse strand. (**A**) Brightfield view of spinal cord area hybridized with Tcp-1. (**B**) Darkfield view of A. Note patterns of hybridization not readily visible under brightfield. (**C**) Background hybridization with reverse strand of Tcp-1. (**D**) Darkfield view of C. Magnification × 500.

9.2 **Staining**

Most staining protocols can be used in conjunction with autoradiography. For most sections, general stains, such as haemalum and eosin, are sufficient to show basic morphology. Haemalum stains the nuclear material purple and eosin stains cytoplasmic material pink. Stain the sections as follows.

(i)   Stain the sections with filtered haemalum for 60 sec.
(ii)   Stain with 0.25% eosin in 25% ethanol for 25 sec.
(iii)   Fix with methanol for 60 sec.

9.3 **Mounting**

To mount the slides, lie them horizontally and place a drop of Spurr resin, or other mountant of choice, above the section. Lower a clean coverslip over the resin with a pair of fine forceps, forcing out any bubbles. Allow to harden for at least 30 min before viewing under the microscope.

10. MICROSCOPY

Examination of tissues and grain counts are carried out most easily under darkfield illumination. Darkfield has a greater depth of field so that silver grains lying in different focal planes can be observed simultaneously. Patterns of grain distribution are more obvious using this illumination (*Figures 2A* and *B*). The illumination can be adjusted so that the morphology of the tissue is visible. Grain counts are calculated by taking a representative area of the section and measuring its area using an eyepiece grid. The number of grains present in the grid are counted and expressed as the number of grains per 100 μm^2. More accurate analysis of grain counts can be made using the SEMPIC Photometric system (20), which measures the intensity of the reflected light of the silver grains. Photographic presentation of the sections can be enhanced by photographing on darkfield using one coloured filter and photographing on brightfield with a different colour filter, and superimposing the two images.

11. CONTROLS AND ARTEFACTS

The *in situ* hybridization technique is particularly vulnerable to misinterpretation due to artefactual results, and therefore rigorous controls must be carried out in each experiment.

Essential control procedures include hybridization with a known non-hybridizing probe, usually for convenience the reverse strand to the complementary RNA strand is used (if the probes have been synthesized off Bluescript vectors the control strand is easily made using the second promoter, see *Figure 1B*). Another essential control is treatment of some sections with RNase prior to application of the probe (*Table 3*, steps 10 and 11) to remove all cellular RNA so that hybridization will not occur. A positive control, using a general probe which hybridizes to the cells under examination, such as actin, should also be used. Genetic controls are useful, such as hybridization with a probe from a gene which is known not to be expressed in that tissue, or, if the

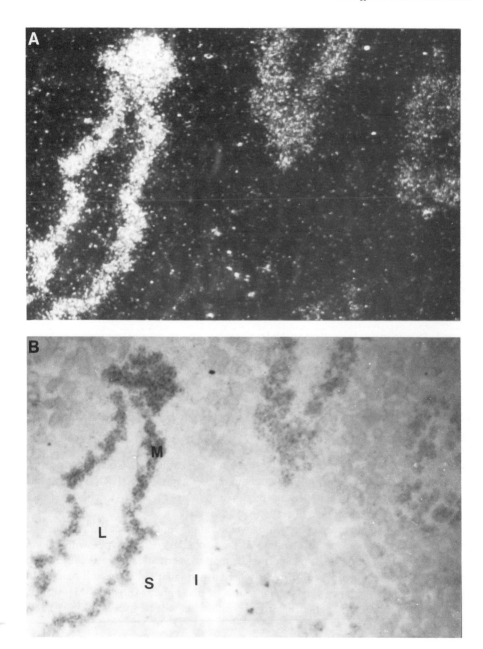

Figure 3. Non-specific binding of control non-hybridizing probe to mature spermatids in frozen testis section. The probe was a ^{35}S-labelled RNA probe and exposure time was 48 h. Hybridization was carried out as described in the text. I: interstitial region; L: lumen; M: maturing spermatids; S: primary and secondary spermatocytes. (**A**) Darkfield exposure. (**B**) Lightfield exposure. Magnification × 800.

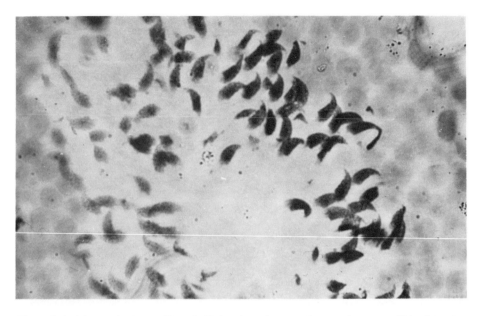

Figure 4. Staining artefact in paraffin embedded testis sections, causing maturing spermatid heads to give the appearance of a positive signal. Magnification × 4000.

gene is regulated developmentally, hybridizations can be carried out to tissues from a developmental stage which is known not to express the gene in question. Animals which carry deletion mutants for the gene provide another useful negative control. Quantitation of results can be aided by test hybridizations with probes from genes which are expressed at different known levels of abundance within the tissue. In our experiments with gene transcription in testis, we find that a reverse strand H-2 Class 1 gene probe is useful since it is expressed at low levels in testis. Stable Class 1 RNA transcripts are at least 50 times more abundant in liver than in testis (*Figure 1A*), whereas Tcp-1 is expressed 22.5 times higher in testis than in liver (21).

The most common artifact observed on *in situ* autoradiographs is the presence of silver grains due to non-specific binding of the probe. Some cell types will show a high level of non-specific binding, such as the mature spermatids in testis (22) shown in *Figure 3*. Staining artifacts also occur where a particular cell type will react with a stain to give an appearance very similar to that of silver grains (*Figure 4*). However, if the above controls are applied incorrect conclusions will be avoided.

12. ACKNOWLEDGEMENTS

The authors wish to thank Martin Burgin for the cutting of tissue sections, Ken Merrifield for assistance in editing and Christine Watson for critical reading of the manuscript. This work is funded by a grant to the Institute of Cancer Research from the MRC/CRC Joint Committee. Rebecca Haffner is an MRC funded graduate student.

13. REFERENCES

1. Gall,J.G. and Pardue,M.L. (1969) *Proc. Natl. Acad. Sci. USA,* **63**, 378.
2. Gall,J.G. and Pardue,M.L. (1971) In *Methods in Enzymology.* Academic Press, New York, Vol. 21, p. 470.
3. Siracusa,L.D., Chapman,V.M., Bennett,K.L., Hastie,N.D., Pietras,D.F. and Rossant,J. (1983) *J. Embryol. Exp. Morphol.,* **73**, 163.
4. Rossant,J., Vijh,M., Siracusa,L.D. and Chapman,V.M. (1983) *J. Embryol. Exp. Morphol.,* **73**, 179.
5. Lo,C.W. (1986) *J. Cell Sci.,* **87**, 143.
6. Hafen,E., Levine,M., Garber,R.L. and Gehring,W.J. (1983) *EMBO J.,* **2**, 617.
7. Awgulewitsch,A., Utset,M.F., Hart,C.P., McGinnis,W. and Ruddle,F.H. (1986) *Nature,* **320**, 328.
8. Pardue,M.L. (1985) In *Nucleic Acid Hybridisation: A Practical Approach.* Hames,B.D. and Higgins,S.J. (eds), IRL Press, Oxford, p. 179
9. Cox,K.H., DeLeon,D.V., Angerer,L.M. and Angerer,R.L. (1984) *Dev. Biol.,* **101**, 485.
10. Akam,M.E. (1983) *EMBO J.,* **2**, 2075.
11. Melton,D.A., Krieg,P.A., Rebagliati,M.R., Maniatis,T., Zinn,K. and Green,M.R. (1984) *Nucleic Acids Res.,* **12**, 7035.
12. Uhl,G.R., Zingg,H.H. and Habener,J.F. (1985) *Proc. Natl. Acad. Sci. USA,* **82**, 5555.
13. Berger,C.N. (1986) *EMBO J.,* **5**, 85.
14. Leary,J.J., Brigati,D.J. and Ward,D.C. (1981) *Proc. Natl. Acad. Sci. USA,* **78**, 6633.
15. Brigati,D.J., Myerson,D., Leary,J.J., Spalholtz,B., Travis,S.J., Fong,C.K., Hsiung,G.D. and Ward,D.C. (1983) *Virology,* **126**, 32.
16. Zinn,K., DiMaio,D. and Maniatis,T. (1983) *Cell,* **34**, 865.
17. Hayashi,S., Gillam,I.C., Delany,A.D. and Tener,G.M. (1978) *J. Histochem.,* **26**, 677.
18. Singh,L. and Jones,K.W. (1984) *Nucleic Acids Res.,* **12**, 5627.
19. Rodgers,A.W. (1979) *Practical Autoradiography.* Review 20, Amersham International plc, Amersham, UK.
20. Dormer,P., Pachmann,K. and Lau,B. (1978) *Histochemistry,* **59**, 17.
21. Willison,K.R., Dudley,K. and Potter,J. (1986) *Cell,* **44**, 727.
22. Gizang-Ginsburg,E. and Wolgemuth,D.J. (1985) *Dev. Biol.,* **111**, 293.

CHAPTER 11

Production of transgenic mice

NICHOLAS D.ALLEN, SHEILA C.BARTON, M.AZIM H.SURANI and WOLF
REIK

1. INTRODUCTION

In a surprisingly short time the production of transgenic mice has almost become a
standard technique. Although the first transgenic mouse was derived by infection of
pre-implantation embryos with a retrovirus (1), most laboratories now use direct microinjection of cloned DNA into one of the pronuclei at the zygote stage. The insertion
of a gene construct into a living organism can markedly extend any analysis hitherto
possible in cell culture systems since gene function can be monitored in a fully
physiological and developmental context. We shall not reiterate what has been said about
the many possibilities that the transgenic mouse system offers, rather we would like
to refer the reader to the recent excellent and comprehensive review by Palmiter and
Brinster (2).

In this chapter we shall focus our discussion on the technique of pronuclear microinjection of DNA. We shall also briefly discuss the use of retroviral-mediated gene
transfer into pre-implantation embryos and embryonic stem cells (ES cells) to derive
transgenic strains.

2. BASIC TECHNIQUES

2.1 Preparation of DNA for microinjection

2.1.1 *Linearized DNA and removal of vector sequences*

Cloned DNA used for microinjection is prepared in a linearized form with as much
vector sequence removed as possible. Linearized DNA has a 5-fold higher integration
frequency compared with circular molecules (3). Although the removal of vector sequences does not alter DNA integration frequency their presence can have a profound
effect on subsequent expression of the transgene with a drop in expression of up to
1000-fold (4,5). The reasons for this are unknown. The means for subsequent easy
removal of the vector sequences should therefore be considered during gene cloning
and construction. Especially suitable vectors are now available that contain sites for
rare cutting enzymes, such as *Not*I and *Sfi*I, in their polylinker sequences, so that even
large inserts can be excised cleanly from the vector in a single step (6,7).

2.1.2 *DNA fragment isolation and purity*

Several methods are available for the preparation of DNA fragments (8). For microinjection purposes, DNA purity is of the utmost importance. In particular, the DNA
sample must be free of organic solvent contaminants which have severe detrimental

effects on the developing zygote. We therefore recommend protocols that specifically avoid the use of phenol, chloroform or ether. The protocol routinely used in our laboratory involves purification of the DNA fragment in low melting point agarose (BRL) followed by direct recovery of the DNA from the gel using a preparative ion-exchange resin chromatography column (NACS-52 Prepac, BRL) as described below.

(i)   Digest the DNA construct ($10-20$ μg) with appropriate restriction enzymes so as to remove as much vector sequence as possible.

(ii) Cast a standard agarose gel of appropriate percentage to resolve the fragment size to be purified (e.g. for a 5-kb fragment cast a 1% agarose gel and electrophorese for 5 h at 50 V).

(iii) Load the DNA sample and electrophorese to achieve a good separation of the desired band from its neighbours.

(iv) Cut out a small well in the gel (0.5 cm wide) immediately in front of the band and fill the slot with molten 0.7% low melting point agarose (BRL). Allow to set.

(v)   Electrophorese the band into the low melting point agarose slice, monitoring its migration by observation under u.v. light. The DNA will first accumulate at the interface of the two gel types before entering the low melting point agarose; this has the beneficial effect of concentrating the DNA band.

(vi) Cut out the DNA band in the low melting point agarose and trim off any excess agarose. Place the gel slice in an appropriate tube and incubate it at 70°C until the agarose has completely melted. Determine the volume of the gel and add 5 vol. of 0.5 M NaCl in TE buffer (10 mM Tris-HCl pH 7.5, 1 mM EDTA). Mix well and incubate for a further 10 min at 70°C.

The DNA/agarose mixture is now put through a disposable ion-exchange column, NACS-52 Prepac (BRL) and purified as follows.

(i)   First hydrate the column by washing it three times with 2.0 M NaCl in TE buffer. Then equilibrate it by washing three times with the DNA binding buffer (for DNA fragments greater than 1 kb use 0.5 M NaCl in TE buffer; for smaller fragments use 0.2 M NaCl in TE buffer).

(ii) Load the warm (40°C) DNA/agarose mixture onto the column and wash the bound DNA with the binding buffer ($3-5$ ml at 42°C) to remove agarose and any gel impurities.

(iii) Elute the bound DNA with 3 vol. of 0.1 ml of 2.0 M NaCl in TE.

(iv) Precipitate the DNA by addition of 0.6 ml cold (-20°C) 95% ethanol. If a small amount of DNA is present (< 10 μg), this should be co-precipitated with 10 μg of carrier tRNA. The presence of tRNA in the final DNA sample does not have any adverse affect on DNA integration frequency or development of the zygote.

(v)   Centrifuge the precipitate for 10 min at 12 000 r.p.m. in a microcentrifuge then wash the pellet well with 70% cold ethanol to remove precipitated salt. Dry the pellet and dissolve it in a small volume of TE buffer.

(vi) Determine the absorbance of the preparation at 260 nm and 280 nm (1 OD_{260} = 50 μg of DNA, OD_{260}/OD_{280} = 1.8 for pure DNA). The preparation can also be checked at this stage by running a small aliquot on a mini agarose gel.

(vii) Dilute the DNA to the desired concentration (see below) with injection buffer

Table 1. Development in culture of injected eggs.

	Day 1 Injected	Day 1 Survived injection	Day 2 2 cell	Day 3 4 – 8 cell	Day 4 Late morula early blastocyst	Day 5 Expanded blastocyst
Injected eggs	130	100	80	70	60	55
Control eggs	(Selected healthy pronuclear 100)		100	95	90	85

Expected survival and development rate of embryos after an average injection session of a construct which is not detrimental to early development.

(10 mM Tris-HCl pH 7.6, 0.1 mM EDTA) and dialyse at 4°C against a large volume of the injection buffer, changing it several times over a 48 h period.

(viii)　Recover the DNA and store it in aliquots at 20 μl at −20°C.

Other methods of DNA fragment preparation from agarose gels that may be used include electroelution, 'squeeze and freeze' method, and isolation by binding of DNA to powdered glass. These methods are described elsewhere (9). Whichever method is chosen extensive dialysis of the sample is recommended.

2.1.3 *DNA concentration*

For most experiments a DNA concentration of $1-2$ ng/μl is used; this corresponds to $200-400$ copies of DNA of a 5-kb fragment in $1-2$ pl of DNA, the volume that can be injected into the pronucleus. This gives an optimal DNA integration frequency of $20-40\%$ (3).

DNA introduced into the mouse germline by pronuclear microinjection tends to integrate into a single site with variable copy number of predominantly head to tail orientation. It is possible that such a molecular organization of the integrated DNA may modulate its expression (10). The integrated copy number can be reduced by reducing the DNA concentration although there are no strict guidelines; for example, mice with single copy integrations have been generated using DNA at a very low concentration (20 molecules/pl). However this strategy must be balanced against a considerable drop in overall integration frequency (4) which would necessitate injection of a larger number of eggs. DNA concentrations greater than 2 ng/μl do not significantly improve DNA integration frequency and at 10 ng/μl embryo survival is markedly decreased (3).

In our laboratory, for each new DNA construct we undertake a pilot experiment in which we monitor development *in vitro* of the injected eggs to the blastocyst stage (*Table 1*). If a significantly high proportion of embryos block at the 1-cell stage (40% compared with 20%) then the DNA is diluted 2-fold or more. Arrest in development is not necessarily due to DNA toxicity *per se*, but could be due to activity of the product from transient expression of the introduced DNA. High transient expression of the gene construct is not necessarily detrimental, indeed, it may serve as a very good assay to test the integrity of a new construct, for example the measurement of indicator genes such as thymidine kinase (11) or β-galactosidase (N.D.Allen, unpublished observations) activity in two-cell embryos.

2.2 **Eggs and embryos**

2.2.1 *Mouse strains*

The choice of mouse strain depends on whether the genetic background of the mouse has a direct bearing on the subsequent use of the transgenic strains. For example, the analysis of particular immunoglobulin gene constructs or major histocompatibility (MHC) constructs will be greatly facilitated in strains of certain genetic backgrounds. In such cases it is wise to inject the DNA directly into eggs from that strain rather than undertaking a breeding programme to cross the transgene into the appropriate strain. In some cases though, such a breeding programme may be necessary as some inbred strains do not respond well to superovulation to provide a sufficient number of eggs per female for microinjection. Furthermore eggs from some strains are more difficult to culture than from others. These different parameters should be established before making a final decision. For most studies the genetic background of the recipient egg is not important and we tend to use hybrid embryos but always with a C57BL maternal genotype for ease of *in vitro* culture, as such embryos are not prone to the two-cell block (Chapter 2, Section 2.2.1).

2.2.2 *Fertilized eggs and embryo culture*

We derive fertilized eggs from superovulated (Chapter 1, Section 5.4) F_1 females (C57BL/6J ♀ × CBA/Ca ♂) mated with CFLP males, an outbred albino strain (all mice are bred from stocks obtained from Bantin and Kingman). Eggs are collected and cultured in M16 culture medium supplemented with 0.4% bovine serum albumin [(BSA, Fraction V, Sigma) Chapter 2, Section 5.2]. Microinjection is carried out in phosphate-buffered medium containing 0.4% BSA [(PB1, 12), Chapter 13, Section 3.1.1].

2.3 **Equipment**

There are two systems commonly used for microinjection, the differences being in the type of microscope and optics used. These are inverted fixed-stage microscopes (e.g. Zeiss IM3S and the Nikon Diaphot) or upright fixed-stage microscopes (e.g. Ergaval), fitted with image-erected optics. The optics of choice are Normarski differential interference contrast optics, which give a very fine resolution on the pronuclear membranes which is essential for easy injection. An alternative less expensive optical system is the Hoffman modulation contrast optics. The type of manipulator used is the same for either type of microscope, the most commonly used one being the standard Leitz manipulator. For use with an inverted microscope the manipulators are on a raised platform which we find uncomfortable and tiring to use compared with the upright microscope, where the manipulators are at bench level and the elbows of the operator can be rested comfortably on the bench.

Microinjection can be carried out in flat culture drops either on the bottom of a plastic Petri dish or in a glass depression slide (13) or in hanging drops in a Leitz manipulation chamber. Either type of chamber can be used with both inverted and upright microscopes. However, hanging drop culture chambers are usually used with the upright microscope. This type of chamber gives better optical results, especially compared with the plastic Petri dish chamber, which cannot be used with Normarski optics.

In our laboratory we use an Ergaval fixed-stage upright microscope with the Normarski optics, standard Leitz manipulators and Leitz manipulation chambers. The microscope is fitted with ×10 and ×25 long working distance objectives and ×10 widefield eye pieces (the image erecting piece introduces a further magnification of 1.5). We will confine our discussion to the use of this equipment. All equipment used for microinjection and its assembly is the same as that described in the following chapter, with a few exceptions which we describe here.

We will describe two types of injection needle, both of which are attached to the standard Leitz instrument holder. In one case the positive pressure required for injection is controlled by a de Fonbrune microsyringe pump (Beaudouin, Paris) connected to the instrument holder by oil lines, as described in Chapter 12, Section 2.3.1. In the other case injection is controlled with an air-filled 50 ml ground glass syringe (Chance); this is connected to the instrument holder by plastic tubing filled with air.

Both types of injection needle to be described are pulled on a Model P-80, Brown-Flaming Micropipette puller (Sutter Instrument Co.), although other makes and designs of pipette puller are commercially available. Needles are given their final shape using a Beaudouin microforge (see Chapter 12, Sections 2.3.2 and 2.4).

2.4 Preparation of micro-instruments

2.4.1 *Holding pipette*

The egg holding pipette is prepared as described in Chapter 12, Section 2.4.1.

2.4.2 *Injection needles*

(i) *Injection needle for use with the de Fonbrune microsyringe.* Glass for this type of needle is standard thick-walled borosilicate glass capillary tubing, outside diameter 1.0 mm, inside diameter 0.58 mm, length 15 cm (obtainable from Leitz, Clark Electromedical Instruments, Drummond Scientific Company).

(1)   Siliconize the glass capillary tubing inside and out by dipping in silicone solution (e.g. Sigmacote, Sigma) and dry in a hot oven.

(2)   Mount the capillary tube in the needle puller and press start. The Model P-80 puller has four possible adjustments, velocity index, pull index, heat index and air flow. The velocity and pull indices relate to the force of pull, the heat index controls the temperature of a platinum filament heating the glass capillary and the air flow cools the filament and glass. All the controls are fully described in the manual which accompanies the instrument.

When the velocity, pull and heat indices are kept constant, control of the air flow allows for a fine adjustment of the instrument to affect the shape of the needle. The required air flow setting may change fractionally from day to day, therefore the first few needles pulled on any one day should be examined under the microscope to check that they have a suitable taper and tip before proceeding further. The shape and dimensions of an optimal needle are illustrated in *Figure 1*.

(3)   Two needles are obtained from each glass capillary. Cut the shaft of the needles to a length of 4−5 cm with a diamond glass cutter. We recommend pulling 20−30 needles for use on the same day. Although a larger batch of needles can be pull-

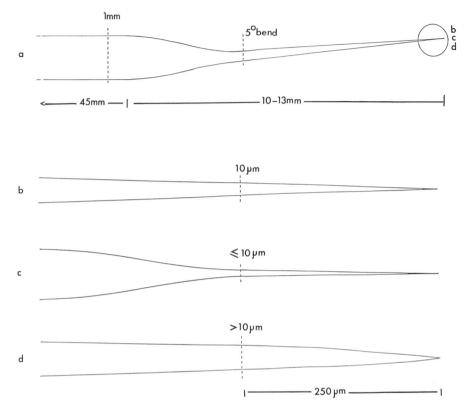

Figure 1. Diagram of an injection needle (not to scale). (**a**) Overall shape and dimensions of an optimal injection needle: the length of the needle from the 1 mm diameter shoulder to the tip should be between 10 and 13 mm. The 5° bend should be made as near to the main shaft of the needle as possible. The length of the mainshaft should be approximately 45 mm. The location of b, c and d on the injection needle is indicated. (**b**) and (**c**) Acceptable shapes: the diameter of the last few microns at the tip of the needle which will enter the egg will be small enough not to damage the membranes irreparably. If the needle is too slender it will not have enough rigidity to respond immediately to the movements of the manipulator. The finer taper at the tip of (**b**) is practicable because of the extra shoulder near to the tip which reduces the thick tapering shaft to a very fine one. (**d**) Unacceptable shape: the needle widens too quickly at the tip.

       ed and stored for use at a later date they tend to pick up dust which can block the needle tips and the tips appear to become less sharp with time.

(4)   Place a needle in a spare Leitz instrument holder. Dip the tip of the needle into the detergent Nonidet P-40 (NP-40, Sigma, use undiluted) and rinse by dipping the tip through two changes of distilled water.

(5)   Mount the needle (in the Leitz instrument holder) horizontally on the microforge (see Chapter 12, Section 2.3.2 for a description of the microforge and its heating filament) and inspect the needle tip. Break the tip of the needle by glancing it against the glass blob on the heating filament. This procedure should open the tip of the needle sufficiently to allow free flow of the DNA solution during injection. The break in the needle tip will not necessarily be visible. However if the tip is again dipped in distilled water and remounted on the microforge, water which has entered by capillarity should be clearly visible in the needle tip.

(6)   Dry the water from the needle tip by warming it very gently with the heating filament. The water will be seen to move quickly out of the needle.

(7)   Turn the needle perpendicular to the heating filament in order to give it a 5° bend; heat the filament to a bright glow and bring it near to the needle at the part where the shaft widens significantly; the needle will bend towards the filament as illustrated in Figure 2b, Chapter 12. See also *Figure 1a* in this chapter.

(ii) *Injection needle for use with a glass syringe.* Glass for this type of needle is thin-walled capillary tubing with an outside diameter of 1.0 mm, an inside diameter of 0.78 mm and length 15 cm. It also contains an internal glass filament which is necessary to enable filling of the needle with DNA (Clark Electromedical Instruments).

(1)   Siliconize the glass capillary tubing as described above.

(2)   The needles are pulled on the Model P-80 puller as described above, although the settings required are slightly different.

(3)   Dip the needle tip into NP-40 (Sigma) and rinse through two changes of distilled water.

(4)   Mount the needle on the microforge and inspect the tip. Warm the tip with a filament to dry off any water. The tip of this type of needle is usually open.

(5)   Move the needle perpendicular to the heating filament and give it a 5° bend as described above.

3. MICROINJECTION

3.1 Setting up the chamber

This chamber is similar to that used for nuclear transfer (Chapter 12, Sections 2.3.1 and 3.1). Microinjection is carried out in drops of PB1 (Chapter 13, Section 3.1.1). A series of up to 15 drops of PB1 (1.5 mm diameter) are placed on the underside of the glass coverslip, within the chamber. If the thick-walled injection needles and de Fonbrune microsyringe are to be used the first drop in the series should be a large drop of the DNA solution (3 mm diameter). Fill the chamber with heavy liquid paraffin oil. Introduce two of the eggs to be injected into each drop of PB1 with a pulled Pasteur pipette. Place the chamber on the operating stage of the microscope.

3.2 Setting up the instruments

The general set up of the instruments is similar to that described for nuclear transplantation (Chapter 12, Section 3.2) except for that of the injection needles, described here.

(i) *Thick-walled needle for use with the de Fonbrune microsyringe.* The needle is filled with heavy liquid paraffin oil before being set up on the manipulator. Back-fill the needle with oil from a plastic syringe fitted with a 5 cm × 30-gauge hypodermic needle inserted down the shaft of the injection needle to where the sides start to narrow. Care must be taken not to introduce any air bubbles into the oil. Attach the needle to the de Fonbrune assembly, again taking care to exclude any air bubbles as these will dampen the response from the de Fonbrune syringe. Bring the needle tip into the microscope field of view and focus with the manipulator height adjustment. The heavy liquid paraffin oil should be seen to fill the rest of the needle gradually and come out of the tip.

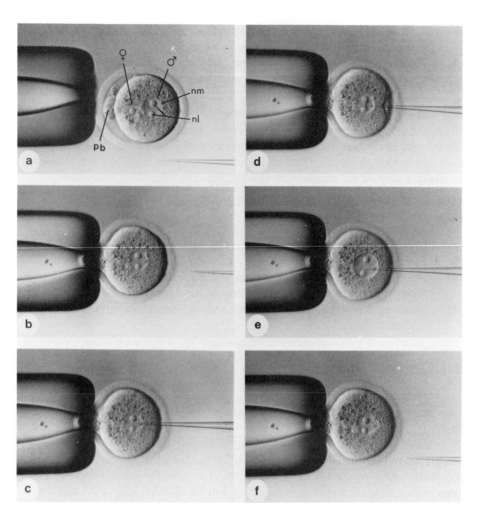

Figure 2. Injection into the male pronucleus. (a) Identify the male and female pronuclei (nm = nuclear membrane; nl = nucleoli; pb = 2nd polar body). (b) Hold the egg by suction on the holding pipette. Align the injection pipette tip with the male pronucleus. (c) Insert needle into pronucleus. (d) Inject DNA. The needle has not yet penetrated the egg and pronuclear membranes; evidence of this is the 'bubble' blown at the tip of the needle and the dimple in the egg membrane caused by extra fluid in the perivitelline space instead of in the pronucleus; the 'bubble' is in fact continuous with the perivitelline space. (e) Jab the needle carefully and inject again; the pronucleus swells, showing successful penetration. (f) After withdrawal of the needle the pronucleus shrinks back to its orginal size.

(ii) *Thin-walled needle for use with 50 ml glass syringe.* The needle tip is filled with DNA before being set up on the manipulator. Simply place the blunt end of the needle shaft into the DNA solution and leave it for 1−2 min. DNA will move up the shaft by capillary action along the internal glass filament and can be seen to collect at the needle tip. Approximately 1 μl of DNA is taken up in this way which is ample for the life of the needle. Place the needle in the instrument holder and assemble it on the manipulator as described above.

3.3 **Pronuclear injection**

(i)   Harvest the fertilized eggs from superovulated, mated females (Chapter 2, Section 2.1) about 18 h after the human chorionic gonadotropin (hCG) injection. After removal of the cumulus cells (Chapter 2, Section 2.1) culture the eggs for 1 h in M16 medium + BSA (Chapter 2, Section 5.2), 37°C in 5% CO_2 in air.

(ii)   Select up to 60 pronuclear eggs and place them in a large drop of PB1 medium (Chapter 13, Section 3.1.1) under oil and culture them in an ungassed incubator at 37°C for at least 15 min before microinjection. Place two eggs in each drop in the manipulation chamber. Restrict the total number of eggs per chamber to a number that can be comfortably operated on over a 20 min period. Prolonged culture out of the incubator may be detrimental to the eggs. Between 15 and 20 eggs is usually appropriate for each group.

(iii)   Bring the microinstruments into the chamber. The thick-walled needle must be filled with DNA from the first drop in the chamber. Bring the needle into the drop and bring the tip into focus by adjusting the manipulator height. Fill the needle by exerting a negative pressure on the de Fonbrune syringe. The oil should be seen to recede and the needle fill with DNA solution until the meniscus is at the edge of the field of view when observed through the ×10 objective. Now equalize the pressure and exert a slight positive pressure to check that the DNA is flowing well. Withdraw the needle from the DNA drop.

(iv)   Move the chamber to bring the first drop containing eggs into the field of view. Adjust the fine focus of the microscope to bring the eggs into focus.

(v)   Bring the microinstruments into the drop and bring them into focus by adjusting the manipulator height. Care must be taken not to crash the two instruments. Switch from the ×10 to the ×25 objective and again make focal adjustments.

(vi)   With the aid of the manipulator joy sticks, orient an egg so that both pronuclei are in focus with the male pronucleus nearest to the injection needle (*Figure 2a*). The male pronucleus is the one further from the polar body and is generally much bigger. Its greater size makes it easier to inject than the female pronucleus.

(vii)   Pick up the egg with the holding pipette by exerting a slight negative pressure on the Agla syringe.

(viii) With the egg firmly held, bring the male pronuclear membrane (and nucleoli) into clear focus by adjusting the manipulator height (*Figure 2b*).

(ix)   Bring the tip of the injection needle into focus and move it with the joy stick into the pronucleus (*Figure 2c*). If the injection needle tip is in the correct plane the pronuclear membrane will be seen to deform. If penetration is not achieved a 'bubble' of DNA solution surrounded by invaginating nuclear and egg membranes will appear when injection of the DNA is attempted (*Figure 2d*). A short, sharp jab is often required to actually penetrate the membranes.

(x)   With the needle tip confidently in the pronucleus exert a positive pressure on the injection needle (de Fonbrune or glass syringe). The pronucleus will be seen to swell if injection is successful (*Figure 2e*). The pronucleus will later shrink back to its original size (*Figure 2f*).

(xi)   Withdraw the injection needle and repeat the procedure for the next egg.

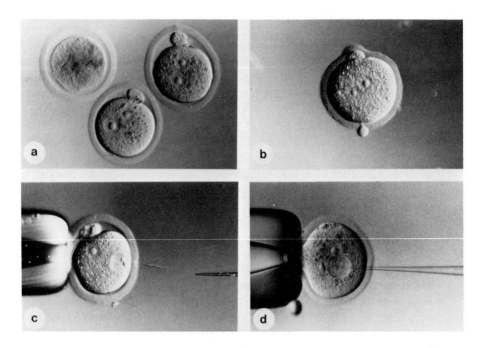

Figure 3. Problems which may be encountered. (**a**) One lysed and two healthy eggs 30 min after injection. (**b**) Egg with cytoplasm herniating through the injection hole in the zona pellucida, 30 min after injection. This egg will not lyse completely and will appear to recover. It will, however, fail to cleave. (**c**) Nucleolus being pulled out of the egg by the injection needle after injection of DNA. (**d**) Nucleolus expelled into the cytoplasm through the second hole in the pronuclear membrane following injection of the DNA.

(xii)   When all of the eggs in the chamber have been injected harvest them into a drop of PB1 and culture them in the incubator.

(xiii) Place fresh eggs from the supply in PB1 in the incubator into the drops in the chamber and repeat the procedure. For each new chamber of eggs it is usually necessary to start with a fresh injection needle.

(xiv)   At the end of the session separate the healthy operated eggs from those that have lysed (*Figure 3a*). Wash them through six drops of M16 medium + BSA and culture them overnight in a drop of M16 + BSA at 37°C in 5% CO_2 in air.

3.4 Problems

Most of the problems that occur during microinjection involve blocking of the injection needle. The needle can become blocked in a number of ways, for example by dust in the DNA solution or, in the case of the thick-walled needle, by small beads of oil that cling to the inner surface of the tip when the needle is filled with DNA. In most cases a needle blocks as a result of proteins it picks up from the egg during injection; this is most likely to occur when an egg has been lysed. In many cases the needle can be unblocked by gently glancing it against the tip of the holding pipette. If this does not clear the blockage, or if the tip breaks to give too wide an opening, then the needle should be exchanged for a fresh one. When a needle cannot be used to penetrate and inject an egg cleanly the consequence is often lysis of the egg. Eggs will lyse to various degrees; many lyse completely (*Figure 3a*) and some appear to recover slowly with culture (*Figure 3b*) although most of the latter fail to divide to the two-cell stage.

As a needle ages, its outer surface also becomes sticky with egg proteins. Although such needles appear to be injecting well, they frequently pull the nucleoli out of the egg when the needle is withdrawn (*Figure 3c*). The nucleoli may also be pushed out of the far side of the nucleus into the cytoplasm; this can happen under the pressure of the incoming DNA solution when the needle has punctured right through the nucleus (*Figure 3d*). Again, such eggs apparently recover but seldom develop.

Few problems are encountered with the holding pipettes, although they can become blocked by large pieces of dust after prolonged storage. Bubbles of air in the oil lines can make the pipette unresponsive to changes in oil pressure. If an egg is picked up too vigorously it may become damaged. Sometimes the zona pellucida of the egg may be broken or have a hole in it; in these cases the egg may be sucked into the holding pipette out of its zona.

3.5 Embryo transfer

In some laboratories operated eggs are transferred to day 1 pseudopregnant foster mothers in the evening after microinjection. We routinely transfer embryos at the two-cell stage the morning after microinjection, to the oviducts of day 1 pseudopregnant foster mothers (see Chapter 13, Section 6.4). This is for three reasons. Firstly, micro-injection is a tiring operation and embryo transfer is not easy to do when tired. Second-ly, and more importantly, a significant percentage of operated embryos block in development at the one-cell stage $(10-20\%)$ and it is of no value to transfer these em-bryos. The block at the one-cell stage can be attributed either to a property of the DNA solution itself as discussed above, or to direct physical damage of the egg during micro-injection. Thirdly, asynchronous transfer accommodates for the slower development of the eggs brought about by their isolation and operation.

4. SCREENING FOR TRANSGENIC OFFSPRING

4.1 Tail autopsy and toe clipping

Following transfer of embryos to the oviduct, pregnancies should be monitored from time to time, for example by weighing the recipient during the mid-gestation period. If by the evening of day 20, successful delivery in an obviously pregnant animal has not occurred, caesarian section should be performed because the number of live fetuses may not be sufficient to induce natural labour (see Chapter 1, Section 5.6).

In our mouse colony healthy young are weaned at 21 days after birth. At this time they are big enough to be checked for the presence of injected DNA sequences by tail biopsy and subsequent DNA analysis. Tail biopsies and numbering of the animal by toe clipping should be done at the same time, since it saves time and possible confu-sion. It is possible, though not advisable, to number animals by ear punching (Chapter 1, Section 3.2). This should be reserved for short-term experiments, since both max-imum number and durability are restricted.

(i)   Lightly anaesthetize the animal with Avertin, hold it with one hand, belly up,

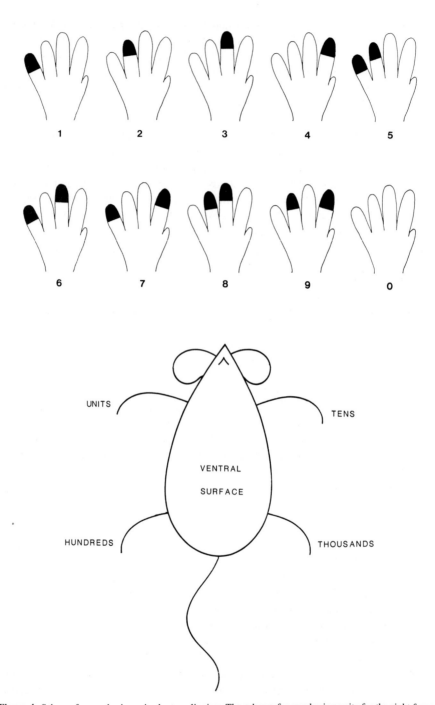

Figure 4. Scheme for numbering mice by toe clipping. The scheme for numbering units for the right forepaw is illustrated. The toe is cut back to half its length to allow unequivocal reading. The pattern is repeated for all four feet in a lateral to medial direction (i.e. units and hundreds are read from left to right while tens and thousands are read from right to left).

and clip individual toes with small and sharp scissors, making sure that enough of the toe is removed to allow unequivocal reading. With a maximum of two toes per foot being removed, a colony of 10 000 animals can be individually marked, as shown in *Figure 4*. Note in this figure that we clip from lateral to medial (starting with the smallest toe and moving toward the bigger ones) with all four feet. We find this system easier to follow than a complete left to right system that is in use in some laboratories.

(ii)   Directly after clipping, remove the terminal third of the tail with sharp scissors and immediately cauterize the stump with a hot razor blade. Cut the tail tissues into small slices using scissors and let the slices drop into 2 ml of phosphate-buffered saline (PBS) in a polypropylene centrifuge tube (10 ml) on ice.

(iii) Using a polytron rotating knife (Kinematica, Luzern, Switzerland) homogenize the tail tissue. Use a setting of $4-5$ on the speed control. Insert the probe into the tube with the sample and slowly move up and down for about $1-2$ min until the tail tissues appear finely minced. Carefully avoid trapping the tissue in the blades and clean the probe thoroughly before homogenizing the next sample.

(iv)   Spin at $1000-2000$ r.p.m. in a bench top centrifuge for 10 min, carefully decant the supernatant, resuspend the pellet in 1 ml of buffer A (75 mM NaCl, 25 mM EDTA, adjust to pH 8) and add 1 ml of buffer B (10 mM Tris-HCl pH 8, 10 mM EDTA, 1% SDS, 400 μg/ml proteinase K). It is essential to resuspend the pellet well in buffer A before adding buffer B to ensure immediate lysis of cells, and hence avoid degradation of the DNA. Incubate the lysates at 37°C overnight, or at 50°C for 3 h. In our hands lysis at 50°C results in better quality DNA. Extract once with phenol and once with chloroform. Bring to 0.3 M NaCl and precipitate by adding an equal volume of isopropanol. Retrieve the precipitate with a small disposable spatula, wash it through 70% isopropanol, and dissolve it in $300-500$ μl of TE buffer. This will give a DNA concentration of $0.5-1$ mg/ml. Provided all the animals are within a particular age group, there will be little variation between individual DNA yields. Because of this, we do not routinely determine the OD of the samples.

4.2 DNA from fetuses

In some instances it is necessary to analyse putative transgenic fetuses. For example, it is possible that a particular gene construction when expressed during embryogenesis exerts a dominant effect that results in pre-natal death. Consequently, when animals are analysed after birth a low frequency of transgenics will be obtained, and those transgenics will not express the introduced DNA.

Enough DNA to genotype individual fetuses can be obtained from day 8 of gestation onwards. For routine purposes we isolate DNA from day $10-11$ onwards when a substantial amount can be obtained (on day 11 up to 100 μg).

(i)   Gently tear apart the uterine musculature with fine forceps and expose the conceptus. Remove decidual tissue and trophoblast if absence of maternal genotype contamination is essential. Isolating DNA from only the embryo proper also has the advantage that cells are dissociated more readily, yielding better quality DNA.

(ii) Put the embryo into an Eppendorf tube containing 0.5 ml of PBS (day 10–day

12) or a 10 ml polypropylene tube (day 13 onwards) containing 2 ml of PBS, and homogenize the tissue by gently pipetting with a Gilson pipettman; use yellow tips for day 10−day 12 stages and blue tips for later stages.

(iii)   Centrifuge at 1000−2000 r.p.m. and proceed as described above for tails, except that a proteinase K concentration of 200 μg/ml in Buffer B should be used for embryos.

4.3 Analysis of DNA

Different protocols exist for the initial analysis of candidate transgenic mice (3,13,14). We routinely run Southern blots (9) although dot blots are quicker to do. However, the additional information provided by a Southern blot will invariably be required at a later stage. We usually digest tail DNA with a restriction enzyme that releases an internal fragment from the integrated DNA construct, or with an enzyme that cuts once in the construct. Because of the predominance of head to tail arrangements, the latter digest will also produce fragments of predictable size. Should the transgene contain sequences that are also present in the mouse genome, a DNA probe can be used that hybridizes with both the transgene and the endogenous gene, provided that the fragments resulting from restriction enzyme digestion are clearly distinct. This offers the additional advantage of precise copy number determination by comparison of the strength of the hybridization signal with that of the endogenous allele. One has to bear in mind though, that a considerable proportion (20−30%) of founder (F_0) mice are mosaics (15), and hence, estimation of copy number per genome equivalent for a particular integration site is only possible in the F_1 or subsequent generations.

Following restriction enzyme digestion and electrophoresis in an agarose gel, the DNA is transferred from the gel to a membrane. The fastest and most reliable blotting and hybridization protocol in our hands is alkaline transfer (16) onto nylon membranes (e.g. Gene screen plus, Dupont, Boston, USA) and hybridization with probes labelled by random oligonucleotide priming (17, Chapter 9, Section 5.2).

4.4 Transgenic mouse husbandry and embryo freezing

For the efficient running of a transgenic mouse house, the importance of clear numbering of mice by toe clipping (described above) and of good records of pairings of mice for breeding, cannot be overstressed. A mouse house can quickly become full to capacity without careful management. Some of the problems of overcrowding can be alleviated by having a programme of embryo freezing to preserve transgenic strains which are not in current use (see Chapter 13). An additional safety factor can be introduced by keeping mice of precious strains in a mouse facility of a collaborating laboratory at a different institute.

5. RETROVIRUS-MEDIATED GENE TRANSFER

As mentioned before, retroviruses have been used to generate transgenic mice by exposing pre-implantation embryos to infectious virus (1). The frequency of germline transformation is similar to that with microinjection of DNA, whilst the technique itself is considerably less demanding. Simply, early morulae are denuded of their zonae, exposed to infectious virus for a brief period in culture and transferred to recipient pseudopregnant females. The various techniques of retroviral gene transfer rely on the availability of high titre virus-producing cell lines. Protocols for infection of embryos and ES cells (see below) by co-cultivation with producing cells or with concentrated supernatant from such cells can be found in original publications (18,19) and in Hogan *et al.* (13).

There are a few important points that distinguish the two routes to transgenesis that are worth mentioning. Transgenic animals generated by integration of retroviruses are invariably mosaics with zero, one or more than one integration site in different cells. This is because embryos are normally infected at the 8−16 cell morula stage (1,20). This will sometimes require a large breeding and screening programme to identify off-spring carrying a particular insertion, in contrast to transgenics generated by pronuclear injection of DNA, which will mostly be non-mosaic and consequently transmit the new gene to 50% of offspring (15). Retroviruses always integrate as single copy genes, at multiple sites of the host genome, and do not create the rearrangements at the site of insertion that are frequently observed with DNA injection (21,22). Hence, they allow for a clearer interpretation of chromosomal position effects (23), and facilitate cloning of the cellular site of insertion, which is of particular importance in the molecular analysis of insertional mutations (24). Whilst the transcriptional control signals of the virus itself function poorly in a variety of tissues and in particular in early embryos, retroviral vectors that carry internal promoters and enhancers have been used to achieve ubiquitous as well as tissue-specific expression of inserted genes (25,26). The general applicability of these principles remains to be evaluated. In particular it will be interesting to see whether the high degree of tissue specificity observed with a variety of DNA constructs can be achieved with retroviral vectors, despite the presence of viral transcription con-trol sequences in the immediate vicinity of the transduced gene. There is also a packag-ing constraint on the length of the viral genomic RNA which is thought to be a maximum of 8−10 kb (27). Clearly some genes, such as for example the gene for Duchenne muscular dystrophy, with an mRNA size of 14 kb (28), cannot be transduced by retro-viral vectors, but the DNA can easily be injected into eggs. It should also be mentioned that the construction of retroviral vectors is certainly less trivial than that of plasmid vectors for microinjection.

The great strength of retroviruses lies in the ability to introduce them into the develop-ing organism much later than at the pre-implantation stage, and into specific tissues in the adult animal (29,30). Upon infection of cells at different stages of differentia-tion, the virus will uniquely mark the descendants of individual cells, and lineage rela-tionships can thereby be deduced (31−34). This approach, although in its infancy, is extremely promising for the analysis of complex lineage relationships in mammalian embryos, and in adult tissue such as the haematopoietic one.

The selective introduction of genes by retroviruses into the haematopoietic tissue of adult organisms has also raised hopes for the correction of single gene defects, such as for example severe combined immunodeficiency syndrome (35). Whilst there are protocols available now for high efficiency transformation of haematopoietic stem cells (32), a number of problems concerning stability of expression and safety have to be solved before clinical application can be considered.

Another approach to germline transformation of mice has recently been developed by Evans and colleagues (18,36). Cultured embryonic stem cells can be infected with retroviral vectors, maintained in culture, and introduced into mice by blastocyst injection (18). The ES cell system offers the great advantage of controlling the number of retroviral insertions and, more importantly, of selecting for specific insertions prior to re-introduction into the germline, for example by applying methods of somatic cell genetics. A mouse strain carrying a mutation in the hypoxanthine phosphoribosyl transferase (HPRT) gene has recently been created by this methodology (36). Future prospects include gene targeting by homologous recombination, selection of insertion sites by physical methods, and saturation of the mouse genome with unique markers for analysis of linkage to classical mutant genes, and eventual cloning of those genes.

Applications of the retrovirus technology can also be envisaged for germline transformation of other mammalian species, in which a lack of embryo culture systems or difficulties with microinjection of zygotes at present hampers the derivation of transgenic animals.

6. ACKNOWLEDGEMENTS

We thank Michael Norris for help with experiments, Sarah Howlett for suggestions and discussion of the manuscript and Mu-Chiou 'Tsu Fun' Huang and the Collicks for entertainment during the writing of this chapter. We also thank Michael Davis and John Hale for expert animal care, and Linda Notton for typing the manuscript. N.D.A. is supported by a AFRC studentship. Part of W.R's work was supported by an EMBO fellowship.

7. REFERENCES

1. Jaenisch,R. (1976) *Proc. Natl. Acad. Sci. USA,* **73**, 1260.
2. Palmiter,R.D. and Brinster,R.L. (1986) *Annu. Rev. Genet.,* **20**, 465.
3. Brinster,R.L., Chen,H.Y., Trumbauer,M.E., Yagle,M.K. and Palmiter,R.D. (1985) *Proc. Natl. Acad. Sci. USA,* **82**, 4438.
4. Chada,K., Magram,J., Raphael,K., Radice,G., Lacy,E. and Costantini,F. (1985) *Nature,* **314**, 377.
5. Townes,T.M., Lingrel,J.B., Chen,H.Y., Brinster,R.L. and Palmiter,R.D. (1985) *EMBO,* **4**, 1715.
6. Michiels,F., Burmeister,M. and Lehrach,H. (1987) *Science,* **236**, 1305.
7. Lathe,R., Vilotte,J.L. and Clark,J. (1987) *Gene,* in press.
8. Maniatis,T., Fritsch,E.F. and Sambrook,J. (1982) *Molecular Cloning : A Laboratory Manual.* Cold Spring Harbor Laboratory Press, New York.
9. Sealey,P.G. and Southern,E.M. (1982) In *Gel Electrophoresis of Nucleic Acids — A Practical Approach.* Rickwood,D. and Hames,B.D. (eds), IRL Press, Oxford, p.39.
10. Reik,W., Williams,G., Barton,S.C., Norris,M.L., Neuberger,M. and Surani,M.A.H. (1987) *Eur. J. Immunol.,* **17**, 465.
11. Brinster,R.L., Chen,H.Y., Warren,R., Sarthy,A. and Palmiter,R.D. (1982) *Nature,* **296**, 39.
12. Whittingham,D.G. and Wales,R.G. (1969) *Aust. J. Biol. Sci.,* **22**, 1065.
13. Hogan,B., Costantini,F. and Lacy,E. (1986) *Manipulating the Mouse Embryo : A Laboratory Manual.* Cold Spring Harbor Laboratory Press, New York.
14. Palmiter,R.D., Chen,H.Y. and Brinster,R.L. (1982) *Cell,* **29**, 701.
15. Wilkie,T., Brinster,R.L. and Palmiter,R.D. (1986) *Dev. Biol.,* **118**, 9.
16. Chomczynski,P. and Qasba,P.K. (1984) *Biochem. Biophys. Res. Commun.,* **122**, 340.
17. Feinberg,A.P. and Vogelstein,B. (1984) *Anal. Biochem.,* **137**, 266.
18. Robertson,E., Bradley,A., Kuehn,M. and Evans,M. (1986) *Nature,* **323**, 445.
19. Jaenisch,R., Fan,H. and Croker,B. (1975) *Proc. Natl. Acad. Sci. USA,* **72**, 4008.
20. Jaehner,D. and Jaenisch,R. (1980) *Nature,* **287**, 456.
21. Panganiban,A.T. (1985) *Cell,* **42**, 5.

22. Covarrubias,L., Nishida,Y. and Mintz,B. (1986) *Proc. Natl. Acad. Sci. USA,* **83**, 6020.
23. Jaenisch,R., Jaehner,D., Nobis,P., Simon,I., Loehler,J., Harbers,K. and Grotkopp,D. (1981) *Cell,* **24**, 519.
24. Gridley,T., Soriano,P. and Jaenisch,R. (1987) *Trends Genet.,* **3**, 162.
25. Stewart,C.L., Schuetze,S., Vanek,M. and Wagner,E.F. (1987) *EMBO J.,* **6**, 383.
26. Soriano,P., Cone,R., Mulligan,R. and Jaenisch,R. (1986) *Science,* **234**, 1409.
27. Shimotohno,K. and Temin,H.M. (1981) *Cell,* **26**, 67.
28. Koenig,M., Hoffman,E.P., Bertelson,C.J., Monaco,A.P., Feener,C. and Kunkel,L.M. (1987) *Cell,* **50**, 509.
29. Jaenisch,R. (1980) *Cell,* **19**, 181.
30. Joyner,A., Keller,G., Phillips,R.A. and Bernstein,A. (1983) *Nature,* **305**, 556.
31. Soriano,P. and Jaenisch,R. (1986) *Cell,* **46**, 19.
32. Sanes,J.R., Rubenstein,J.L.R. and Nicolas,J.F. (1986) *EMBO J.,* **5**, 3133.
33. Turner,D.L. and Cepko,C.L. (1987) *Nature,* **328**, 131.
34. Keller,G., Paige,C., Gilboa,E. and Wagner,E.F. (1985) *Nature,* **318**, 149.
35. Belmont,J.W., Henkel-Tigges,J., Chang,S.M.W., Wager-Smith,K., Kellems,R.E., Dick,J.E., Magli,M.C., Phillips,R.A., Bernstein,A. and Caskey,C.T. (1986) *Nature,* **322**, 385.
36. Kuehn,M., Bradley,A., Robertson,E. and Evans,M. (1987) *Nature,* **326**, 295.

CHAPTER 12

Nuclear transplantation in fertilized and parthenogenetically activated eggs

S.C.BARTON, M.L.NORRIS and M.A.H.SURANI

1. INTRODUCTION

Recent major advances in techniques of nuclear transplantation allow manipulation of the genetic constitution of mouse eggs or embryos by the removal or addition of pronuclei or nuclei. Such procedures have made possible the investigation of a number of important areas in development, for instance:

(i)   the developmental contribution of the maternal and paternal pronuclei;
(ii)   nuclear-cytoplasmic interactions;
(iii)   the developmental potential of nuclei transferred to eggs from later stage embryos, and the feasibility of 'cloning';
(iv)   reproduction in mammals by parthenogenesis.

These studies have provided a new understanding of the functional differences between parental genomes though the precise reasons for the limited developmental potential of eggs receiving advanced donor nuclei, and of parthenogenetic eggs, are as yet unknown.

Since the pronuclei and nuclei of eggs and early embryos are relatively large, piercing the egg membrane with a pipette large enough to extract the nucleus without damaging it usually leads to lysis of the egg. An efficient method has been developed (1) whereby the nucleus is removed in a small pocket of cytoplasm surrounded by intact egg membrane. This nucleoplast can then be fused to another egg or cell with the aid of Sendai virus. In this way over 100 enucleated eggs or 40−50 reconstituted eggs can be produced in a working session of about 4 h.

2. BASIC TECHNIQUES

2.1 Culture

Eggs are cultured in a standard carbonate-buffered culture medium [M16, (2), Chapter 2, Section 5.2] with 0.4% bovine serum albumin (BSA) (Fraction V, Sigma) in plastic Petri dishes (Sterilin) under oil in a humidified atmosphere of 5% CO_2 in air at 37.5−38.5°C (see Chapter 2, Section 1). Manipulations are carried out in phosphate-buffered medium with 0.4% BSA [PB1, (3), Chapter 13, Section 3.1.1] to which is added, for most of the procedures, 1 μg/ml Cytochalasin D (CCD) (Sigma) and 0.1 μg/ml Nocodazole (Sigma), both dissolved in dimethylsulphoxide (DMSO). The disruption of microfilaments (by CCD) and of microtubules (by Nocodazole) renders the eggs more malleable and resistant to damage by manipulation. Stock solutions of drugs, the CCD at a strength of 2 mg/ml DMSO and the Nocodazole at 0.2 mg/ml DMSO, are dispensed in small aliquots and stored at $-20\,^{\circ}$C for up to six months. Immediately before use 1 μl of each is added to 2 ml PB1 (PNC). Eggs are transferred in small groups to PNC for manipulation and washed as soon as possible afterwards because the effects of exposure to Nocodazole for more than 4 h may be harmful.

Oil for culture is light liquid paraffin oil (weight per ml at $20\,^{\circ}$C ~ 0.846 g) from BDH. Oil for the micromanipulation chambers and instrument lines is heavy liquid paraffin oil (weight per ml at $20\,^{\circ}$C ~ 0.875 g) from BDH. Each batch of both light and heavy oil is tested for toxicity before general use. It is filtered through Whatman's No. 1 filter paper into clean glass bottles and kept covered but we have not found it necessary to sterilize it.

2.2 Preparation of Sendai virus

Sendai virus, or haemagglutinating virus (Japan HVJ strain), is used as the fusogen. The procedures used for growing the virus in embryonated chicken eggs are described in detail elsewhere (4,5) and will be dealt with only briefly here.

(i)   Incubate virus-free fertilized chicken eggs on their sides at $30\,^{\circ}$C for 11 days.

(ii)   Turn the eggs on a candling machine to reveal the embryo, which appears as a dark shadow in the centre of the egg, and mark the injection site of each egg between two main veins near the air sac.

(iii)   Lightly swab an egg with 70% alcohol and file the shell at the site of injection.

(iv)   Using a sterile syringe fitted with a 25 gauge needle, inject 0.1 ml virus diluted in Hanks' balanced salt solution without glucose (HBSS-G) at a concentration of 0.1 haemagglutination unit (HAU) per egg. To do this insert the needle to a depth of about 1 cm, direct it towards the pointed end of the egg and slowly inject the virus suspension into the egg.

(v)   Seal the opening in the shell by applying melted paraffin wax and incubate the eggs upright on their pointed ends at $30\,^{\circ}$C for 3 days.

(vi)   After this time cool the eggs at $4\,^{\circ}$C overnight to kill the embryo and constrict the blood vessels to prevent haemorrhaging when the allantoic fluid is collected.

(vii)   Harvest the virus quickly at $4\,^{\circ}$C to obtain a high titre of virus. Using sterile conditions, cut the top of the egg shell and lift the flap. Cut the allantoic membrane and draw off the allantoic fluid with a pipette. This fluid should be evenly cloudy and of a pale yellow colour; discard samples contaminated with yolk or blood.

(viii) Pool the fluid from all the eggs and keep on ice.

(ix)   Centrifuge at 3000 r.p.m. for 10 min.

(x)   Take the supernatant and centrifuge again at 15000 r.p.m. for 1 h at $4\,^{\circ}$C. Disperse the resulting white pellet in 1 ml of cold HBSS-G.

(xi)   Take a small sample to determine HAU [see established procedures using human or guinea pig erythrocytes (4)] and dilute the virus preparation with HBSS-G to approximately 20 000 HAU/ml.

(xii)   Freeze 1 ml aliquots on dry ice and store at $-70\,^{\circ}$C. The Sendai virus preparation can be used for cell fusion for several years.

Two methods are available for inactivation of Sendai virus prior to use in cell fusion experiments. U.v.-inactivated Sendai virus can contain traces of infectious virus highly undesirable near a colony of mice. The method of choice is therefore a chemical one using beta propiolactone (BPL) (6). Great care should be exercised in handling BPL because it is a potent carcinogen and also because it is highly labile at temperatures greater than 4°C. The following procedure is used.

(i)   Thaw one or two aliquots of the stock solution of virus and vortex vigorously.
(ii)   Centrifuge at 35 000 g for 30 min. Resuspend the resulting pellet in 0.9 ml HBSS-G and disperse by vortexing.
(iii)   Transfer the solution to a tube with a tightly fitting cap and add 5 mg BSA.

The rest of the procedure is usually carried out in a cold room to avoid premature loss of activity of the highly labile BPL. Because of the toxicity of BPL it is advisable to wear two pairs of gloves and other protective clothing, including a mask. All solutions should be prepared and checked prior to transfer to the cold room so that the steps involving BPL can be carried out rapidly. All the solutions are kept on ice.

(iv)   In the cold room, first add 20 μl of BPL to 162 μl of glass-distilled water.
(v)   Quickly add 50 μl of this solution to 0.95 ml of a saline bicarbonate solution (1.68 g $NaHCO_3$ + 0.85 g NaCl + 0.2 ml of 0.5% phenol red + 100 ml 2 × glass-distilled water).
(vi)   Quickly add 100 μl of BPL + saline bicarbonate solution to the 0.9 ml of viral preparation from step (iii) above.
(vii)   Keep the tube on ice and mix the contents by vortexing at frequent intervals for 10−15 min to ensure thorough mixing in the tightly closed tube.
(viii) Place the tube on a gently rotating wheel at 37°C for 2 h and keep overnight at 4°C to ensure complete hydrolysis of BPL.
(ix)   Aliquot 20 μl volumes of the inactivated Sendai virus in Eppendorf tubes, freeze on dry-ice and store at −70°C. The inactivated Sendai virus remains fusogenic when stored at −70°C for at least a year.
(x)   When required for use, thaw an aliquot of the Sendai virus and dilute with HBSS-G to yield Sendai virus at a concentration of 1000−10 000 HAU/ml. Since there is great variation in the potency of different batches of Sendai virus and because no fixed relationship between the HAU titre and fusogenic potency can be assumed, the dilution factor is determined by pilot experiments. High concentration of a potent fusogenic viral preparation will cause extensive lysis of cells whereas excessive dilution will reduce the fusion to unacceptably low levels. We have found that the diluted preparation of the Sendai virus can be used for at least one week without loss of fusogenic activity if stored at 4°C.

2.3 Equipment

2.3.1 *Micromanipulator assembly*

In this laboratory we use the standard Leitz micromanipulator as described by Gardner (7) with an Ergaval fixed-stage microscope (Carl Zeiss, Jena) with image-erected and Nomarski interference optics. The microscope is fitted with ×10 and ×32 long-working-distance objectives and ×10 wide-field eyepieces. (See also Chapter 3, Section 7.1.)

Positive and negative pressure in the operating instrument is controlled by a de Fonbrune microsyringe pump (Beaudouin, Paris) and in the holding instrument by a simple 2 ml Wellcome Agla syringe to which has been added a coiled wire return spring before assembly. The connecting lines are of plastic tubing which is transparent to enable the detection of air bubbles in the system and which is rigid enough for all the operating pressures to be transmitted to the pipette tips and not be absorbed by the tube walls.

The glass operating instruments are attached to standard Leitz instrument tubes and held in single instrument heads on the manipulators. The syringe connecting lines and instrument tubes are filled with heavy liquid paraffin oil, great care being taken to exclude air bubbles from the system.

Manipulations are carried out in hanging drops in a Leitz (Puliv) micromanipulation chamber (internal height 3 mm) filled with heavy liquid paraffin oil. The chambers are cleaned immediately after use with hot water and then alcohol, and kept wrapped in dust-free tissue. The coverslips which form the upper surface of the manipulation chamber must fit the width of the chamber exactly and are obtainable from Leitz. Before use they are cleaned with alcohol, dipped in a silicone solution (e.g. Sigmacote), air dried, cleaned with alcohol again, and placed in a hot oven to discharge the static electricity.

The manipulation assembly should be set up on a firm workbench at a height comfortable for the operator. Good air conditioning is desirable as manipulation becomes more difficult and liable to mishaps when the ambient temperature rises above about 20°C (71°F), when the cell membranes become more sticky and the heavy liquid paraffin becomes less viscous.

2.3.2 *Other equipment*

A standard dissecting microscope such as the Wild M5 is required for normal handling of eggs and for loading and unloading the manipulation chambers.

A simple needle or electrode puller is needed to produce finely drawn micropipettes. Various designs are available commercially or can be made in the laboratory workshop. To break, bend and polish the tips of micropipettes, a microforge is required (e.g. de Fonbrune, Beaudouin, Paris) which provides a magnification of about ×100 and an accuracy of below a micron. The eyepiece of the microscope of the forge should contain a graticule, for instance a millimetre cross bar marked in one hundredths. The heating filament of the forge is a 2.5 cm length of platinum or platinum/iridium wire (diameter $0.25 - 0.3$ mm) bent to a V-shape and mounted horizontally. For fashioning the bevelled nuclear transfer needles two blobs of glass are fused onto the filament, one on the tip and one on the upper surface near the tip (see also Chapter 3, Section 9.2).

Also, some means of bevelling the tips of nuclear transfer needles is required. For this purpose machines are available commercially, or can be made in the laboratory workshop. The grinding surface of the bevelling machine used in this laboratory is a glass plate covered with a very fine diamond grinding disc which turns at a variable speed up to 200 r.p.m. (see *Figure 1*).

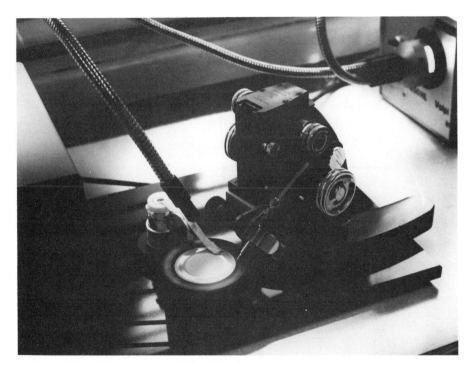

Figure 1. Bevelling machine made in the laboratory workshop. The pipette is held in a clamp attached to a modified microscope stand. Lighting is from a fibre optic lamp. The pipette is positioned over the grinding surface with the aid of a hand lens. The grinding surface is kept wet with a fabric wick dipped in distilled water.

2.4 Preparation of micro-instruments

In the following text the orientation of the instruments on the microforge will be described as it actually is and not as it appears through the microscope of the forge because the inversions vary on different microscopes. Glass micro-instruments are prepared from standard thick-walled borosilicate glass capillary tubing, outside diameter 1.0 mm, inside diameter 0.58 mm, length 15 cm, obtainable from Leitz, Clark Electromedical Instruments or Drummond Scientific Company. Other types of glass, for instance Pyrex, will have different handling characteristics regarding pulling and bevelling.

2.4.1 *Holding pipettes*

(i)   Draw the capillary tubing by hand over a small Bunsen flame to an outside diameter of between 80 and 100 μm and break off 1.5−2 cm from the narrowing of the tubing. Leave the main shaft at about 4 cm.

(ii)   Clamp the pipette onto the microforge in a horizontal position and inspect the tip for size and squareness of the break.

(iii)   If the tip is absolutely square, heat the filament to a dull red glow and bring it close to the tip to heat polish it and close it down to an internal diameter of about 20 μm (*Figure 2a*).

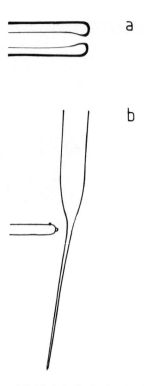

Figure 2.a. Holding pipette: diagram of finished tip. **b.** Putting bend on shaft of pipette.

(iv)   The holding pipette needs a slight bend at the place of narrowing in order to bring the tip above the height of the shoulder of the pipette when held in the horizontal position. To do this turn the pipette to the vertical, the main shaft upwards, heat the filament to a bright glow and bring it near to the pipette at the part where it begins to widen significantly; without being touched the pipette will bend slowly towards the filament, as in *Figure 2b*.

2.4.2 *Nuclear transfer needles*

(i)   Siliconize the glass capillary tubing inside and out by dipping in silicone solution and drying off in a hot oven.

(ii)   Adjust the needle puller to draw needles that come to a fine point about 8−9 mm from the first narrowing of the tube. Leave the thick shaft of the needle about 4.5 cm long.

(iii)   Mount the needle horizontally on the microforge and inspect the final half centimetre to check that the taper is very gradual.

(iv)   Break off square where the outside diameter is 25−30 microns. To do this heat the filament until it just begins to change colour and gently fuse the blob of glass on the upper surface of the tip of the filament to the shaft at the desired spot. When the glass is firmly attached, but without being narrowed or bent, switch

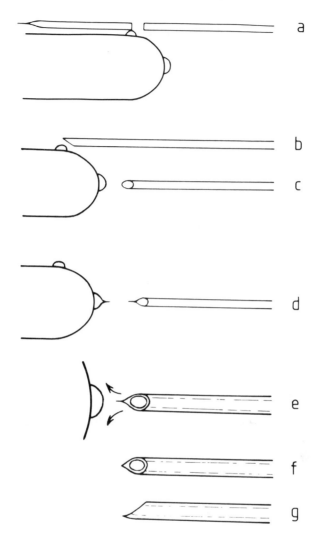

Figure 3. Nuclear transfer needle. **a.** Breaking pulled capillary at an outside diameter of 25−30 μm. The position of the two blobs of glass fused onto the end of the microforge filament is shown. **b.** Orientation of the tip for heat polishing. **c.** Orientation of the tip for drawing point. **d.** Point drawn on tip. **e.** Removing point to make sharp edge on tip. **f.** Finished tip from side. **g.** Finished tip from above.

     off the heat; the sudden contraction of the filament on cooling should break the glass cleanly (*Figure 3a*).

(v)   Bevel the pipette for 30−60 sec on a grinding machine to an angle of about 35°.

(vi)   Wash carefully with sulphuric acid, rinse with distilled water (DW), treat with the detergent NP-40 (Sigma, use undiluted) (the NP-40 enhances the non-stick properties of the siliconized surface), and finally rinse thoroughly with distilled water. These washings can be performed by attaching the pipette to the needle of a small syringe with a short length of plastic tubing and drawing the liquids in and out of the tip of the pipette. The washing regime used in this laboratory is:

1.	DW	three times	5.	DW	three times
2.	H_2SO_4	three times	6.	NP-40	three times
3.	DW	three times	7.	DW	12 times
4.	H_2SO_4	three times	8.	DW — second vessel	12 times

A final jet of air is pushed through the pipette to clear it of water. It is as well to check that the bevel is satisfactory before embarking on the washing procedure.

(vii) After washing remount the pipette horizontally in the microforge and orient the tip as in *Figure 3b*.

(viii) Heat the filament until it just begins to change colour and bring the tip close enough to the glass blob to heat polish the bevelled surface without narrowing the orifice.

(ix) Turn the needle so that the bevelled orifice faces away from the viewer (*Figure 3c*).

(x) Now a sharp edge is made on the tip of the pipette to allow it to penetrate the zona pellucida without damaging the egg membrane. Turn the heat down very slightly, bring the tip of the pipette into contact with the blob of glass on the filament tip, and then withdraw it to draw out a fine point on the smooth tip of the pipette (*Figure 3d*). Turn off the heat and reduce the point to a sharp edge by grazing it against the upper and lower surfaces of the blob of glass on the end of the filament (*Figure 3e* and *f*).

(xi) Finally, still with the orifice facing away from the operator, turn the pipette to the vertical position in the microforge, heat the filament to a bright glow and make a bend of about 5° on the shaft at the part where it begins to widen significantly (as in *Figure 2b*). This is to bring the tip above the level of the shoulder of the pipette when held in the horizontal position to allow vertical movement in the microdrop without the shoulder hitting the upper surface of the manipulation chamber. Viewed from the side (as it will be oriented when mounted in the micromanipulator) the finished pipette tip is as in *Figure 3f* and viewed from above as in *Figure 3g*.

(xii) Thoroughly dry the pipettes in a warm oven and store in a Petri dish immobilized on a strip of Blu-tac across the base.

2.5 Eggs and embryos

2.5.1 *Mouse strains*

Two main considerations determine the choice of strains of mouse from which to obtain eggs.

(i) It is often required to culture the manipulated 1-cell eggs for a few days before the next stage of the experiment. However, the development of eggs of most mouse strains *in vitro* is blocked to a greater or lesser extent at the 2-cell stage (see Chapter 2, Section 2.2.1). Hybrid embryos with C57BL in the genotype are an exception to this and we use F_1 females from a mating of C57BL/6J females crossed with CBA/Ca males (both from Bantin and Kingman) to provide the eggs for manipulation.

(ii)   In order to be able to check the provenance of the nuclei when assessing the results of the experiments, both coat colour and some easily assessed enzyme differences are desirable. For example the C57BL × CBA F_1 animals are black agouti in coat colour (AB/aB) and are homozygous for the B form of the isozyme glucose phosphate isomerase (GPI-1B) (8). To provide a distinguishable contrast, mice from an outbred line of albino animals (CFLP, obtained from Bantin and Kingman) which have been carefully screened and bred to be homozygous for the A form of GPI are used. GPI analysis is carried out on cellulose acetate plates using the Titan system supplied by Helena Laboratories. For instance, eggs from F_1 females mated with CFLP males are used to make androgenetic or gynogenetic eggs.

2.5.2 *Fertilized eggs*

To obtain maximum numbers of eggs, F_1 females (~ 28 days old) are superovulated by injecting them intra-peritoneally with 7.5 iu PMS followed 46 − 48 h later with 7.5 iu hCG and placing them with the males, one female to one male, immediately after the hCG injection (see Chapter 1, Section 5.4). The time of injection varies according to what stage of pronucleus is required but, for instance, injecting at 16.30 h will usually provide eggs with pronuclei beginning to migrate from the surface of the egg by mid-day of the following day. The females are checked the following morning for copulation plugs.

    Pronuclear eggs are retrieved from the females on the morning of the first day of pregnancy and cleared of the cumulus cells with hyaluronidase as described in Chapter 2, Section 2.1.

2.5.3 *Activated eggs*

Eggs for activation are obtained from F_1 C57BL/CBA females, both because of the absence of the 2-cell block in culture and because conditions for achieving reliable and consistent activation in this cross are now well established. Activation of the unfertilized egg can be brought about through many different agencies and various methods have been described in detail by Kaufman (9,10). We will describe only our current method of producing haploid and diploid parthenogenetic mouse eggs which is that of Cuthbertson using ethanol as the activating agent (11,12).

(i)   Collect eggs from the superovulated, but unmated, females 17 − 18 h after the hCG injection.

(ii)   Remove the cumulus cells with hyaluronidase (Chapter 2, Section 2.1) in PBI (Chapter 13, Section 3.1.1) and wash the eggs through three large drops of M16 + BSA (Chapter 2, Section 5.2) and then through three large drops of 7% ethanol in M16 + BSA.

(iii)   Leave the eggs in the third alcohol drop until 4.5 min have elapsed from their entry into the first alcohol drop. A minute before the end of the period pick up the eggs in the alcohol medium in the handling pipette and hold them there until the end of the 4.5 min, then gently expel them into a large wash drop of M16 + BSA. The mixing of the two media that occurs can cause the eggs to swirl up to the surface of the drop and stick there if care is not taken. Wash the eggs through a further six drops of M16 + BSA.

(iv)   If *haploid parthenogenones* are required culture the eggs in M16 + BSA under oil at 37.5°C and they will start to form the second polar body. Within 4 h a haploid pronucleus will have developed and the second polar body will be fully extruded.

(v)   *Diploid parthenogenones* can be produced by suppressing formation of the second polar body and so keeping both products of the second meiotic division within the developing egg. This is achieved by incubation in medium containing the cytoskeletal inhibitor Cytochalasin B (CCB). Immediately after activation and washing, transfer the eggs to a large drop under oil of M16 + BSA + 5 μg/ml CCB (stock CCB is dissolved in DMSO at 2 mg/ml) and incubate for 3−4 h. By this time the two pronuclei should be clearly visible within the cytoplasm. Very carefully wash the eggs through nine large drops of M16 + BSA to remove the CCB and return them to culture. After another 2 h, when they have had time to recover from removal from CCB, the eggs should be carefully screened and any anomalous ones discarded.

2.5.4 *Preimplantation embryos*

Cleavage stage embryos for nuclear transplantation experiments can be obtained either by flushing from the oviduct or uterus or by culturing from the 1- or 2-cell stage *in vitro* (see Chapter 2, Section 2).

3. NUCLEAR TRANSFER

3.1 **Setting up the chamber**

(i)   Fix a coverslip across the manipulator chamber with a little of a Vaseline/wax mixture (~5% hard paraffin wax) dispensed from a syringe onto the uprights of the chamber.

(ii)   Using the lowest power of the dissecting microscope place drops of Sendai virus solution and PNC medium on the underside of the coverslip using a pulled Pasteur pipette. First place a large drop of Sendai virus (a hemisphere of diameter ~3 mm) near to the rear upright of the chamber, then ~12 smaller drops of PNC (hemisphere diameter ~1.5 mm) across the centre of the chamber (see *Figure 4a*).

(iii)   Carefully fill the chamber to the lateral edges of the coverslip with heavy liquid paraffin using a shortened Pasteur pipette.

(iv)   Introduce the eggs to be manipulated into the drops with a pulled Pasteur pipette. It is a good safeguard always to keep to the same pattern of egg placing, with a visual marker as well as a positional marker when donor and recipient eggs are not obviously different: for instance, donor eggs in groups of four in the four drops nearest the Sendai virus drop, then an empty drop to receive discarded nuclei and any debris, and finally the recipient eggs in pairs in the remaining drops. At this stage also have ready in the incubator a culture dish with a series of large drops of M16 + BSA under oil to receive the operated eggs.

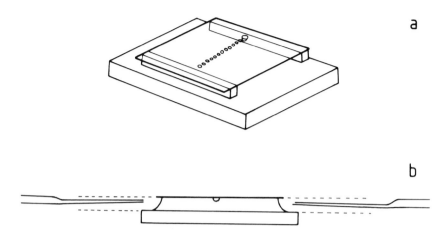

Figure 4.a. Manipulation chamber with drops placed on underside of coverslip. **b.** Manipulation chamber with instruments positioned ready to enter, seen from the front.

3.2 Setting up the instruments

(i)   Place the loaded chamber on the microscope stage and focus with the lower power objective on an egg. Then move the chamber to the rear of the stage. The plane of focus is now a little above the bottom of the operating drops and to avoid mishaps later the focussing knob of the microscope should not be touched again whilst setting up the instruments, focussing of the instruments being achieved by raising and lowering the manipulators.

(ii)   Set the two manipulators to move horizontally — mark 0 on the dial (see Chapter 3, Figure 8, left-hand manipulator).

(iii)   Fit a holding pipette into the instrument tube attached to the Agla syringe, after all air bubbles have been expelled from the system, then fix the instrument tube into the instrument head of one of the manipulators (generally the left-hand manipulator).

(iv)   Arrange the instrument head so that the holding pipette slopes upwards a little towards the tip — less than 5°.

(v)   Bring the pipette tip into the microscope field of view and focus with the manipulator height adjustment. Push oil through to the end of the pipette with the Agla syringe and then withdraw the pipette assembly to the side of the microscope.

(vi)   In the same way attach a nuclear transfer pipette to the de Fonbrune assembly on the right-hand manipulator and fill with oil. Take care that the bevel of the tip is absolutely vertical as seen down the microscope — see *Figure 3g*. Withdraw the nuclear transfer assembly to the side.

(vii)   Finally, bring the manipulation chamber back into the centre of the microscope stage with an operating drop in the microscope field. The relative angles and positions of the instruments should appear from the front as in *Figure 4b*.

(viii) Gently move the instruments, one at a time, into the chamber until they appear in the drop. Each instrument should always be moved to its own rear or front position in the drop when not in use to avoid clashes.

3.3 **Nuclear removal and transfer**

The steps involved in making heterozygous diploid androgenetic eggs will be described in detail as an example of the technique. $F_1 ♀ \times$ CFLP♂ pronuclear eggs are used as both donors and recipients.

(i)   Harvest the eggs about 18 h after hCG injection and after removal of the cumulus cells (Chapter 2, Section 2.1) culture for an hour in M16 + BSA (Chapter 2, Section 5.2).

(ii)   Select early pronuclear eggs with the pronuclei still fairly peripheral as recipient eggs and later stage eggs with the pronuclei larger and closer together as donor eggs. Put about 30 of each type into large drops of PNC medium under oil and culture in an ungassed incubator for at least 15 min before manipulation.

(iii)   First prepare the recipient eggs by removing the female pronucleus. Orient an egg so that when it is held by the polar body region both pronuclei will be in focus and so that when the bevelled pipette is pressed into the egg the male pronucleus will lie behind the pipette and the female pronucleus in front of it. This is in order to keep the identity of the pronuclei clear once the egg has been disturbed by the pipette.

In distinguishing between the pronuclei the female pronucleus is usually clearly the smaller of the two and in the early stages lies near to the second polar body (see Chapter 2, Figure 1). Since the sperm may enter the egg at any point on the surface the male pronucleus can in some cases be as near to the polar body as the female pronucleus. The absolute size of the pronuclei at any stage varies from egg to egg but the female pronucleus is always smaller. However in older eggs, where the pronuclei have migrated to the centre and enlarged, the difference in size is sometimes not obvious. If there is any doubt about the identity of the pronuclei the egg is discarded. (The number of nucleoli is no indication of their identity.)

(iv)   Manoeuvre the egg into the appropriate position using both the enucleation and holding pipettes and hold it gently but firmly by suction so that the zona pellucida and part of the polar body deform into the holding pipette (*Figure 5a*).

(v)   Raise the egg a little from the bottom of the drop, focus the microscope on the outermost edge of the zona pellucida, and bring the tip of the enucleation pipette into focus.

(vi)   Press the enucleation pipette gently through the zona pellucida and into the egg, which will deform round it without the membrane breaking (*Figure 5b*).

(vii)   Take the tip close to the female pronucleus, and apply gentle suction. The egg membrane will deform into the pipette and the pronucleus will follow.

(viii) When the pronucleus is inside the tip of the pipette gently withdraw the pipette through the zona pellucida (*Figure 5c*). The egg membrane will stretch out into a narrow tube. Pinch the extended membranes against the edge of the hole in the zona pellucida to break them. They will usually seal again on both sides of the break (*Figure 5d*).

(ix)   Repeat steps (iii) to (viii) for all the recipient eggs and evacuate the female pronuclei into an empty drop.

(x)   Now move the chamber to bring a drop with donor eggs into view. Manoeuvre

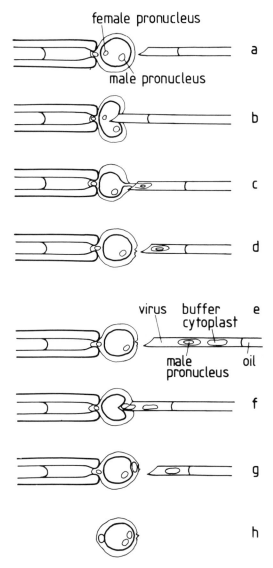

Figure 5. Making heterozygous diploid androgenetic eggs. **a–d.** Removing the female pronucleus from a fertilized egg to make a haploid androgenetic egg. **e–g.** Transferring second male pronucleus to haploid androgenetic egg. **h.** After fusion.

an egg to bring the male pronucleus into an appropriate position and hold it firmly with the holding pipette. This time it is not necessary to hold the egg by the polar body region.

(xi) Penetrate the zona pellucida as before, and first take a small cytoplast from the egg and pinch it off. This will act as a buffer later on.

(xii) Then take out the male pronucleus. (The nuclear membrane may deform a little on entering the enucleation pipette. This will not hurt it, but if the pipette is too

247

narrow or the tip restricted by adherent oil blobs or debris some of the nucleoli may shift forward in the nucleus, press against the membrane and then squeeze through. If this is seen to happen discard the nucleus and try to clear the tip of the pipette or fit a new one.)

(xiii) Withdraw both pipettes from the drop, the enucleation pipette still containing the male pronucleus.

(xiv) Move the chamber to bring the Sendai virus drop into view and bring the enucleation pipette into this drop.

(xv) Push the nucleoplast so that it protrudes a little from the end of the pipette to bathe it in the virus, and then draw it about 100 μm into the pipette so that a small volume of virus preparation occupies the end of the pipette.

(xvi) Withdraw the pipette from the drop, move the chamber to a recipient egg drop and pick up an egg by the same region as when it was enucleated [this can be easily identified by the polar body and the deformation of the zona pellucida (*Figure 5c*)].

(xvii) Probe the transfer pipette into the original hole in the zona pellucida and gently expel the virus solution and the nucleoplast into the perivitelline space (*Figure 5f* and *g*). The extra cytoplast can be useful in nudging the end of the karyoplast out of the pipette and also as a buffer if the oil moves suddenly. If there is a large volume of cytoplasm following the pronucleus in the karyoplast, adjust the size of the karyoplast by withdrawing the pipette as soon as the pronucleus is inside the zona pellucida and nipping off a cytoplast as before.

Fusion will occur at any time within the next hour or so, sometimes within minutes of transfer (*Figure 5h*).

(xviii) Repeat the transfer procedure for all the recipient eggs.

(xix) Move the chamber to the dissecting microscope and gather the eggs together, wash them briefly and culture them in M16 + BSA.

(xx) At the end of the session wash all the eggs through nine drops of fresh M16 + BSA, check carefully for fusion and continue culture in M16 + BSA.

3.4 **Other nuclear transfers**

Heterozygous diploid gynogenetic eggs can be made in the same way as in Section 3.3, or by using haploid parthenogenetic eggs for either or both partners (see *Figure 6*), but we have found that these fully heterozygous eggs with only maternal genomes do not develop differently from parthenogenetic eggs which have been diploidized by suppressing second polar body formation. Similarly a variety of aneuploid eggs can be produced. More than one karyoplast will fuse with the egg membrane provided they are transferred into the perivitelline space within a short time of one another, so it is possible to make triploid or tetraploid eggs with different numbers of maternal and paternal genomes.

The techniques for handling nuclei from later stage embryos are basically the same as in Section 3.3 (13,14). The size of the transfer pipette will need to be adjusted a little to suit the size of the nucleus. The nuclei of 8-cell and later stage embryos are often difficult to see clearly and great care is needed in monitoring the eggs for fusion. It is also possible to transfer nuclei into 2-cell blastomeres. In this case, however carefully

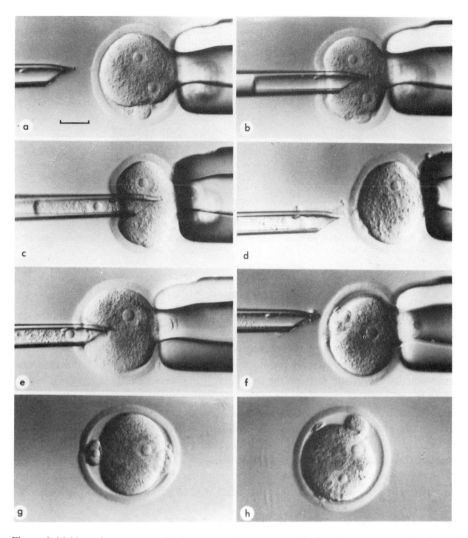

Figure 6. Making a heterozygous diploid gynogenetic egg using a fertilized egg as donor and a haploid parthenogenetic egg as recipient. Scale bar = 30 μm. **a−d.** Removing female pronucleus from fertilized egg. **e−f.** Transferring female pronucleus to haploid parthenogenetic egg. **g.** Before fusion. **h.** After fusion.

it is placed, the karyoplast has a tendency to slip into the cleft between the two cells and it is advisable to leave the manipulated embryos in the manipulation chamber until fusion has been observed. In this way we have made 2-cell chimaeras, removing the nucleus from one cell of a 2-cell $F_1 \times F_1$ (F_2) embryo and replacing it with a 2-cell nucleus from a CFLP × CFLP embryo. As the chimaeric embryos are still in their zonae they can be transferred at the 2-cell or 4-cell stage to the oviducts of day 1 recipients (Chapter 13, Section 6.4.5). Of six such embryos transferred at the 4-cell stage six were born, two overtly chimaeric (judged by coat colour and blood GPI type), one wholly of the CFLP type and three wholly of the F_2 type.

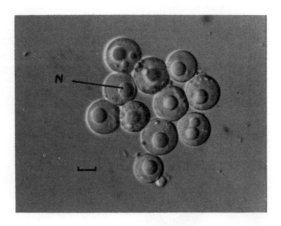

Figure 7. Pronucleoplasts. Male pronuclei with minimum cytoplasm removed from fertilized eggs. N = nucleolus. Scale bar = 10 μm.

3.5 Karyoplasts with minimum cytoplasm

If nuclei or pronuclei virtually free of cytoplasm are required, for instance for biochemical analysis, they can be prepared as karyoplasts with a minimum volume of cytoplasm. The volume of cytoplasm behind the nucleus in the enucleation pipette can be adjusted either (i) by making the nucleus protrude from the tip of the pipette and then lowering the pipette through the medium/oil interface until the karyoplast breaks off, or (ii) by inserting part of the tip of the enucleation pipette into the tip of the holding pipette, pushing the nucleus out of the enucleation pipette a little, and then applying very gentle suction to both pipettes. The karyoplast should break off immediately behind the nucleus. With both these methods it is easier to adjust the amount of cytoplasm when there is a large excess of cytoplasm than when the excess is only moderate. It is however possible to make karyoplasts with cytoplasm less than 10% of total volume, but the smaller the nucleus the more difficult it is to achieve this ratio (*Figure 7*).

3.6 Bisection of eggs

If for some reason it is undesirable to expose the eggs to Cytochalasin or other drugs and large karyoplasts are required these may be prepared by operating on zona-free eggs (Chapter 2, Section 3.2.1) in hanging drops of PB1 (Chapter 13, Section 3.1.1). A large holding-type pipette, internal orifice about 40 μm diameter, and a solid glass needle of diameter $10-12$ μm with a blunt flamed tip (see Chapter 3, Section 9.3), both made from siliconized capillary tube, treated with NP-40 and well washed, are used. The egg is drawn gently into the holding pipette and expelled again to elongate it and the solid needle is taken across the egg at the desired point and down through the medium/oil interface (15). The polar body or bodies can be removed in the same way. The survival rate for this procedure is lower in PB1 than if it were performed in PNC but not unworkably so.

4. ASSESSMENT OF RESULTS

The precise conditions for optimum fusion efficiency vary from one Sendai virus culture preparation to another but a fusion rate of more than 90% is to be expected once these are established. When the reconstitution has been achieved the eggs can be treated as any other eggs. They may be cultured *in vitro* for preimplantation studies (see Chapter 2), or they may be transferred to pseudopregnant recipients to observe *in vivo* development. Embryos at the 1-, 2- or even 4-cell stage are transferred to the oviducts of day 1 recipients and at the late morula/early blastocyst stage to the uteri of day 3 recipients (see Chapter 13, Section 6.4.6). (In this respect it is helpful as a rule to transfer experimental embryos to one oviduct or uterine horn and normal embryos to the other in order both to check the effectiveness of the pseudopregnancy and to assess individual variations in successful recipients.) The reconstituted eggs can be used in aggregation chimaeras or blastocyst reconstructions (e.g. 16) and analysed for their proteins and enzymes (see Chapters 6−8) or for their chromosomes (see Chapter 5).

Diploid parthenogenetic, gynogenetic or androgenetic embryos seldom continue development beyond day 10 or 11 after transfer to pseudopregnant recipients (17−20). At this stage their gross morphology is abnormal to a greater or lesser extent but their tissues are still alive and healthy enough to be subjected to GPI or other enzyme analysis. Dissection and assessment of mid-gestation stage embryos are dealt with in Chapter 3. Combination chimaeras between abnormal and normal embryos may be analysed later in gestation or continued to term and adulthood to allow further tissue analysis or test breeding.

5. CURRENT LIMITATIONS AND FUTURE PROSPECTS OF FUSION-ASSISTED NUCLEAR TRANSPLANTATION TECHNIQUES IN MAMMALIAN EGGS

While both viral fusogens such as the one described here and chemical fusogens such as polyethylene glycol (PEG) (21,22) have been used successfully in studies involving mammalian eggs, these methods have some major limitations. We have found that PEG, the most widely used chemical fusogen for somatic cells, is very unreliable for most purposes when employed on early embryos. It tends to be variable in its activity and can cause extensive lysis and prevent development because it is difficult to remove from the plasma membrane.

Sendai virus has proved much more successful as a fusogen for mammalian eggs but again this agent has several drawbacks. There is always an element of risk, however small, that some viral particles will escape inactivation and this could have catastrophic consequences for a mouse colony. We have found that Sendai virus is an effective fusogenic agent only so long as karyoplasts from eggs and very early embryos are used. In attempts to fuse cells from more advanced embryos or adult tissues the fusion rate dropped to 0−10%. This clearly limits the questions which can be investigated. Further, once Sendai virus has been used to fuse a karyoplast to an egg, blastomeres derived from this reconstituted egg cannot be used later to provide donor nuclei, since Sendai virus is ineffective when employed a second time. This presumably results from the loss of cell-surface receptors for the virus. Hence this rules out the possibility of serial nuclear transplantation when Sendai virus is used alone. Also Sendai virus suffers from the same batch variability as PEG.

A cell fusion method of promise for the future is Zimmermann's technique for electro-fusion (23). This is a highly versatile and effective fusogenic method which has the advantage that the conditions can be varied easily to suit different cell types and also that it does not restrict the serial transplantation of nuclei. The technique involves firstly the application of a pulse of alternating current to align the cells to be fused and achieve close contact or 'agglutination', followed by a pulse of direct current to induce breakdown of the membranes at the site of cell contact, which eventually leads to fusion. The intensity of the current, the duration of its application and the number of pulses, as well as the fusion and post-fusion media, temperature and pre-treatment of the cells, can all be varied, with the prospect of eventually arriving at the optimum conditions needed for fusion of a particular cell type. Mammalian eggs and blastomeres have been fused using this technique (24,25) so that some of the parameters have already been established, but where there is a significant difference in the size of the cells to be fused, as with a large egg and a small karyoplast in the nuclear transfer technique described here, the precise alignment of the cells necessary for the electrofusion method to proceed is as yet problematical. Also where there is a big difference in the stage or type of cells to be fused the particular conditions necessary for fusion are again different. However it is generally thought that once the physical parameters are established for a given system, fusion using this method can be achieved very reliably and without many of the problems associated with viral and chemical fusogens.

6. ACKNOWLEDGEMENTS

We wish to thank Sarah Howlett for helpful discussions, Katharine Stroud for drawing the diagrams and Paul Miles and Geoff Leeson for making the needle puller and bevelling machine.

7. REFERENCES

1. McGrath,J. and Solter,D. (1983) *Science,* **220**, 1300.
2. Whittingham,D.G. (1971) *J. Reprod. Fert. (Suppl.),* **14**, 7.
3. Whittingham,D.G. and Wales,R.G. (1969) *Aust. J. Biol. Sci.,* **22**, 1065.
4. Harris,H. and Watkins,J.F. (1965) *Nature,* **205**, 640.
5. Giles,R.E. and Ruddle,R.H. (1973) *In Vitro,* **9**, 103.
6. Neff,J.M. and Enders,J.F. (1968) *Proc. Soc. Exp. Biol. Med.,* **127**, 260.
7. Gardner,R.L. (1978) In *Methods in Mammalian Reproduction.* Daniel,J.C. (ed.), Academic Press, New York, p. 137.
8. Chapman,V.M., Whitten,W.K. and Ruddle,F.H. (1971) *Dev. Biol.,* **26**, 153.
9. Kaufman,M.H. (1978) In *Methods in Mammalian Reproduction.* Daniel,J.C. (ed.), Academic Press, New York, p. 21.
10. Kaufman,M.H. (1983) *Early Mammalian Development: Parthenogenetic Studies.* Cambridge University Press, Cambridge.
11. Cuthbertson,K.S.R. (1983) *J. Exp. Zool.,* **226**, 311.
12. Kaufman,M.H. (1982) *J. Embryol. Exp. Morphol.,* **71**, 139.
13. McGrath,J. and Solter,D. (1984) *Science,* **226**, 1317.
14. Surani,M.A.H., Barton,S.C. and Norris,M.L. (1986) *Cell,* **45**, 127.
15. Barton,S.C. and Surani,M.A.H. (1983) *Exp. Cell Res.,* **146**, 187.

16. Barton,S.C., Adams,C.A., Norris,M.L. and Surani,M.A.H. (1985) *J. Embryol. Exp. Morphol.*, **90**, 267.
17. Surani,M.A.H. and Barton,S.C. (1983) *Science,* **222**, 1034.
18. Surani,M.A.H., Barton,S.C. and Norris,M.L. (1984) *Nature,* **308**, 548.
19. Mann,J.R. and Lovell-Badge,R.H. (1984) *Nature,* **310**, 66.
20. Barton,S.C., Surani,M.A.H. and Norris,M.L. (1984) *Nature,* **311**, 374.
21. Eglitis,M.A. (1980) *J. Exp. Zool.,* **213**, 309.
22. Spindle,A. (1981) *Exp. Cell Res.,* **131**, 465.
23. Zimmermann,V. and Vienken,J. (1982) *J. Membr. Biol.,* **67**, 165.
24. Kubiak,J.Z. and Tarkowski,A.K. (1985) *Exp. Cell Res.,* **157**, 561.
25. Willadsen,S.M. (1986) *Nature,* **320**, 63.

CHAPTER 13

The low temperature preservation of mouse oocytes and embryos

MAUREEN J.WOOD, DAVID G.WHITTINGHAM and WILLIAM F.RALL

1. INTRODUCTION

The preservation of oocytes or embryos in liquid nitrogen (LN$_2$) at $-196\,^{\circ}$C for extensive periods without genetic change and subsequent recovery of normal offspring from the stored material has important implications in animal breeding, biomedical research and medicine. The research worker must have access to a vast array of mutant, inbred and congenic lines of mice. However, specialized stocks are costly and time consuming to develop and maintain. A bank of frozen embryos from these stocks provides insurance against the loss of valuable genetic material by breeding failure, genetic contamination or disease, and obviates the expensive maintenance of animals not in current use. Furthermore the international distribution of embryos is simpler, cheaper and more humane than the transport of animals and fewer diseases are likely to be carried by embryos, thus requiring less stringent quarantine restrictions.

Frozen human embryos are increasingly important in the management of infertility. After stimulated ovulation and fertilization *in vitro*, more embryos may be obtained than can be returned immediately to the mother. By freezing these additional embryos for transfer in subsequent cycles, the chance of a successful pregnancy from a single egg collection is increased. The storage of unfertilized eggs circumvents the legal and ethical objections to the preservation of human embryos, and may become important for the woman at risk of losing normal ovarian function. It must be emphasized, however, that whilst the safety of storing embryos is well documented, the same is not yet true for oocytes preserved at $-196\,^{\circ}$C.

The methods developed by Whittingham, Leibo and Mazur (1) and Wilmut (2) that resulted in the successful freezing of mouse embryos, i.e. slow cooling ($0.2-2.0\,^{\circ}$C/min) in the presence of dimethylsulphoxide (DMSO) and slow warming ($4-25\,^{\circ}$C/min), were soon adapted for freezing mouse oocytes (3) and the embryos of other species including man (4,5). Further modifications were made as embryo banking became part of the routine in many laboratories and with the increasing use of frozen embryos by commercial cattle breeders. Nowadays, several distinct protocols employing a variety of cryoprotectants, shorter cooling procedures and rapid warming rates are in use worldwide. However, contemporary comparisons of different freezing protocols have not been made. Thus it is impossible to judge the superiority of any procedure. Recently, live offspring were obtained from mouse embryos preserved at $-196\,^{\circ}$C without the formation of ice, i.e. by vitrification rather than freezing (6); and reports of alternative vitrification procedures are appearing (7,8).

In this chapter we shall restrict ourselves to a description of the following.

(i)   The methods used routinely for oocyte and embryo freezing in our laboratory. These procedures were devised using simple, inexpensive apparatus but can be modified for use with commercially available biological freezers.

(ii)   The vitrification method that in our hands yields viable embryos, albeit with a lower rate of survival than after conventional freezing techniques.

Comprehensive treatments of the theoretical aspects of embryo freezing can be found in references 4, 9 and 10.

2. DEVELOPMENTAL STAGES

All of the pre-implantation stages of the mouse give rise to viable offspring after freezing at $-196°C$. To date, after vitrification, only the 8-cell (6), morula and early blastocyst (7) stages have continued development to form normal late stage fetuses or liveborn progeny.

The 8-cell stage embryo is preferred for storage for several practical reasons.

(i)   It is less sensitive than 1- or 2-cell embryos to handling *in vitro*.

(ii)   Its morphological appearance and potential for further development correlate well. Early abnormalities are eliminated by discarding retarded or fragmenting embryos.

(iii)   The freezing procedure is slightly simpler for 8-cell embryos compared to blastocysts.

(iv)   If one or two blastomeres are damaged the remaining cells have the potential to form a viable fetus.

(v)   After thawing, a period of 24 h culture *in vitro* allows the selection of those embryos that resume cleavage for transfer to pseudopregnant recipients.

Unfertilized mouse eggs are frozen for three major purposes, namely:

(i)   as a model for the development of procedures for human oocyte preservation;

(ii)   as a model to study cellular responses to low temperature;

(iii)   to retain the option to make specific genetic combinations by fertilization *in vitro* at a later date.

2.1 Source of oocytes and embryos

Unfertilized or fertilized eggs and cleavage stage embryos are obtained from prepubertal or mature mice induced to ovulate by administration of pregnant mares' serum gonadotrophin (PMSG) and human chorionic gonadotrophin (hCG) 44−52 h apart (see Chapter 1, Section 5.4). There is no evidence that hormone priming of the donor female alters the sensitivity of the resultant embryos to cooling and warming. However, some mouse strains are refractory to gonadotrophin stimulation and in these cases fertilized eggs and embryos may be collected from mature naturally ovulating females. Most F1 hybrid mice superovulate reliably. Thus, for the preservation of particular genes when the genetic background is unimportant, the problems of superovulation may be circumvented by pairing gonadotrophin-primed F1 hybrid females with males carrying the appropriate gene.

2.2 Collection of oocytes and embryos

Newly ovulated oocytes are released from the oviduct between 13 and 14.5 h post hCG by tearing the walls of the ampulla with Watchmakers' forceps (see Chapter 2, Section 2.1.1). The oocytes are frozen either in the intact cumulus masses or after removal of the cumulus cells by treatment with hyaluronidase (Chapter 2, Section 2.1.1). In the latter instance they are pooled and washed free of the enzyme before distribution to the freezing tubes.

Fertilized eggs and cleavage stage embryos are flushed from the reproductive tract at appropriate intervals after the estimated time of ovulation or the injection of hCG (Chapter 2, Section 2.2). All stages are collected in PB1 or M2 (Section 3.1) for freezing and in PB1, M2 or HB1 (Section 4.1) for vitrification.

3. FREEZING

The formation of large quantities of ice in any mammalian cell is accepted as the major factor leading to cell death after thawing. Either the mechanical stress imposed by the ice crystals or the osmotic stress that occurs when the ice forms or melts may be lethal. Thus, the successful freezing of eggs and embryos depends upon adequate dehydration of the cells to prevent the formation of excessive amounts of intra-cellular ice during cooling or warming.

The removal of water from cells may be effected in one of two ways.

(i) At sub-zero temperatures: as ice forms in the suspending medium the concentration of extra-cellular solutes increases, the vapour pressure of water inside the cell is higher than that outside the cell, and water flows from the cell to restore equilibrium across the membrane.

(ii) Before cooling to sub-zero temperatures: the addition of a non-permeating solute, such as sucrose, to the suspending medium results in partial dehydration of the blastomeres. This method of water removal may be combined with a holding period at a relatively high sub-zero temperature (typically around $-30°C$) to permit further water loss before rapid cooling to $-196°C$ (11) or the embryos may be cooled rapidly from room temperature by direct immersion into LN_2 (12).

The procedures employing a non-permeating solute to dehydrate the embryo before cooling are relatively new and untested. It remains to be seen whether they will be adopted for routine embryo banking. By contrast, procedures in which gradual dehydration occurs during continuous slow cooling have been used successfully in many laboratories over the past 14 years. The methods described here are based on the cooling rate procedures devised for embryo preservation by Whittingham, Leibo and Mazur (1) and Wilmut (2) and on the modifications of these procedures described by Whittingham, Wood, Farrant, Lee and Halsey (13).

The essential features of the procedures are as follows:

(i) choice of suspending medium;

(ii) presence of a cryoprotective compound during cooling;

(iii) induction of ice ('seeding') at a temperature just below the freezing point of the freezing medium;

Table 1. Summary of procedures for freezing and thawing mouse oocytes and embryos.

See Section 3 of text for details.

1. Mark and heat sterilize the ampoule (Section 3.1.2).
2. Introduce 0.15 ml medium (Section 3.1.1) into the ampoule (Section 3.1.3.i).
3. Pipette the oocytes/embryos into the ampoule (Section 3.1.3.ii).
4. Cool the sample to 0°C.
5. After 10 min, add 0.15 ml 3 M DMSO to the sample (Section 3.2.3).
6. After 15 min, transfer the ampoule to the seeding bath (Section 3.3.1) at −6°C.
7. After 2 min, seed the sample (Section 3.3.iii).
8. After 5 min, transfer the seeded samples to the cooling bath (Section 3.4.2) at −6°C.
9. Cool at ~0.5°C/min to either −40°C or below −70°C before immersing the ampoule in LN$_2$ (Section 3.4.3).
10. Warm the samples at the appropriate rate as described in Section 3.5.
11. Dilute the DMSO (Section 3.6).

(iv)   slow cooling (~0.5°C/min) terminated at −40°C or at temperatures below −70°C by immersion in LN$_2$;

(v)   storage at −196°C in LN$_2$;

(vi)   careful removal of the cryoprotectant after thawing to avoid osmotic shock.

The factors affecting survival are interdependent and a change in one variable may necessitate changes in all other variables to maintain the same level of survival. We emphasize that for successful egg or embryo freezing it is essential to follow precisely a single protocol. Here we shall describe the procedures used by two of us (MW and DW) for freezing oocytes and embryos. Our protocol is summarized in *Table 1* and described in detail below.

3.1 Sample preparation

3.1.1 *Choice of medium for freezing*

To avoid damage during prolonged exposure to the uncontrolled pH changes of bicarbonate buffered media, eggs and embryos are collected and frozen in media with a stable pH of between 7.2 and 7.4 in air. Commonly used media are:

(i)   a simple physiological saline (PBS);
(ii)   a modified Dulbecco's phosphate-buffered saline (PB1: 14; *Table 2*);
(iii) a Hepes-buffered embryo culture medium (M2: 15; *Table 2*; see also Chapter 2, Section 5.2).

PB1 and M2 are supplemented with pyruvate, glucose and bovine serum albumin (BSA); M2 also contains lactate. These energy sources help to prevent depletion of metabolic pools in the embryos during collection.

3.1.2 *Choice and preparation of sample container*

Oocytes and embryos can be frozen in plastic screw-top phials (Nunc, Kamstrup, Denmark), glass test tubes (10 mm o.d., 75 mm long; Arthur H.Thomas Co., Philadelphia, USA), borosilicate glass ampoules (1−2 ml capacity; Wheaton Scientific, Millville, USA) or plastic insemination straws (IMV, L'Aigle, France). For ease of loading, intact cumulus masses are often frozen in glass test tubes. A reliable procedure for freez-

Table 2. Composition of media for oocyte and embryo freezing.

Component	PB1		M2	
	g/l	mM	g/l	mM
NaCl	8.000	136.87	5.533	94.66
KCl	0.200	2.68	0.356	4.78
$CaCl_2.2H_2O$	0.132	0.90	0.252	1.71
KH_2PO_4	0.200	1.47	0.162	1.19
$MgCl_2.6H_2O$	0.100	0.49	–	–
$Na_2HPO_4.12H_2O$	2.898	8.09	–	–
$MgSO_4.7H_2O$	–	–	0.293	1.19
$NaHCO_3$	–	–	0.349	4.15
Hepes[a]	–	–	4.969	20.85
Na pyruvate[b]	0.036	0.33	0.036	0.33
Na lactate[c]	–	–	2.610	23.28
Glucose	1.000	5.56	1.000	5.56
BSA[d]	3.000	–	4.000	–
Penicillin G (potassium salt)	0.060	–	0.060	–
Streptomycin sulphate	–	–	0.050	–
Phenol red	0.010	–	0.010	–

[a]Calbiochem-Boehring: Ultrol.
[b]Calbiochem-Boehring.
[c]Sigma grade DL-V, 60% syrup.
[d]Crystalline: Miles-Serovac.

ing embryos in straws is described by Renard and Babinet (16). We prefer 2 ml glass ampoules for freezing denuded oocytes and embryos for the reasons listed below.

(i)   Survival after rapid warming is higher for embryos frozen in glass tubes or ampoules compared to plastic phials (this probably reflects the superior heat transmission properties of glass).

(ii)   Glass ampoules can be heat sealed to prevent contamination with debris that accumulates during storage in the LN_2 refrigerator. The silicone bungs used to seal test tubes are easily lost.

(iii)   There is less chance of accidentally warming the relatively large sample contained in an ampoule compared to a straw (0.2−0.3 ml versus 20−40 μl).

(iv)   The large size of the ampoule (compared to a straw) simplifies the labelling and later identification of frost covered samples.
   The disadvantages of glass ampoules are:

(i)   during storage, LN_2 enters poorly sealed ampoules: when the LN_2 vapourizes during thawing the ampoule may explode;

(ii)   compared to straws, relatively few ampoules can be stored in a LN_2 refrigerator.
   In our experience, washing the ampoule is unnecessary and any traces of detergent left on the glass may be lethal to the embryos. Before sterilization, mark each ampoule using glass marking ink (Gurr). A simple code number is sufficient. Allow the ink to dry and then heat sterilize the ampoules at 150°C for 90 min.
   The permanent code numbers are essential for long-term storage but can be difficult

to read when the glass is frosted. To simplify retrieval, immediately before loading the ampoule we attach an adhesive label of the type used by electricians to identify wires (wire-markers; W.H.Brady and Co. Ltd, Banbury, UK).

3.1.3 *Preparing the sample for cooling*

(i)   Using a disposable hypodermic needle (19 gauge, 50 mm long), introduce 0.15 ml of M2 into the labelled ampoule.

(ii)   Pipette the embryos (~ 30 per sample) into the medium. To avoid adding a large volume of medium with the embryos, introduce one or two air bubbles into a finely drawn mouth-operated Pasteur pipette before picking up the embryos in the smallest possible volume of M2. Then place the tip of the pipette into the M2 in the ampoule and gently expel the contents until one air bubble is expelled. Under the microscope check that no embryos remain in the pipette.

(iii)   Cool the sample to 0°C in an ice bath before adding the cryoprotectant (Section 3.2.3).

3.2 Cryoprotectant

The presence of a cryoprotective compound during cooling is essential for embryo survival. The way in which such compounds act is not fully understood, but it is probable that they protect the embryo from the damaging effects of the high concentrations of solutes generated as the water freezes in the suspending medium.

3.2.1 *Choice of cryoprotectant*

A variety of permeating compounds including DMSO, glycerol, methanol, propylene glycol and ethylene glycol afford protection to fertilized mouse eggs and embryos during freezing (1,16,17,18,19). Unfertilized eggs are protected by DMSO or glycerol (3,20). In our laboratory DMSO is preferred for all routine embryo and oocyte freezing for several reasons.

(i)   To date, live mice have been recovered only from oocytes preserved using DMSO as the cryoprotectant (3,21).

(ii)   After freezing in the presence of DMSO all the pre-implantation stages of the mouse develop normally.

(iii)   The procedures for addition and removal of DMSO are relatively simple. This is of particular importance to novices and in laboratories where freezing is not commonplace.

(iv)   Survival after thawing is high. Typically, 50% of random-bred, and up to 30% of inbred, embryos frozen at the 8-cell stage develop normally *in vivo*. The proportion of thawed oocytes that are fertilized *in vitro* depends in part on the genotype of the sperm and varies between 1% and 60%. After embryo transfer approximately 50% of the fertilized eggs form normal fetuses.

(v)   The safety of long-term storage in DMSO is proven (Section 5.1).
        Embryo survival may be affected by:

(i)   the concentration of cryoprotectant;

(ii)   the rate and temperature at which the cryoprotectant is added to the sample and

Table 3. Preparation and addition of DMSO before freezing.

Preparation of 3.0 M DMSO

BDH Spectroscopic grade DMSO is preferred.
1.   Pipette 7.87 ml medium (PB1 or M2) into a Falcon plastic test tube.
2.   Add 2.13 ml DMSO.
3.   Mix gently to avoid denaturing the BSA in the medium.
4.   Sterilize using a 0.22 μm Millipore filter.

A. Addition of DMSO to mouse oocytes and 1-cell to morula stage embryos

1.   Cool the 3.0 M DMSO to 0°C.
2.   Pipette the embryos into an ampoule containing 0.15 ml of M2 and cool to 0°C (~ 10 min).
3.   Add 0.15 ml of 3.0 M DMSO.
4.   Mix thoroughly.
5.   Wait 15 min before transferring the sample to the seeding bath.

B. Addition of DMSO to mouse blastocysts

1.   Bring the 3.0 M DMSO to room temperature.
2.   Pipette the embryos into an ampoule containing 0.15 ml of M2 at room temperature.
3.   Add 0.05 ml of 3.0 M DMSO.
4.   Mix thoroughly.
5.   Wait 5 min.
6.   Repeat steps 3, 4 and 5.
7.   Repeat steps 3 and 4.
8.   Cool the sample to 0°C.
9.   Wait 15 min before transferring the sample to the seeding bath.

the time of exposure before cooling to sub-zero temperatures;
(iii)   the rate and temperature at which the protectant is removed after thawing (Section 3.6).

3.2.2 *Concentration of DMSO*

The survival of mouse embryos is high after freezing in medium containing DMSO at concentrations between 1.0 and 2.0 M. Less than 1.0 M DMSO affords insufficient protection against cooling damage, whilst at concentrations greater than 2.0 M the DMSO itself may be toxic. Practically, a concentration of 1.5 M is used.

3.2.3 *Addition of DMSO*

A considerable heat of solution is generated when DMSO is mixed with the freezing medium. Consequently, undiluted DMSO is not added directly to the sample containing embryos. Instead a 3.0 M solution of DMSO is prepared (*Table 3*) and an aliquot is added to the sample to give a final concentration of 1.5 M DMSO. The addition is made in a single step at 0°C to samples of unfertilized eggs or embryos up to the morula stage (*Table 3A*). The survival of blastocysts is improved if the concentration of DMSO is increased gradually at room temperature (*Table 3B*).

The effect on embryonic survival of permeation by the cryoprotectant is not understood fully; but conditions that increase the penetration of glycerol before freezing improve the survival of mouse ova (9). The amount of additive entering the embryo is influenced by the permeability of the embryo and the time and temperature of exposure before

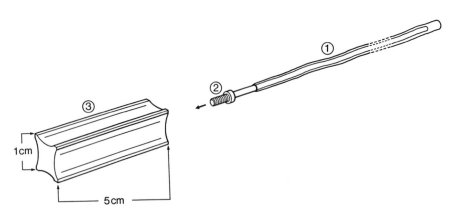

Figure 1. Ampoule holder made from a length (25 cm) of plastic coated copper wire (**1**) brazed onto a copper screw (**2**) which screws into a moulded Perspex bar (**3**). Up to four ampoules are attached to the Perspex bar with a rubber band.

freezing. The time and temperature of exposure appear less critical for DMSO than for glycerol probably because DMSO is more freely permeable. Optimal levels of survival of the 8-cell stage of the mouse are achieved after exposure at 0°C for only 5 min. Prolonged exposure at 0°C (up to 30 min) does not reduce survival and is more practical for routine embryo freezing. We hold samples on ice for 15 min after the addition of DMSO. During this time:

(i)   the ampoule is removed briefly from the ice bath, held with long forceps and heat sealed with a twin jet ampoule sealer;
(ii)   using rubber bands, the samples are attached to holders (*Figure 1*) in preparation for transfer to the seeding bath;
(iii)   the temperatures of the seeding and cooling baths are adjusted.

3.3 Induction of ice in the suspending medium: 'seeding'

During cooling at rates compatible with embryo survival the suspending medium rarely freezes spontaneously at its freezing point (about −3.5°C) and the sample can supercool to temperatures as low as −21°C. As ice nucleation occurs and the latent heat of fusion is released, the temperature of the sample rises to the melting point then falls rapidly until thermal equilibrium with the cooling bath is restored. The precise rate of cooling depends on the difference in temperature between sample and cooling bath and may be of the order of 10°C/min. The dehydration necessary for cell survival can occur only after ice is formed in the suspending medium. If the rate of cooling after nucleation is too rapid, then the amount of water left in the cell when intra-cellular freezing occurs is incompatible with survival.

To avoid supercooling, the sample is induced to freeze (seeded) at a temperature just below the freezing point of the suspending medium.

(i)   After equilibration with DMSO, transfer the sealed ampoule from the ice bath at 0°C to a constant temperature bath (seeding bath; Section 3.3.1) at −5 to −6°C.
(ii)   Wait 2 min.

(iii)   Seed the sample by touching the ampoule just above the meniscus with the tips of fine forceps (Watchmakers' No. 5 are ideal) previously cooled in LN_2. As soon as frost appears on the glass remove the forceps. NB Take care not to cool the sample excessively.

(iv)   Return the ampoule to the seeding bath to allow the latent heat of fusion to disperse.

(v)   After 5 min, check that ice has formed in the sample and transfer the ampoule to a cooling bath at the same temperature as the seeding bath.

Alternative methods of seeding are described elsewhere (9,10).

If the sample does not seed, try again. If you are still unsuccessful the most likely reasons are that the temperature of the bath is too high or the concentration of DMSO used is too high. Lower the temperature of the bath by 1°C and try to seed again.

3.3.1 *Seeding bath*

Any bath that can be held readily at −5 or −6°C for longer than 15 min may be used for seeding. We use constantly stirred industrial methylated spirit or ethanol cooled in a commercially available thermoelectric bath (Virtis minifreezer: Technation, Middlesex, UK). A simple alternative is a mixture of common salt and ice. Other suitable equipment is described by Leibo and Mazur (9).

3.4 **Cooling**

Embryonic survival is dependent on the removal of some intra-cellular water before temperatures are reached at which the blastomeres freeze internally. When dehydration is achieved during continuous cooling the rate of cooling is critical. After the induction of ice in the suspending medium, water leaves the embryo in response to the increasing concentration of extra-cellular solutes. As the temperature is lowered, progressively more water is lost and the blastomeres shrink. The rate of water loss is determined by the permeability of the cell membrane to water and by the ratio of surface area:volume of the cell. Mouse embryos seem to respond as a single unit despite the change in the surface area:volume ratio of the individual blastomeres at each cleavage division. If cooling is too rapid, the embryo has insufficient time to respond before the temperature is reached where intra-cellular ice nucleation can occur; and eventually the cell contents freeze with the formation of lethal amounts of intra-cellular ice. Conversely, if cooling is too slow, the cells are exposed for an excessive length of time to high concentrations of extra-cellular solutes, pH changes and reduced cell volume. Ideally embryos are cooled at a rate just slow enough for them to remain in osmotic equilibrium with the extra-cellular solution.

3.4.1 *Rate of cooling*

Compared to other cell types, mammalian embryos require relatively slow rates of cooling (0.2−2.0°C/min) for optimal survival. In practice, mouse embryos in 1.5 M DMSO are cooled at rates of 0.3−0.5°C/min. The rate of cooling is measured over the range −10 to −60°C. Slow cooling is terminated at −40°C or at temperatures below −70°C by immersing the sample in LN_2. The survival of 8-cell embryos in 1.5 M DMSO is poor if slow cooling is terminated at temperatures higher than −35°C or between −45

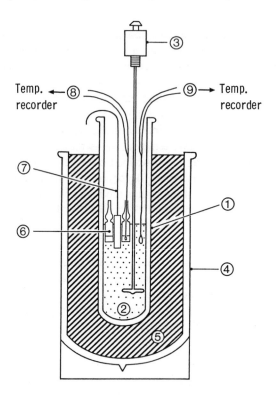

Figure 2. Cross-section of a cooling bath: (**1**) partially evacuated, unsilvered dewar; (**2**) coolant; (**3**) variable speed stirrer; (**4**) evacuated, silvered Dewar; (**5**) LN$_2$; (**6**) ampoule containing embryos; (**7**) pliable ampoule holder (see *Figure 1*); (**8**) and (**9**) copper-constantan thermocouples monitoring the temperature of a blank sample and the coolant.

and −70°C. Mouse blastocysts will tolerate the termination of slow cooling over a wider range of temperatures (13), but are usually cooled slowly to −40 or −70°C. Oocytes are cooled slowly to below −70°C before storage in LN$_2$.

3.4.2 *Cooling bath*

A cooling bath similar to that used in our laboratory (*Figure 2*) is described in detail by Leibo and Mazur (9). Briefly, it consists of a partially evacuated (10−100 mm of mercury) unsilvered Dewar (capacity 950 ml; i.d. 70 mm; internal height 300 mm; e.g. Dilvac: Day-Impex Ltd, Colchester, UK) containing 500 ml of the coolant (industrial methylated spirit or ethanol). The coolant is stirred constantly with a variable speed stirrer. For cooling, the dewar is immersed in an outer dewar (capacity ∼4 l) containing LN$_2$. The rate of cooling depends on the volume of coolant, the degree of evacuation of the inner dewar and the depth to which it is immersed in LN$_2$. Ideally the temperature is monitored using copper-constantan thermocouples and a low temperature recorder (e.g. Speedomax 2500 multipoint: Leeds and Northrup Ltd, Birmingham UK). One thermocouple records the temperature of the coolant whilst a second is located in a blank sample (remember to seed the blank sample). The temperature of the coolant and the sample usually differ by approximately 1°C. If no recorder is

264

available the cooling characteristics of a dewar may be determined in advance using a low temperature thermometer and a stopwatch. Keeping a constant volume of coolant and immersing the inner dewar to a given depth in LN_2 yields a highly reproducible rate of cooling. If the depth of immersion in LN_2 is constant throughout then the cooling curve is slightly curvilinear. To maintain a constant rate of cooling between -10 and $-60°C$ the inner dewar may be gradually lowered into the LN_2 or the outer dewar may be raised on an adjustable laboratory jack. A comparison of survival after cooling at a constant rate or at a gradually decreasing rate has not been made. We maintain a constant rate; but Leibo and Mazur (9) consider that a gradual reduction in rate may be desirable to allow sufficient time for osmotic equilibrium to be reached at lower temperatures.

3.4.3 *Cooling the seeded sample*

After seeding the sample carry out the following procedure.

(i)   Check that the temperature of the coolant in the cooling bath is within $0.5°C$ of the seeding bath temperature.

(ii)   When the latent heat of fusion has dissipated (~ 5 min; observe the temperature of the blank sample), suspend the ampoule in the cooling bath so that the contents are submerged.

(iii)   Adjust the outer dewar containing LN_2 to give a cooling rate of $0.3-0.5°C/min$.

(iv)   Terminate slow cooling when the temperature of the sample reaches $-40°C$ or continue cooling to below $-70°C$. At the chosen temperature, remove the ampoule from the cooling bath and immerse it (still attached to the holder) in LN_2.

(v)   Wait 15 min before further handling.

(vi)   Cool the aluminium ampoule storage cane in LN_2.

(vii)   Keeping the ampoule immersed in LN_2, cut the rubber band attaching it to the holder.

(viii) Using long forceps with the tips cooled previously in LN_2, load the ampoule onto the storage cane.

(ix)   Carry the canes submerged in LN_2 to the LN_2 refrigerator.

3.5 **Warming**

The temperature at which slow cooling is terminated determines the rate of warming for optimal survival of embryos. When slow cooling is terminated at relatively high sub-zero temperatures (around $-40°C$), the embryos contain some intra-cellular water which vitrifies during rapid cooling to $-196°C$ (22). To avoid devitrification and the growth of lethal amounts of intra-cellular ice, rapid warming is essential. Conversely, embryos cooled slowly to temperatures below $-70°C$, survive slow but not rapid warming. For most mammalian cells, optimal survival is achieved after rapid warming irrespective of the cooling procedure. It is unclear why these very dehydrated embryos demand slow warming. Perhaps slow rehydration is necessary to allow the gradual re-assembly of sub-cellular structures.

In practice, for mouse embryos in 1.5 M DMSO cooled slowly ($0.3-0.5°C/min$) before rapid cooling to $-196°C$:

(i)   When slow cooling is terminated at $-40°C$ warm the sample rapidly at about $500°C/min$ by agitating the sample in water at $40°C$.

(ii)   When slow cooling is continued to below $-70°C$ warm the sample slowly at a rate between 4 and $20°C/min$. Take the sample from LN_2 and stand it on the bench where the rate of warming is approximately $20°C/min$. Alternatively, cool a boiling tube (i.d. 35 mm; external height 200 mm) in LN_2 then suspend it in air at room temperature and transfer the sample from LN_2 to the tube. This latter method yields a warming rate of about $10°C/min$.

Irrespective of the rate of warming, after the last of the ice disappears hold the sample at room temperature for 5 min before removing the DMSO.

3.6 Removal of the DMSO after thawing

The presence of a penetrating cryoprotectant during cooling necessitates the careful removal of the additive after thawing to avoid damage by osmotic shock. The response of embryos to a return to physiological conditions (i.e. from medium with an osmolality of 1.5 osmol to culture medium with an osmolality of about 0.3 osmol) depends on:

(i)   the amount of cryoprotectant that penetrated the blastomeres before cooling;

(ii)   their relative permeability to water and the protective compound;

(iii)   temperature.

Water is usually more freely permeable than the protectant with the result that when the suspending medium is diluted, water enters more rapidly than the cryoprotectant leaves and the blastomeres swell. Traditionally, to control the degree of swelling, the cryoprotectant was removed from mouse embryos in a gradual stepwise manner at $0°C$ or $20°C$ (1, 2). More recently it was demonstrated that mouse embryos tolerate a rapid two-step dilution from 1.5 M DMSO at room temperature (13).

After thawing DMSO is removed from oocytes and embryos alike as follows.

(i)   After the ice has melted in the sample, score the neck of the ampoule with a glass cutter and break off the top.

(ii)   Wait 5 min to allow the sample to reach room temperature.

(iii)   Add 1 ml of M2 (at room temperature) to the 0.3 ml sample and mix thoroughly.

(iv)   After 1 min, using a 1 ml Falcon plastic pipette transfer the contents of the ampoule to an embryological watchglass.

(v)   Rinse the ampoule twice using 1 ml of M2 per rinse and pipette the washings into the watchglass.

(vi)   Pick up the embryos in a finely drawn Pasteur pipette and wash through two changes of M2 (2 ml per wash).

After removal of the DMSO, oocytes are fertilized *in vitro* (Section 6.2) before transfer to pseudopregnant foster mothers (Section 6.4). Fertilized eggs and embryos may be transferred immediately to recipient but survival is improved by a period in culture (Section 6.3) before transfer.

3.7 Safety

It is vital to read a standard safety manual that explains the proper handling of liquid gases. Then obey the rules!

When freezing embryos be especially aware of the following points.

(i)   LN$_2$ will boil and splash when the outer dewar of the cooling equipment is filled with LN$_2$ for cooling. Protect your eyes and skin.

(ii)   The aluminium storage canes will stick to the skin after cooling to $-196°C$. Wear a loose-fitting glove.

(iii) Badly sealed ampoules will contain LN$_2$ after storage and may explode. Take no chances. ALWAYS wear a safety visor or at least goggles when removing ampoules from LN$_2$ and during the initial stages of thawing. Protect others in the vicinity.

4. VITRIFICATION

The vitrification approach seeks to simplify embryo cryopreservation procedures by taking advantage of the ability of highly concentrated solutions of cryoprotectants to avoid crystallization when cooled to low temperatures (23). Instead, these solutions supercool and become so viscous that they form a glassy solid by a physical process called vitrification. In the vitrification approach, embryos are equilibrated in a mixture of cryoprotectants (the vitrification solution) and, when the cytoplasm has dehydrated

Table 4. Composition of the stock vitrification solution (VS1) and the Hepes-buffered saline (HB1) used to prepare the various dilutions of VS1.

Component	VS1[a]		HB1	
	g/l	mM	g/l	mM
Buffered saline				
NaCl	8.000	136.90	8.000	136.90
KCl	0.200	2.68	0.200	2.68
MgCl$_2$.6H$_2$0	0.050	0.25	0.100	0.50
KH$_2$PO$_4$	0.120	0.88	0.136	1.00
CaCl$_2$.2H$_2$0	0.010	0.07	0.120	0.90
Na pyruvate	–	–	0.036	0.33
Hepes[b]	4.766	20.00	4.766	20.00
Glucose	1.000	5.56	1.000	5.56
BSA[c]	0.750	–	3.000	–
Penicillin G (potassium salt)	–	–	0.060	–
Phenol red	0.010	–	0.010	–
Cryoprotectants				
DMSO	20.502	2.62	–	–
Acetamide	154.970	2.62	–	–
Propylene glycol	100.00	1.30	–	–
Polyethylene glycol (mol. wt 8000)	60.00	6.0%w/v	–	–

Vitrification solutions pH = 7.8−8.0.
HB1 pH = 7.2.
[a]Store in glass.
[b]Calbiochem-Boehring: Ultrol.
[c]Crystalline: Miles-Serovac.

Low temperature preservation of mouse oocytes and embryos

Table 5. Equilibration of 8-cell mouse embryos in the vitrification solution.

A. Solutions

1.　Prepare three dilutions of the stock VS1 using HB1 (*Table 4*):
　　(a) 25% VS1 : (1 part VS1 : 3 parts HB1)
　　(b) 50% VS1 : (1 part VS1 : 1 part HB1)
　　(c) 90% VS1 : (9 parts VS1 : 1 part HB1)
2.　Cool the 50% VS1 and 90% VS1 to 4°C.

B. Preparation of the straw

1.　Aspirate a 1 cm column of 90% VS1 into a 0.25 cc plastic insemination straw until the cotton/PVA plug is wetted.
2.　Using a 20 gauge, 3.5″ long, hypodermic needle and disposable *glass* syringe, introduce a 2 cm column (~40 μl) of 90% VS1 into the centre of the straw.
3.　Cool the straw to 4°C.

C. Stepwise equilibration procedure

1.　Transfer the embryos with a mouth operated micropipette into 3 ml of 25% VS1 and mix thoroughly by gently swirling the dish.
2.　Wait 15 min.
3.　Place the embryo suspension in a cold room (2−4°C). Wait 10 min and then perform the remaining steps of the procedure in the cold room.
4.　Transfer the embryos in a small volume of 25% VS1 into 3 ml of cold 50% VS1. Mix thoroughly by swirling the suspension with a micropipette. (If the embryos float, rinse the pipette with 50% VS1 and pipette the embryos to the bottom of the dish.)
5.　After 10 min, transfer the embryos into 3 ml of cold 90% VS1 and mix thoroughly.
6.　Approximately 5 min after transfer into 90% VS1, load the shrunken embryos into the 2 cm column of 90% VS1 in the straw.
7.　Heat seal the straw at both ends (use a plastic bag sealer). NB Do not warm the embryo suspension with the fingers.

---

sufficiently, the embryo suspension (contained in a plastic insemination straw) is vitrified by immersion in LN_2. The rate of cooling is relatively unimportant provided it is rapid enough to prevent crystallization. Warming must be rapid enough to prevent devitrification (crystallization) during the temperature interval of −100 to −25°C. Thus, vitrification provides a rapid and simple method for embryo cryopreservation by eliminating the need for seeding the embryo suspension and the controlled period of slow cooling. Perhaps the two most important factors that influence the success of vitrification are:

(i)　the composition of the vitrification solution;
(ii)　the procedures used to equilibrate embryos in the solution.

4.1 Vitrification solution

The vitrification solution must be sufficiently concentrated to avoid crystallization when cooled at practicable rates and yet must not produce chemical toxicity or osmotic injury during equilibration or dilution. The solution (VS1: *Table 4*) developed by Rall and Fahy (23) is a mixture of the following cryoprotectants dissolved in modified Dulbecco's saline.

(i)　Equimolar concentrations of DMSO and acetamide to take advantage of the ability of certain amides to reduce the chemical toxicity of concentrated DMSO solutions (24,25).

268

(ii)   Propylene glycol because of its glass forming tendency.
(iii) The non-penetrating polymer, polyethylene glycol, to increase the ability of the
       solution to vitrify (26).

The development *in vitro* and *in vivo* of 8-cell mouse embryos is reduced after exposure to the stock VS1 (6). For vitrification the embryos are suspended in a 90% VS1
solution (see *Table 5*) because this concentration of cryoprotectants appears to be less
toxic.

4.2 Equilibrating embryos in the vitrification solution

The equilibration procedure is designed to yield a concentrated, dehydrated cytoplasm
that is capable of vitrifying on subsequent cooling. The procedure must not lead to
chemical toxicity or excessive osmotic stresses during either equilibration or subsequent dilution (23). Whether embryos are injured or not depends partly on the extent
to which the cryoprotectants in the vitrification solution permeate the cytoplasm.
Although some permeation appears necessary for successful preservation, complete
permeation increases the likelihood of injury due to chemical toxicity and osmotic swelling during dilution. In practice, embryos are equilibrated at room temperature in 25%
VS1 and then exposed in two steps to increasing concentrations of the vitrification solution at 4°C. The procedure is given in *Table 5*.

4.3 Cooling

Ten minutes after transferring the embryos into 90% VS1, vitrify the embryo suspension by cooling the straw in LN_2. The only requirement is that the rate of cooling is
rapid enough to prevent crystallization ($>20°C/min$). Three convenient cooling methods
are:

(i)   direct transfer into LN_2 ($\sim2500°C/min$);
(ii)   transfer into cold N_2 vapour ($-150°C$), e.g. in the neck of a LN_2 refrigerator
       ($\sim200°C/min$);
(iii) transfer to a gradually cooling isopentane bath ($\sim20°C/min$) to approximately
       $-120°C$, followed by direct transfer into LN_2.

Rapid cooling in LN_2 may in some cases lead to fracture of the glassy suspension
and result in damage to the zonae and blastomeres of some embryos.

4.4 Storage of vitrified samples

Vitrified embryo suspensions must be stored at temperatures below the glass transition
temperature of the solution ($-120°C$) to prevent injury resulting from devitrification.
Long-term storage is best accomplished by immersing the straws in LN_2 (see Section 5).

4.5 Warming

After storage, the vitrified suspension must be warmed rapidly ($>200°C/min$) to prevent the deleterious effects of devitrification (8, 23). One convenient method is to transfer
the sample into cold (4°C) water. The rate of warming obtained is approximately
$2000°C/min$. Do not use a warm water bath because even a transient increase in the

Table 6. Diluting embryos out of the vitrification solution.

Warm the embryos and carry out the initial dilution at 4°C. Immediately after warming, dry the straw with a tissue and cut off the heat sealed ends. Remove the cryoprotectants in one of the following ways.

A. Stepwise dilution procedure

1. Expel the contents of the straw into a small Petri dish containing 3 ml of cold (4°C) 50% VS1 solution and rinse the straw with 50% VS1. Mix contents of dish thoroughly and wait 10 min.
2. Transfer embryos with a micropipette into 3 ml of cold 25% VS1. Mix thoroughly and wait 10 min.
3. Remove the embryo suspension from the cold room and wait 10 min.
4. Transfer the embryos into 3 ml of 12.5% VS1 (a 1 to 8 dilution of stock VS1 with HB1; *Table 4*). Mix and wait 10 min.
5. Transfer the embryos into PB1 or M2 and wait 10 min before further handling.

B. Sucrose dilution procedure

1. Expel the contents of the straw into a small Petri dish containing 3 ml of cold (4°C) 1.08 M sucrose in PB1. Mix thoroughly (the embryos will float) and wait 5 min.
2. Remove the embryo suspension from the cold room and wait 10 min.
3. Transfer the shrunken embryos into PB1 or M2.

temperature of the suspension may result in chemical toxicity and further permeation of the cryoprotectants.

4.6 Dilution of the vitrification solution

Careful attention to the method used to dilute embryos from the vitrification solution is important because of the toxic nature of the concentrated solution and the presence of permeating cryoprotectants in the cytoplasm. It is important to ensure that the temperature of the embryo suspension does not rise above 4°C until the cryoprotectants are diluted to at least 25% of the stock VS1 concentrations. Rapid dilution by direct transfer of the embryos into isotonic saline must also be avoided to prevent injury due to osmotic swelling as water flows into the cells to restore osmotic equilibrium. Excessive osmotic stress is prevented by diluting the cryoprotectant in one of two ways.

(i)   Slowly using a stepwise procedure (23; *Table 6A*) in which the concentration of the vitrification solution is reduced in two steps at 4°C to 25% of the stock VS1 concentration. Then the remaining intra- and extra-cellular cryoprotectants are diluted slowly in several steps at room temperature.

(ii) Rapidly using a 'sucrose' procedure (8; *Table 6B*) based on the method described by Leibo and Mazur (9). The vitrification solution is replaced at 4°C with a saline solution containing 1.08 M sucrose. The sucrose does not penetrate the blastomeres and thus acts as an osmotic buffer to prevent excessive swelling of the embryos as the cryoprotectants leave the cytoplasm.

4.7 Safety

(i)   Be aware of the manufacturer's recommendations for the handling of acetamide.
(ii) Obey the rules for handling liquid nitrogen (see Section 3.7).

5. STORAGE OF SAMPLES

Frozen and vitrified embryos are stored at $-196\,°C$ in any of the commercially available LN_2 refrigerators. Provision should be made to ensure that the level of LN_2 is maintained. Ideally, an alarm system is fitted to monitor either a rise in temperature or a loss in weight in the event of the LN_2 reaching a dangerously low level. As an added precaution, we keep samples of embryos from each mouse stock in more than one refrigerator.

5.1 **Length of storage**

At $-196\,°C$ only photophysical reactions, e.g. ionization due to radiation, occur. In the absence of normal enzymatic repair processes, it is possible that lethal amounts of damage could accumulate in embryos stored for long periods. To date there is no evidence to support this theory. Normal fertile mice have been recovered from 8-cell embryos stored in LN_2 for up to 13 years. Furthermore, no genetic damage was induced when prolonged storage was simulated by exposing frozen embryos to the equivalent of approximately 2000 years of normal background radiation (27).

5.2 **Records**

It is essential to keep full records for every stock preserved. Remember that someone else may recover the embryos. The minimum amount of information required is:

Date of preservation
Details of the stock preserved
Ampoule/straw number
Number of oocytes/embryos per ampoule/straw
Developmental stage
Method of preservation, i.e. freezing or vitrification
Location in LN_2 refrigerator
Appropriate method of thawing and dilution procedure.

For frozen embryos the following should also be recorded: freezing medium, sample volume, final concentration of DMSO, cooling rate, final temperature before immersion in LN_2.

For vitrified embryos include details of the vitrification solution.

5.3 **Nomenclature**

There is no evidence that mice derived from frozen embryos are altered either genotypically or phenotypically by the processes of freezing and thawing. It remains to be demonstrated that embryos are genetically unchanged after vitrification. However, phenotype may be modified by the intra-uterine environment, so embryos recovered in the parental strain may differ from those recovered in another strain. It is important to know the history of experimental animals and all relevant information should be included in the stock description. International rules govern the designation of inbred strains of mice recovered from frozen embryos (28). So far the question of the nomenclature of stocks preserved by vitrification has not been considered.

6. VIABILITY ASSAYS AND RECOVERY OF PRESERVED STOCKS

The number of stored eggs or embryos representing each mouse stock will vary according to the requirements of the individual laboratory. In most cases it is advisable to
preserve sufficient material to permit multiple recoveries before it becomes necessary
to replenish the bank. The precise number of eggs/embryos to be stored can be determined only after assessing the viability of each batch preserved. The procedures for
assessing viability are identical to those for the recovery of live mice except that the
foster mother may be killed in late gestation (Day 15−16) and the number of normal
fetuses counted. The possibility of post-natal loss should not be overlooked when
estimating the number of embryos needed to recover a stock.

6.1 Oocytes

After removal of the DMSO, thawed oocytes or cumulus masses may be transferred
to the oviducts of females mated 2−4 h previously with fertile males (3). Survival is
higher if the oocytes are fertilized *in vitro* and transferred at the 2-cell stage to the
oviducts of Day-1 pseudopregnant recipients (Section 6.4.5).

6.2 Fertilization *in vitro*

The procedures we find successful for the fertilization *in vitro* of thawed oocytes (21)
are summarized in *Table 7* and details of the steps are given below.

6.2.1 *Media*

The sperm is collected and dispersed in a modified Tyrode's solution (T6: 15: *Table*

Table 7. Summary of procedures for fertilization *in vitro* (IVF) of thawed mouse oocytes.

See Section 6.2 of text for details.
Carry out all incubations at 37°C in a humidified atmosphere of 5% CO_2 in air.

A. <u>On the day before IVF</u>

1.   Pipette 1 ml drops of T6 (*Table 8*) into two Falcon plastic petri dishes or two embryological wat-
      chglasses. Overlay with paraffin oil. Incubate overnight. These will be used for the initial disper-
      sion of the sperm (See B3 below).
2.   Add BSA (Sigma Fraction V; 15 mg/ml) to T6 (T6 + BSA), filter sterilize and incubate overnight.

B. <u>On the day of IVF</u>

1.   Correct the pH of the T6 + BSA medium to 7.5−7.6, prepare the fertilization drops (Section 6.2.2)
      and incubate.
2.   Collect the sperm in the T6 drops prepared in A1 above and incubate (Section 6.2.3).
3.   After 25 min, pipette an aliquot of well dispersed, motile sperm into each fertilization drop to give
      a final concentration of $1-2 \times 10^6$ sperm/ml. Incubate.
4.   Thaw the oocytes, dilute the DMSO (Section 3.6) and wash in M2. Hold in air on a hotblock at 37°C.
5.   2 h after collection of the sperm, wash the oocytes in T6 + BSA and transfer to the fertilization
      drops (Section 6.2.4). Incubate.
6.   After about 4 h, collect the oocytes, wash in medium M16 + BSA (see Chapter 2, Section 5.2)
      and incubate overnight in drops of medium M16 + BSA under oil (Section 6.2.5).
7.   After about 20 h incubation, transfer 2-cell stage embryos in M2 to Day 1 pseudopregnant reci-
      pients (see Section 6.4.5).

Table 8. Composition of medium T6 used for fertilization *in vitro*.

Component	g/l	mM
NaCl	5.719	97.84
KCl	0.106	1.42
$MgCl_2.6H_2O$	0.096	0.47
$Na_2HPO.12H_2O$	0.129	0.36
$CaCl_2.2H_2O$	0.262	1.78
$NaHCO_3$	2.101	25.00
Na lactate[a]	2.791	24.90
Na pyruvate[b]	0.052	0.47
Glucose	1.000	5.56
Penicillin G	0.060	–
(potassium salt)		
Streptomycin sulphate	0.050	–
Phenol red	0.010	–

T6 is a modified Tyrode's solution
[a]Sigma grade DL-V, 60% syrup.
[b]Calbiochem-Boehring.

Table 9. Preparation of Sørensen's phosphate buffer for pH colour standards.

1.	Prepare *Stock A* (1/15 M KH_2PO_4): 9.08 g KH_2PO_4, 10 mg phenol red, double glass distilled H_2O to 1 litre.
2.	Prepare *Stock B* (1/15 M Na_2HPO_4): 11.8 g $Na_2HPO_4.2H_2O$, 10 mg phenol red, double glass distilled H_2O to 1 litre.
3.	Prepare solutions of known pH by mixing aliquots of stocks A and B.

pH (at 18°C)	Stock A (ml)	Stock B (ml)
6.6	62.7	37.3
6.8	50.8	49.2
7.0	39.2	60.8
7.2	28.5	71.5
7.4	19.6	80.4
7.6	13.2	86.8

4.	Check pH using a meter and if necessary make minor adjustments by adding aliquots of the appropriate stock (A to lower pH, B to raise pH).
5.	Sterilize aliquots of each solution by Millipore filtration into Falcon test tubes.
6.	Store at 4°C.
7.	Solutions keep uniform pH for ~6 months.

8). For sperm capacitation and fertilization the medium is supplemented with 15 mg/ml Sigma Fraction V BSA (T6+BSA).

6.2.2 *Preparation of the fertilization drops*

For fertilization, cumulus free oocytes or intact cumulus masses are incubated with fully capacitated sperm in drops of T6 + BSA overlaid with paraffin oil. Prepare the fertilization drops as follows.

(i)   Incubate the T6+ BSA overnight at 37°C in 5% CO_2 in air.
(ii)   Next day, using 0.1 M NaOH bring the pH of the medium to 7.5−7.6. A precise

measure of pH is not necessary (and difficult to obtain from a rapidly cooling sample). An adequate colour reference is provided by Sorensen's phosphate buffer with phenol red as indicator (*Table 9*).

(iii)   Pipette 100 μl of medium on to a 60 mm Falcon plastic petri dish.

(iv)   Surround but do not cover the drop with warm, pre-gassed, paraffin oil.

(v)   Add 300 μl of T6 + BSA to the initial drop.

(vi)   Overlay the drop with paraffin oil.

(vii)   Incubate at 37°C in 5% CO_2 in air whilst preparing the sperm suspension.

6.2.3 *Source and collection of sperm*

The fertilizing capacity of sperm is strain dependent and whenever appropriate C57BL/6 × CBA/Ca F1 hybrid males are preferred as sperm donors. Fully mature, proven males that have been separated from females for about two weeks are used and two sperm suspensions are prepared. Collect and disperse the sperm as detailed below. NB Work quickly and use a 37°C hot-block to maintain the temperature of the medium.

(i)   Dissect one cauda epididymis and one vas deferens from each of two donors.

(ii)   Blot the dissected tissues on a laboratory wipe to remove blood.

(iii)   Place the tissues in 1 ml of *warm* T6 under oil.

(iv)   Dissect the remaining caudae + vasa deferentia into a second aliquot of T6.

(v)   Under a dissecting microscope, gently ease the sperm from the vasa deferentia using two pairs of Watchmakers' forceps. Puncture the caudae with the tips of the forceps and express the sperm. Remove the tissue.

(vi)   Incubate the sperm suspension at 37°C in a humidified atmosphere of 5% CO_2 in air for 25 min.

(vii)   Select the better dispersed more motile sperm suspension and using a haemocytometer estimate the concentration of sperm.

(viii)   Dilute the sperm suspension into the prepared fertilization drop (Section 6.2.2) to give a final concentration of $1-2 \times 10^6$/ml for fertilization.

6.2.4 *Incubation of thawed oocytes with capacitated sperm*

The thawed oocytes are incubated for 4 h with capacitated sperm. Under the conditions described here mouse sperm are fully capacitated approximately 2 h after collection. We adopt the following procedure.

(i)   After removal of the DMSO (see Section 3.6) hold the thawed oocytes in M2 at 37°C in air.

(ii)   About 2 h after collection of the sperm, wash the oocytes in T6 + BSA to remove all traces of Hepes buffer.

(iii)   Transfer the oocytes in a minimum volume of medium to the fertilization drop ($\sim$ 100 oocytes per drop).

(iv)   Incubate for 4 h at 37°C in an atmosphere of 5% CO_2 in air.

6.2.5 *Culture of fertilized oocytes and embryo transfer*

(i)   After 4 h incubation, pick up the oocytes from the fertilization drop and wash through two changes of medium M16 + BSA (Chapter 2, Section 5.2) to remove excess sperm.

(ii)   Culture in drops of medium M16 + BSA under oil.

(iii) On the next day, transfer 2-cell embryos to the oviducts of Day-1 pseudopregnant recipients (Section 6.4.5).

6.3 Embryos

The recovery of live mice is the ultimate assay of the viability of embryos preserved in LN$_2$. However, the morphological appearance of the thawed embryo and its ability to resume cleavage *in vitro* correlate well with survival after embryo transfer. By making these intermediate assessments the time and cost of transferring non-viable embryos is spared. Furthermore, the resumption of embryonic development seems to be delayed after thawing. A period of culture *in vitro*, to allow metabolic recovery before embryo transfer, increases the proportion of normal fetuses recovered. Alternatively, the developmental delay can be overcome by transferring 8-cell embryos immediately after thawing to Day-1 pseudopregnant recipients (16) (Day 1 is the day of detection of the copulation plug). For blastocysts, however, asynchronous transfer without a period of culture after thawing does not overcome developmental delay (29,30).

6.3.1 *1-cell to 8-cell stage*

After removal of the DMSO examine the embryos under a dissecting microscope and classify them according to the following criteria.

(i)   Normal, i.e. similar to unfrozen embryos of the same developmental stage.
(ii)   Blastomeres swollen.
(iii) One or more blastomeres lysed.
(iv)   Degenerate.
(v)   Zona pellucida lost or broken.

The likelihood that normal embryos will resume development in culture is high. Swelling of the blastomeres suggests that some DMSO remains in the cells or that transport across the membrane has been disturbed. Swollen embryos may recover but some will lyse. The potential for normal development of embryos with damaged blastomeres depends on the stage at which the embryo was frozen and on the number of blastomeres damaged. The loss of one blastomere from a 2-cell embryo is usually lethal and few damaged 4-cell embryos form normal blastocysts. After losing one or two blastomeres, an 8-cell embryo will form a blastocyst with a reduced number of cells in the inner cell mass and may develop normally after embryo transfer. Early cleavage stages (< 8-cell) develop poorly in culture without a zona pellucida and cannot be transferred to the oviduct until after compaction. Later stage embryos develop normally without a zona as do all stages retained within a cracked zona. However, the possibility of infection by viruses arises in the absence of an intact zona (31).

6.3.2 *Blastocysts*

Immediately after thawing blastocysts may be collapsed and morphological assessment is difficult. Totally degenerate embryos can be eliminated.

6.3.3 *Development in vitro*

Thawed embryos are washed in medium M16 + BSA (Chapter 2, Section 5.2) and cultured overnight in drops of the same medium under oil. Embryos that resume normal cleavage and blastocysts that re-expand are transferred to the oviducts of Day-1 (2-cell to blastocyst stages) or the uterine horns of Day-3 (morula and blastocyst stages) pseudopregnant recipients (Section 6.4).

6.4 **Embryo transfer**

For the completion of development and recovery of live mice, pre-implantation stage embryos must be transferred to the reproductive tract of pseudopregnant foster mothers. The transfer can be made surgically to the oviduct or uterus by a dorsal route or non-surgically through the cervix. The non-surgical procedure is described elsewhere (32,33). It has the disadvantage that there is no way of knowing in which uterine horn the embryos have been deposited. Also, it is a difficult technique to master. Here we describe the surgical methods of transfer to the oviduct (Section 6.4.5) or uterus (Section 6.4.6).

6.4.1 *Foster mothers*

For routine embryo transfer we prefer F1 hybrid (e.g. C57BL/6J × CBA/Ca or C57BL/6 × CBA/H) foster mothers because over 90% establish pseudopregnancy when mated with sterile males and become pregnant after embryo transfer. Outbred mice (e.g. MF1; Harlan-Olac Ltd, Bicester, UK) may be used but avoid inbred strains. The mice should be at least 12 weeks of age and weigh 20 g or more.

6.4.2 *Induction of pseudopregnancy*

Pseudopregnancy is induced by mating naturally cycling females with vasectomized males (Chapter 1, Section 5.3) or with naturally sterile males e.g. males carrying the translocation T145H (34). To obtain sufficient recipients on a specific day, select females in pro-oestrus for mating or make use of the Whitten effect (see Chapter 1, Section 5.2). Synchronization of females with exogenous gonadotrophins reduces the proportion of mated females that maintain pseudopregnancy.

6.4.3 *Number of embryos*

Between three and six embryos are transferred to each oviduct or uterine horn (i.e. a total of 6–12 embryos per recipient). When fewer embryos are transferred pregnancy may not be established and there is a reduced chance of survival for very small litters. The transfer of too many embryos results in crowding and retarded growth.

6.4.4 *Anaesthetic*

The recipients are anaesthetized a few minutes before surgery with a 1.2% solution of Avertin (Chapter 1, Section 5.5) injected intraperitoneally at a dose of 0.02 ml per g of body weight.

6.4.5 *Transfer to the oviduct*

All of the pre-implantation stages (pro-nucleate egg to expanded blastocyst) can be

Table 10. Preparation of a pseudopregnant recipient for embryo transfer to the oviduct or uterus.

1.	Anaesthetize the mouse (Section 6.4.4).
2.	Place in the prone position (head at 12 o'clock), swab the back with 70% ethanol and part the hair in the mid-line just posterior to the thorax.
3.	Using fine pointed scissors (10.5 cm long), make a 2 cm long mid-line incision.
4.	Using fine serrated forceps (10.5 cm long) deflect the skin to the left and identify the ovary through the body wall in the region of the lumbar fossa. Use the scissors to incise the abdominal muscles over this area.
5.	Bring the ovary, oviduct and part of the uterine horn to the exterior towards the mouse's head and place on a small gauze pad.
6.	Proceed as described in Section 6.4.5 (transfer to the oviduct) or in Section 6.5.6 (transfer to the uterus).

transferred to the ampulla of a recipient on Day 1 of pseudopregnancy (the day on which the copulation plug is found). Prepare the recipient as outlined in *Table 10* and proceed as follows:

(i)   Under a dissecting microscope (e.g. Wild M5), examine the ovarian capsule and locate the fimbrial end of the oviduct (usually lying free beneath the ovary and pointing towards the midline).

(ii) Using two pairs of Watchmakers' forceps, tear the ovarian capsule where it is least vascularized and adjust the position of the mouse to facilitate entry to the oviduct (head at about 10 o'clock for a right-handed operator).

(iii) Under a second microscope, introduce several 'marker' air bubbles into a finely drawn Pasteur pipette and pick up the embryos in a small volume of medium M2 (*Table 2*).

(iv) Using Watchmakers' forceps (held in the left hand) as a retractor, expose the fimbrial end of the oviduct, introduce the tip of the pipette and hold it in place with the forceps. Expel the eggs and air bubbles (observe the bubbles through the wall of the oviduct), gently withdraw the pipette and hold the fimbria closed with the forceps for several seconds.

(v)   Return the ovary and oviduct to the abdomen.

(vi) Deflect the skin incision to the right flank, expose the oviduct, turn the mouse round (head at 5 o'clock) and repeat the procedure.

(vii) After replacing the right ovary and uterine horn in the abdominal cavity, suture the skin with one or two Michel clips. It is unnecessary to suture the abdominal muscles.

6.4.6 *Transfer to the uterus*

Only 8-cell, morula and blastocyst stage embryos can be transferred to the uterine horns of Day 3 or Day 4 pseudopregnant recipients. (Day 1 is the day of detection of the copulation plug). Earlier pre-implantation stage embryos degenerate soon after transfer to the uterine environment.

For freshly collected embryos, synchronous or asynchronous transfer with the embryo more advanced than the endometrium are successful (*Table 11*). Embryos that are less advanced than the uterus (e.g. 8-cell embryos in a Day 4 uterus) rarely implant. After manipulation *in vitro*, e.g. freezing or culture, embryos may be delayed developmentally and survival is increased by asynchronous transfer (*Table 11*) to allow

Table 11. Synchronous and asynchronous embryo transfer to the uterus.

Type of transfer	Developmental stage of embryo	Stage of recipient pseudopregnancy[a]
Synchronous	8-cell	Day 3
	morula/ early blastocyst	Day 4
Asynchronous	morula/ early blastocyst	Day 3
	fully expanded/ hatched blastocyst	Day 3 or Day 4

[a]Day 1 is the day of detection of the copulation plug.

Table 12. The survival of mouse oocytes and embryos preserved by freezing or vitrification.

Method of preservation	Developmental stage	Strain	Survival[a]	Reference
Freezing				
1.5 M DMSO Slow cool	Oocyte	F1 hybrid[b]	16[c]	(21)
to −70°C Slow warm	8-cell	F1 hybrid[d]	42	(13)
		Random-bred MF1	50	(13)
		F2 hybrid[e]	20−30	(35)
		Inbred CBA/CaH	21	(36)
		XO	14	(36)
1.5 M DMSO Slow cool	8-cell	F1 hybrid[d]	36	(13)
to −40°C Rapid warm		Random-bred MF1	54	(13)
		F2 hybrid[e]	35	(27)
Vitrification				
90% VS1	8-cell	Random-bred MF1	18	(6)

[a]Expressed as the % of oocytes or embryos preserved that give rise to normal fetuses or liveborn young.
[b]Collected from F1 hybrid (C57BL/6JLac × CBA/CaLac) donors.
[c]After fertilization with CBA/CaH-T6 sperm.
[d]Collected from randombred MF1 females mated with C57BL males.
[e]Collected from F1 hybrid (C3H/HeH × 101H) females mated with males of the same hybrid cross.

time for recovery before the onset of implantation.

After preparing the recipient mouse as described in *Table 10*, proceed as follows:

(i) Using a low power dissecting microscope with a long working distance (e.g. Prior Stereomaster 240; BDH), check that the ovary has active (red) corpora lutea.

(ii) Under a second microscope, introduce several marker bubbles to a finely drawn Pasteur pipette and pick up the embryos.

(iii) Hold the pipette and a needle (25-gauge hypodermic or No 7 darning needle) in the right-hand and use fine blunt forceps placed across the oviduct to hold up the uterus (do not grip the uterine horn).

(iv)   With the needle, puncture the uterine horn on the anti-mesometrial side close to the utero-tubal junction. Withdraw the needle (keep your eye on the hole), introduce the tip of the pipette (avoid jabbing the wall of the uterus) and expel the embryos in a small volume of medium (observe the movement of the air bubbles but do not introduce air to the uterus).

(v)   Withdraw the pipette and check that the embryos have been expelled.

(vi)   Return the ovary and uterus to the abdominal cavity without handling the uterus.

(vii)   Deflect the skin incision to the right flank and repeat the procedure.

(viii)   Close the skin with one or two Michel clips (it is unnecessary to suture the abdominal muscles).

6.4.7 *Post-operative management*

After surgery keep the mice warm until they recover from the anaesthetic. They can be housed in small groups until just before parturition.

6.5 Overall survival

The level of survival that may be expected after preserving eggs and embryos using the methods described here is shown in *Table 12*.

7. THE PRESERVATION OF HUMAN OOCYTES AND EMBRYOS

The methods used for freezing human embryos are derived directly from those developed for mouse and cattle embryos. All of the pre-implantation stages of the human (pronucleate egg to blastocyst) have given rise to normal pregnancies after freeze preservation. The methods used are diverse and details can be found elsewhere (5, 37, 38, 39). Overall survival is low, being estimated as less than 5% (40) of the embryos frozen.

There is a single report of a pregnancy following the fertilization *in vitro* of frozen-thawed human oocytes (41).

The successful vitrification of human embryos has not been reported. Furthermore, we feel strongly that the potential toxicity of the vitrification solution described here makes it unsuitable for the preservation of human oocytes and embryos.

8. REFERENCES

1. Whittingham,D.G., Leibo,S.P. and Mazur,P. (1972) *Science*, **178**, 411.
2. Wilmut,I. (1972) *Life Sci.*, **II**, part 2, 1071.
3. Whittingham,D.G. (1977) *J. Reprod. Fert.*, **49**, 89.
4. Wilmut,I. (1986) In *Manipulation of Mammalian Development.* Gwatkin,R.B.L. (ed.) Plenum Press, New York.
5. Trounson,A. and Mohr,L. (1983) *Nature*, **305**, 707.
6. Rall,W.F., Wood,M.J., Kirby,C. and Whittingham,D.G. (1987) *J. Reprod. Fert.*, **80**, 499.
7. Scheffen,B., Van Der Zwalmen,P. and Massip,A. (1986) *Cryo-Letters*, **7**, 260.
8. Rall (1986) *Cryobiology*, **23**, 548.
9. Leibo,S.P. and Mazur,P. (1978) In *Methods In Mammalian Reproduction.* Daniel,J.C.,Jr (ed.), Academic Press, New York, p. 179.
10. Whittingham,D.G. (1980) In *Low Temperature Preservation in Medicine and Biology.* Ashwood-Smith,M.J. and Farrant,J. (eds), Pitman Medical, Tunbridge Wells, UK p. 65.
11. Renard,J-P., Bui-Xuan-Nguyen and Garnier,V. (1984) *J. Reprod. Fert.*, **71**, 573.
12. Takeda,T., Elsden,R.P. and Seidel,G.E.,Jr (1984) *Theriogenology*, **21**, 266.

13. Whittingham,D.G., Wood,M.J., Farrant,J., Lee,H. and Halsey,J.A. (1979) *J. Reprod. Fert.*, **56**, 11.
14. Whittingham,D.G. (1974) *Genetics*, **78**, 395.
15. Quinn,P., Barros,C. and Whittingham,D.G. (1982) *J. Reprod. Fert.*, **66**, 161.
16. Renard,J-P. and Babinet,C. (1984) *J. Exp. Zool.*, **230**, 443.
17. Rall,W.F. and Polge,C. (1984) *J. Reprod. Fert.*, **70**, 285.
18. Rall,W.F., Czlonkowska,M., Barton,S.C. and Polge,C. (1984) *J. Reprod. Fert.*, **70**, 293.
19. Miyamoto,H. and Ishibashi,T. (1983) *J. Exp. Zool.*, **226**, 123.
20. Fuller,B.J. and Bernard,A. (1984) *Cryo-Letters*, **5**, 307.
21. Glenister,P.H., Wood,M.J., Kirby,C. and Whittingham,D.G. (1986) *Gamete Res.*, **16**, 205.
22. Rall,W.F., Reid,D.S. and Polge,C. (1984) *Cryobiology*, **21**, 106.
23. Rall,W.F. and Fahy,G.M. (1985) *Nature*, **313**, 573.
24. Baxter,S.J. and Lathe,G.H. (1971) *Biochem. Pharmacol.*, **30**, 1079.
25. Fahy,G.M., MacFarlane,D.R., Angell,C.A. and Meryman,H.T. (1984) *Cryobiology*, **21**, 407.
26. MacKenzie,A.P. (1977) *Phil. Trans. R. Soc. Lond. Ser. B*, **278**, 167.
27. Glenister,P.H., Whittingham,D.G. and Lyon,M.F. (1984) *J. Reprod. Fert.*, **70**, 229.
28. Staats,J. (1976) *Cancer Res.*, **36**, 4333.
29. Whittingham,D.G. (1977) In *The Freezing of Mammalian Embryos*. Ciba Foundation Symposium 52, Elliott,K. and Whelan,J. (eds), Elsevier/North Holland, Amsterdam, p. 97.
30. Wood,M.J. (1986) In *Proceedings of the Fourth World Congress on In Vitro Fertilization*. Johnston,I. (ed.), Plenum Press, New York.
31. Carthew,P., Wood,M.J. and Kirby,C. (1985) *J. Reprod. Fert.*, **73**, 207.
32. Marsk,L. and Larson,K.S. (1974) *J. Reprod. Fert.*, **37**, 393.
33. Moler,T.L., Donahue,S.E. and Anderson,G.B. (1979) *Lab. Anim. Sci.*, **29**, 353.
34. Lyon,M.F. and Meredith,R. (1966) *Cytogen.*, **5**, 335.
35. Whittingham,D.G., Lyon,M.F. and Glenister,P.H. (1977) *Genet. Res.*, **29**, 171.
36. Whittingham,D.G., Lyon,M.F. and Glenister,P.H. (1977) *Genet. Res.*, **30**, 287.
37. Zeilmaker,G.H., Alberda,A.T., van Gent,I., Rijkmans,C.M.P.M. and Drogendijk,A.C. (1984) *Fertil. Steril.*, **42**, 293.
38. Cohen,J., Simons,R.F., Edwards,R.G., Fehilly,C.B. and Fishel,S.B. (1985) *J. In Vitro Fert. Embryo Transfer*, **2**, 59.
39. Lassalle,B., Testart,J. and Renard,J-P. (1985) *Fertil. Steril.*, **44**, 645.
40. Whittingham,D.G. (1985) In *In Vitro Fertilization and Donor Insemination*. Thompson,W., Joyce,D.N. and Newton,J.R. (eds.) Royal College of Obstetricians and Gynaecologists, London, p. 269.
41. Chen,C. (1986) *Lancet*, No. 8486, 884.

CHAPTER 14

Fertilization of human oocytes and culture of human pre-implantation embryos *in vitro*

PETER R.BRAUDE

1. INTRODUCTION

The technique of *in vitro* fertilization (IVF) and embryo replacement has enabled many previously infertile couples to have children of their own. However although pregnancies are possible, many improvements to the procedure remain to be made, the majority of these being likely to result from an improved understanding of the mechanisms of fertilization in humans, and the reasons for failure of development of the human embryo *in vitro* and after transfer.

This chapter will concentrate on the fundamental techniques for spermatozoal preparation, oocyte retrieval, fertilization of oocytes and the culture and development of human embryos *in vitro*. Although the procedures given are appropriate for running a clinical service, they are oriented towards research on human material *in vitro* which is required to improve our understanding of the therapeutic process (1).

2. SOURCES OF HUMAN OOCYTES AND EMBRYOS FOR RESEARCH

2.1 Spare oocytes and embryos

2.1.1 Therapeutic IVF

The chance of successfully establishing a pregnancy following fertilization *in vitro* increases significantly when more than one embryo is replaced into the uterus (2). In order to maximize the number of embryos available for replacement, regimes for superovulation have been developed to enable the recovery of multiple oocytes at laparoscopy. These regimes use human menopausal gonadotropins, a mixture of luteinizing hormone (LH) and follicle stimulating hormone (FSH) prepared from post menopausal urine (Pergonal, Serono), on their own (3), or in combination with clomiphene, an estrogen receptor antagonist (Clomid, Merrell; Serophene, Serono) (4). A pre-ovulation dose of 5000 – 10 000 i.u. of human chorionic gonadotropin (hCG; Profasi, Serono Labs; Pregnyl, Organon Labs), given 34 – 36 h prior to follicular aspiration, initiates germinal vesicle breakdown and the final maturation of the oocytes.

Since the incidence of multiple pregnancy also increases with increasing numbers of embryos replaced, a balance must be found between the improved chance of establishing a pregnancy and the increased risks associated with multiple gestations such as spontaneous abortion and prematurity (5). For this reason, it is usually recommended that the number of embryos replaced be limited to three. This inevitably means that where embryo cryopreservation facilities are not available, a number of 'spare' embryos will be generated which, with the approval of the couple, could be used for research (see Section 8, Ethical and Legal considerations).

2.1.2 *Gamete intra-fallopian transfer*

Gamete intra-fallopian transfer (GIFT) is a relatively new technique which has shown promise as a means of therapy for infertile couples where patency of the fallopian tubes can be demonstrated (i.e. in unexplained infertility, or where there are low numbers of sperm, or there is a defect in sperm transport) (6). The oocytes are retrieved from the follicles as for IVF, but instead of fertilizing the oocytes *in vitro*, they are replaced into the fallopian tubes with a small aliquot of washed spermatozoa using a fine catheter during the same surgical procedure. As with IVF the risks of multiple pregnancy suggest that the number of oocytes transferred to the fallopian tubes should be limited to between two and three oocytes. Thus spare oocytes may be available from this clinical procedure for research purposes.

2.2 **Donated oocytes**

Patients undergoing laparoscopic sterilization may agree to be superovulated for the purposes of donating oocytes for embryological research (7). As these women are volunteers, superovulation regimes must be safe and hence are milder than those used for therapeutic IVF and GIFT. The egg retrieval procedure should not prolong the duration of anaesthesia unduly beyond the time normally required for a sterilization procedure. Two regimes for superovulation of sterilization patients have been described (*Figure 1*). Either a fixed schedule of stimulation is used where the day of menses is altered by manipulating the length of the preceding menstrual cycle with progesterone (8), or a variable schedule is used where the day of administration of hCG is adjusted to the availability of operating lists (9). Oocytes capable of being fertilized *in vitro* are obtained using both procedures but the yield of oocytes is low especially when the patients have been on oral contraceptives for a prolonged period.

3. OOCYTE RETRIEVAL

Oocyte retrieval can be performed either under direct vision by laparoscopy or transvesically, transvaginally or transurethrally, using ultrasound visualization of the ovaries (10). Laparoscopic oocyte retrieval is used more frequently than ultrasound probably because of the familiarity of gynaecologists with this operative technique in their general practice. However, as general anaesthesia is not required for methods using ultrasound, patients can be treated as day-cases more easily and with fewer staff and facilities. Not surprisingly, the balance is changing in favour of ultrasound-guided aspiration.

3.1 **Laparoscopic oocyte retrieval**

The detailed procedure for laparoscopy and aspiration of follicles is beyond the scope of this book but a brief resumé is included so that embryologists involved in therapeutic

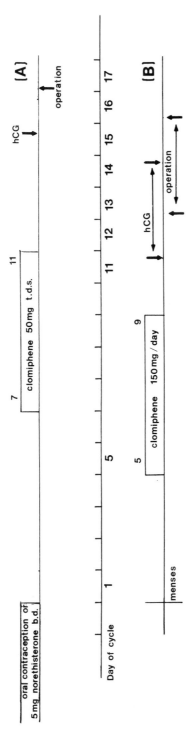

Figure 1. Two regimes described for superovulation of oocyte donors undergoing laparoscopic sterilization. (**A**) Fixed operation day – day of withdrawal bleed regulated by use of hormones. (**B**) Variable day of operation based on day of menses.

IVF will understand the surgical procedure and their role in the preparations for surgery. A more detailed account is given by Craft (11).

Under general anaesthesia a small skin incision is made below the umbilicus and a Veress needle with the gas tap in the open position is inserted into the peritoneal cavity. The gas mixture (usually 100% CO_2 see Section 3.4) is insufflated through the needle at 1 l/min. The intra-abdominal pressure is checked to see that it is not more than 5 mm Hg above the resting pressure for the needle, confirming that the needle is placed correctly within the peritoneal cavity. The abdomen is then filled with about 2.5 l of gas. The Veress needle is withdrawn, the skin incision enlarged to 1.5 cm and the laparoscope trocar and cannula inserted into the peritoneal cavity. The trocar is removed and the laparoscope, with fibre optic light cable attached is inserted through the cannula under direct vision. If in the correct place, the intraperitoneal contents will be seen. The pelvic organs are assessed for ovarian access before a second 1 cm incision is made in the midline, just above the pubic hair-line for the trocar and cannula for the grasping forceps. The ovarian ligament is located and held with these forceps thereby enabling the position of the ovary to be manipulated for optimum access to the follicles. If the ovary is mobile, twisting the ovarian ligament enables the inferior surface of the ovary to be accessed.

Once follicles have been identified and an access route chosen, the trocar and cannula for the aspirating needle are inserted laterally between the two previous incisions. An assistant holds the ovary in position with the grasping forceps and the aspirating needle is passed down the cannula and the appropriate follicle pierced. The follicle must not be pierced at its thinnest point as the sudden release of pressure may cause a rent in the wall forming a poor seal around the needle. Suction is applied with a special vacuum pump (Craft Suction Unit, Rocket of London) at 80 – 100 mm Hg, or by hand using a 20 ml syringe. Once the follicle is seen to collapse, suction is ceased and the test tube containing the follicle aspirate is passed to the embryologist for scrutiny. If an oocyte is found the needle is withdrawn and the next follicle aspirated. If the oocyte is not found in the first aspirate, the follicle is re-inflated with flushing medium (*Table 2*) and suction re-applied to empty the follicle. During this manoeuvre the tip of the needle is moved about within the follicle to try and curette the granulosa cells to which the egg may still be attached. Irrigation and aspiration is repeated until the oocyte is obtained. When all available follicles have been aspirated, any fluid in the Pouch of Douglas (the utero-rectal fossa) is aspirated in case an egg has escaped with leaking follicular fluid. Finally, the ovary is checked for any bleeding, the gas is released from the abdomen and the skin incision repaired.

3.2 Instruments for laparoscopic oocyte retrieval

The preparation of instruments for laparoscopy should be as meticulous as for any other kind of tissue culture work. Difficulty is often experienced in explaining to medical staff that although great care is taken routinely to ensure instruments are sterile, this is not sufficient to ensure that they are prepared adequately for oocyte retrieval for IVF and embryo culture. It is for this reason that we suggest that the preparation of the instruments should be supervised by the embryologist in conjunction with the theatre sister.

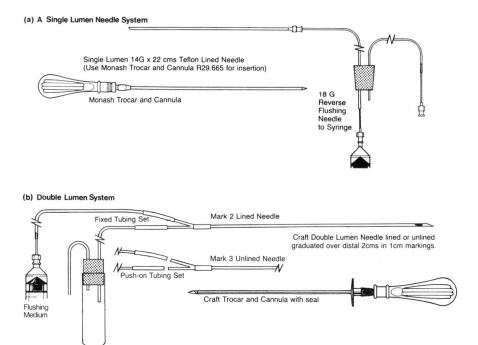

(a) A Single Lumen Needle System

Single Lumen 14G x 22 cms Teflon Lined Needle
(Use Monash Trocar and Cannula R29.665 for insertion)

Monash Trocar and Cannula

18 G
Reverse
Flushing
Needle
to Syringe

(b) Double Lumen System

Fixed Tubing Set

Mark 2 Lined Needle

Craft Double Lumen Needle lined or unlined
graduated over distal 2cms in 1cm markings.

Mark 3 Unlined Needle

Push-on Tubing Set

Craft Trocar and Cannula with seal

Flushing
Medium

Figure 2. Types of needles used for laparoscopic oocyte retrieval.

3.2.1 *Types of needles for oocyte retrieval*

A number of different needles are available for follicle aspiration. In general, these contain either a single or a double lumen. With single channel needles, after the follicular fluid has been aspirated, the bung is removed from the collection tube and flushing medium is passed back down the needle into the follicle. Once the follicle is distended the bung is replaced in the collection tube and the follicle re-aspirated (see *Figure 2a*). Double channel needles allow for continuous flushing of the follicle. Aspiration takes place through the channel with the large bore, and the fine bore channel allows the follicle to be irrigated without withdrawing the needle (*Figure 2b*). The only disadvantage we have found with this system is that the needle is oval and hence a firm seal through the ovarian capsule may not always be obtained. Round double channel needles have been developed (Go Medical Industries, Castlemed, UK). In some types of needle (Renou, Craft Mk1 and Mk2; Rocket Ltd) the aspiration channel is lined with Teflon tubing in order to create a continuous smooth walled tube all the way from the needle tip to the collection vessel. This should decrease the likelihood of the sticky cumulus mass being lost in the aspiration system. However, unless care is taken in the preparation of these needles and they are relined frequently, then the lining tubing may kink with repeated sterilization. Newer models (Craft Mk3) no longer have this lining.

3.2.2 *Lining a needle*

We reline our needles with Teflon tubing (0.9 mm I.D.; Rocket of London, Cat. No.

R57.532) after one or two uses. The old lining is pulled from the needle and the bevel of the needle checked under a dissecting microscope. Any barbs or rough edges are taken off with a fine grinding stone. The needle is then rinsed thoroughly with tissue culture grade water. The new lining tubing is stretched until it is thin enough for a length to be passed down the tube from the top and to appear at the bevel. The stretched portion is pulled through the needle until the normal thickness tubing now lines the needle. As the tubing tends to shrink with heating, the end is cut with about 2 cm sticking out of the needle. After sterilizing this end is cut flush with the bevel. Before sterilizing, the tubing is rinsed thoroughly with tissue culture grade water. Rather than autoclaving, we prefer to double bag the completed needle set and sterilize in a hot dry oven at 160°C for 2 h.

3.3 **Preparation of laparoscopy instruments**

We would advise against the commonly used method of disinfecting instruments for laparoscopy, i.e. soaking for 10 min in a glutaraldehyde solution (Cidex, Surgikos Ltd). This procedure does not ensure full sterilization (i.e. of spores), although it is viricidal (e.g. for hepatitis virus and HIV) and bactericidal, and provides adequate decontamination of instruments between patients. However, the solution is highly toxic to all cells, and besides being volatile, it is very adherent to surfaces with which it comes into contact. It is thus impossible to remove the fixative completely and it is likely to contaminate the gloves of the operator or scrub nurse, or the instruments. Autoclaving as a method of sterilization is preferred. Autoclavable instruments and laparoscopes are available (Stortz).

3.4 **Gases for laparoscopy**

It is essential that adequate measures are taken to ensure as little fluctuations as possible in the pH of the fluid surrounding the egg. In standard laparoscopy procedures, 100% CO_2 is used to insufflate the abdomen. Because of its rapid reabsorption, this is the safest mixture to be used should inadvertent penetration of a major vessel occur. Theoretically the prolonged contact of the thin walled follicles with this gas might cause a decrease in the intrafollicular pH although direct evidence for this assumption is lacking. However, it is clear that the aspiration of gas with the follicular contents can lower the pH of the aspirated fluid, especially if the gas is trapped above the follicular aspirate in capped tubes. For these reasons mixtures containing a physiological CO_2 level (e.g 5% CO_2:5% O_2:90% N_2) have been used in preference to 100% CO_2 for oocyte aspiration (12). The abdomen is initially gassed with CO_2 for safety, and once the position of the cannula within the peritoneal cavity has been verified, the reduced CO_2 mixture is insufflated.

As follicular fluid has poor buffering capacity, the oocyte needs to be retrieved as quickly as possible from the aspirated fluid and transferred to adequately buffered medium. The proximity of the laboratory is thus one important factor in maintaining ideal conditions for the oocyte. Should a laboratory not be available immediately adjacent to the theatre, a trolley containing a stereo-microscope with a heated stage and a thermostatically controlled hot-block (BT1; Grant Instruments) maintained at 37°C will suffice. Once the oocyte is located it is transferred to a test tube, containing Hepes-buffered flushing medium (*Table 2*) kept at 37°C, for transport to the main laboratory.

4. FLUSHING AND CULTURE MEDIA

The principles for making up media for embryo culture have been elucidated in detail by Pratt in Chapter 2 (Section 5) which should be read in conjunction with this section. Only modifications to the general principles which are specific to human fertilization and culture will be dealt with here.

4.1 **Ionic constituents**

A number of media have been used successfully for human *in vitro* fertilization and embryo culture. These media have not been designed specifically for that purpose but have been modified from media used for growth of animal pre-implantation embryos *in vitro* or for the growth of cell lines e.g. Earles Balanced Salts (EBS) (13), Whittingham's T6 (14), Menezo's B2 (15), Ham's F10 (16). No real differences in fertilization or cleavage rates among these various media have been demonstrated to date, and acceptable pregnancy rates have been achieved with all of them.

Some basic media can be purchased ready-made, in liquid ($10\times$ strength), or in powdered form (EBS, Ham's F10; Flow Labs). However, many groups prefer to make up their own media, thereby having the advantage of personal quality control over each ingredient (17). Recipes for media based on EBS are given in *Tables 1−4*. The osmolarity of the media must be assiduously checked and adjusted to fall within the range of 280−285 mOsmol/l otherwise fertilization may fail.

4.2 **Macromolecules**

Most laboratories performing therapeutic IVF add serum from the patient to the media for fertilization and growth. The necessity for the presence of macromolecules in the

Table 1. Ingredients required to make up Earles balanced salt solution for human IVF and embryo culture.

EBS, $10\times$ concentrate	Earles balanced salt solution, $10\times$ strength, without sodium bicarbonate (FLOW Labs)
Water	Suitable for tissue culture, ANALAR (BDH), or Water for injection, B.P. (Boots Hospital Products)
Sodium bicarbonate	(BDH: ANALAR grade), keep for tissue culture only
Sodium pyruvate	(SIGMA Chemical Co.) stock solution 1.1 mg/ml; freeze in 1 ml and 4 ml aliquots
Penicillin	Crystapen (Glaxo Ltd) 600 mg vials. Store in fridge. Make up to 600 mg/ml with 1 ml tissue culture grade water
Gentamicin	Garamycin (Schering). Supplied by Flow Labs as 10 mg/ml stock
Heparin	Mucous Heparin WITHOUT preservative. (Paynes and Byrne Ltd). 0.2 ml ampoules containing 5000 i.u. Store in fridge
BSA Fr. V	Bovine Serum Albumin Powder Fraction V (SIGMA Cat. No. A9647)
Hepes	Free acid; (Ultrol, Calbiochem). 21 mM stock solution (5.96 g/100 ml). pH adjusted to 7.35−7.4 with 0.2 M NaOH. Can be stored for up to 3 months after preparation

Fertilization of human oocytes

Table 2. Follicle flushing medium and embryo replacement medium (Hepes-buffered EBS for use in air).

To make 400 ml antibiotic-free Hepes-buffered EBS (HEBS):
1. Thaw one 4 ml aliquot sodium pyruvate (*Table 1*).
2. Weigh 0.1348 g of sodium bicarbonate. Add 20 ml of water and dissolve.
3. Measure 40 ml of EBS 10× concentrate into conical flask. Add 295 ml of water and the thawed aliquot of sodium pyruvate.
4. SLOWLY add the sodium bicarbonate solution, using a pipette and shaking gently. (Doing this step VERY slowly prevents precipitation of calcium carbonate which gives the solution a cloudy appearance. If this occurs − discard and start again).
5. Add 33.6 ml of Hepes stock solution (*Table 1*).
6. Check the osmolarity (it should be 280−285 mOsmol/l). If too high add water. If too low add 10× EBS concentrate.

For flushing medium:
7. Add 350 μl of heparin (*Table 1*) to 350 ml of the Hepes-buffered EBS. Sterilize by Millipore filtration. Store refrigerated in 50 ml aliquots in tissue culture flasks.
8. Millipore filter the remaining 50 ml of the stock HEBS for making embryo replacement medium.

For embryo replacement medium (containing 10% heat-inactivated maternal serum):
9. Add 2 ml of heat-inactivated serum (*Table 4*) to 18 ml of Hepes-buffered EBS.
10. Optional step. Filter sterilize (Millipore) discarding the first 0.5−1 ml of the filtered solution (as the filters can contain wetting agents which may inhibit normal embryonic development).
11. Store in a fridge in sterile tubes or flasks for the replacement procedure.

Table 3. *In vitro* fertilization medium and embryo culture medium (bicarbonate-buffered EBS)

To make 100 ml of bicarbonate-buffered EBS stock:
1. Thaw one 1 ml aliquot of sodium pyruvate (*Table 1*).
2. Weight out 0.21 g of sodium bicarbonate. Add 10 ml of water and dissolve.
3. Measure 10 ml of EBS 10× concentrate into a conical flask. Add 77 ml of water and the thawed 1 ml aliquot of sodium pyruvate.
4. SLOWLY add sodium bicarbonate solution, using pipette and shaking gently. (Doing this step VERY slowly prevents precipitation of calcium carbonate which gives the solution a cloudy appearance. If this occurs − discard and start again).
5. Add 10 μl of penicillin solution (*Table 1*) and 0.2 ml of gentamicin solution (*Table 1*).
6. Check osmolarity (should be 280−285 mOsmol/l). If too high add water, if too low add 10× EBS concentrate.

For fertilization medium:
7. Add 0.4 g of BSA (*Table 1*) to 80 ml of the EBS and leave to dissolve slowly by itself. DO NOT shake otherwise it will froth and form lumps which are difficult to dissolve. Millipore filter and store in aliquots of 10 ml in Falcon tubes (no. 2001) in fridge.
8. Millipore filter the remaining 20 ml of the stock and store in aliquots of 4 ml in Falcon tubes (no. 2002) in the fridge (for making embryo culture medium − see below).

For embryo culture medium (containing 10% heat-inactivated maternal serum = EBS+HIS[a]:
9. Remove 0.4 ml of EBS from one of the 4 ml aliquots stored in fridge and replace with 0.4 ml of heat-inactivated serum.
10. Optional step. Millipore filter, discarding the first 0.5−1 ml.
11. Use in drops of 50−100 μl under oil for culture of pronucleate stage embryos and onwards.

[a]For 15% maternal serum, remove 0.6 ml from the 4 ml aliquot in step 9 and replace with 0.6 ml of serum.

Table 4. Preparation of patients' serum for IVF culture and replacement media.

1.	Aseptically, take $10-15$ ml of blood using a tissue culture tested non-toxic syringe (e.g. Steriseral, NI Medical).
2.	Remove the needle and empty the syringe into a 15 ml sterile plastic centrifuge tube (Falcon 2095, Becton Dickinson).
3.	Centrifuge IMMEDIATELY in bench-top centrifuge ($5-10$ min, at 1000 g − maximum speed recommended for no. 2095 tube is 3600 r.p.m.).
4.	*As soon as centrifuge cycle is complete*, transfer the plasma to a sterile plastic tube (Falcon no. 2001) using a sterile Pasteur pipette.
5.	Leave the tube of plasma on a bench to clot. [This may take a few hours in the smooth walled plastic tube. Clotting may be accelerated if glass tubes (washed as for tissue culture) are used, or if the end of a sterile glass Pasteur pipette is placed into the plastic tube].
6.	Remove the fibrin coagulum by stirring with a sterile glass Pasteur pipette.
7.	Heat-inactivate by incubating for 30 min in a water-bath at 56°C.
8.	Filter the sample using a disposable sterile Millex-GS 0.22 μm filter (Millipore Ltd) remembering to discard the first few drops (~ 0.5 ml) of the sample.
9.	Aliquot and dispense the plasma into sterile plastic tubes. Freeze for future use.

culture medium has been questioned recently (18) although the experiments only demonstrated the requirements for the culture of mouse embryos to the blastocyst stage. Comparable controlled experiments on human embryos have not yet been reported.

The concentration of serum used for fertilization in humans varies from 7.5% to 10%, and for culture, from 10% to 15% ($13-17$). The protocol recommended for the preparation of serum for IVF, culture and embryo replacement is shown in *Table 4*. It will be appreciated that the addition of patient's serum to each individual batch of medium makes it difficult to compare the results for human embryonic development *in vitro* in different types of medium, and to perform strictly controlled experiments aimed at elucidating mechanisms of fertilization and embryonic development. In an attempt to overcome this problem for our own work on fertilization *in vitro*, and for our therapeutic IVF programme, we have standardized our fertilization medium replacing the serum with 5 mg/ml Fraction V bovine serum albumin (BSA; Sigma A9647). The BSA can be purchased in a large batch and will last for a number of years if kept frozen. Consistent fertilization rates can be achieved and pregnancies established wtih oocytes fertilized in this medium.

The media for growth of post-fertilization embryos are difficult to standardize as BSA may not be adequate for growth to the blastocyst stage (19). Both pooled human serum and pooled human cord serum have been used successfully, but because of the risk of transmission of disease, it is inadvisable to use these embryos for replacement unless the serum has been adequately screened.

4.3 Water

The most troublesome ingredients in *in vitro* culture systems for pre-implantation embryos are the water and the oil. Our belief is that no matter how good the specification of the water purification system available in the laboratory, unless the system is used regularly and is maintained efficiently, it is preferable to buy water. Analar water (BDH) is chemically pure and adequate for IVF and culture, but being non-sterile and stored in large volumes may not be free of bacterial toxins. Alternatively, water for injections (BP) is sterile and non-pyrogenic, is packed in smaller quantitites, and the standards for its preparation are controlled strictly. Water for injection or irrigation (Boots) appears to be satisfactory for the culture of human embryos *in vitro*.

4.4 Oil

We have used BDH light paraffin oil for a number of years with little difficulty. The method for the preparation of the stock is described by Pratt (Chapter 2, Section 5.1). Oil from stock for immediate use is first washed with tissue culture medium by vigorous mixing of about 200 ml of EBS and 500 ml of oil in a one litre sterile tissue culture flask (Nunc). The emulsion is then left to separate into phases. The oil is stored for use at room temperature in this container (i.e. floating on the medium). As with all tissue culture ingredients, strict quality control must be observed. Zona-intact and zona-free mouse embryos are cultured to the blastocyst stage to ensure non-toxicity of reagents (see Chapter 2, Section 6).

The potential risk of toxicity of the oil (20), and the time taken to ensure its correct preparation (13), have led a number of groups to seek alternative means of working in small volumes of medium. The use of oil is to prevent changes in the osmolarity occurring by evaporation of water during incubation. However, fertilization and culture can be achieved in test tubes (e.g. Falcon 2001, Becton Dickinson) in volumes of about 1 ml in a humidified atmosphere.

4.5 Antibiotics

Antibiotics are added to most tissue culture media to inhibit the growth of micro-organisms. These organisms may be either environmental contaminants or pathogenic organisms present in the same sample (21). As penicillin and streptomycin are the most frequently used antibacterials in tissue culture media, two points are worth noting about the use of these antibiotics in human IVF protocols.

Penicillin allergies are not infrequent, and we advise either that penicillin should be left out of all media for flushing follicles and embryo replacement procedures, or that specific batches of penicillin-free medium are kept aside for use in atopic individuals. The use of a filter (Millex-GV, Millipore) in the flushing line will protect the patient at operation should inadvertent contamination of the media with bacteria have occurred.

Gram-negative organisms are the most likely to interfere with fertilization *in vitro* (22) or to cause infection if injected into the female genital tract at replacement. *Escherichia coli* is known to bind firmly to the spermatozoa and may not be removed during preparation for *in vitro* fertilization (21). Although streptomycin is effective against most Gram-negative pathogens, the incidence of resistance is increasing. A more effective alternative is gentamicin (Sigma/Schering) used at a concentration of 20 μg/ml.

4.6 Gases

Most media for *in vitro* fertilization and embryo culture contain 25 mM bicarbonate which should ensure a pH of 7.2−7.4 when gassed with 5% CO_2. There is still uncertainty about the ideal oxygen concentration for human embryo development *in vitro*. Although early work on murine and bovine embryos suggested that development *in vitro* to the blastocyst stage was better in 5% oxygen (23), it is clear that high rates of blastocyst formation are achieved routinely for mouse embryos cultured in media gassed with 5% CO_2 in air (21% O_2). No reliable comparative data are available for human embryos as the rate of blastocyst formation *in vitro* is poor in both mixtures. However, comparable pregnancy rates occur with embryos grown either in 5% CO_2 in air or in a mixture of 5% O_2:5% CO_2:90% N_2.

4.7 Temperature

Incubators for the fertilization of human oocytes and the culture of pre-implantation embryos should be maintained between 37°C and 37.5°C. As exposure to abnormally low temperatures can disrupt microtubule polymerization and hence have effects on the meiotic spindle, the medium for flushing follicles should be kept warm and the oocytes transferred to a warm environment as soon as possible after location in the aspirate.

5. OOCYTES

5.1 Manipulation of oocytes

Once the oocyte has been located in the aspirate from the follicle, it should be transferred as soon as possible to the environment in which it will be cultured for fertilization *in vitro* (e.g. 1 ml of EBS+BSA/serum under oil equilibrated overnight with 5% CO_2 in air). The oocyte is transferred using a sterile glass Pasteur pipette which has been flamed to remove sharp edges from the tip. If a large number of red blood cells are still present, the oocyte should be washed through a clean drop of medium. The oocyte is incubated for 4−5 h prior to insemination as it has been suggested that this improves the fertilization rate (24).

5.2 Grading of oocytes

Current superovulation regimes create a population of follicles which are not totally synchronous in development. Although the pre-operative injection of hCG may start germinal vesicle breakdown, the cytoplasmic maturation of the oocyte may not be complete and appropriate for the initiation of normal embryonic development (25).

A mature pre-ovulatory oocyte (*Figure 3a*) at aspiration will have reached metaphase II of its second meiotic division. The pre-ovulatory follicle itself will be large, with an average diameter of 2.2−2.8 cm, and contain 3−10 ml of fluid (26). The cumulus mass surrounding the oocyte will be well expanded with a consistency of mid-cycle cervical mucus. Usually, immature oocytes (*Figure 3b*) are retrieved from small follicles (<2 ml) and may not have undergone germinal vesicle breakdown. The surrounding cumulus cells are well compacted around the oocyte and difficult to remove even after being in the presence of spermatozoa for 18 h or more. However, evidence is now accumulating that maturation *in vitro* can be achieved for a proportion of these immature oocytes by leaving them in culture for between 12 and 36 h (27). Although the fertilization rate of these *in vitro* matured oocytes is reduced, full development to the blastocyst stage *in vitro* can be achieved. The competence of embryos derived from oocytes matured *in vitro* to establish a pregnancy is reduced (28).

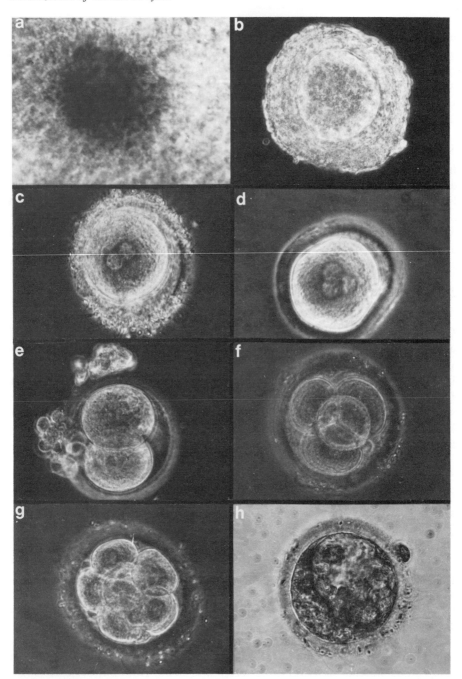

Figure 3. a. A mature human pre-ovulatory oocyte. Note the large expanded cumulus mass occupying the entire frame and obscuring detailed examination of the oocyte. **b**. An immature human oocyte with tightly packed cumulus cells. **c**. A fertilized human oocyte showing two pronuclei (20 h after insemination). **d**. A tripronucleate oocyte (19 h after insemination). **e**. A human two-cell embryo (40 h after insemination). **f**. A human four cell embryo (46 h after insemination). **g**. A human eight cell embryo (74 h after insemination). **h**. A human expanding blastocyst showing the blastocoelic cavity and the area of the inner cell mass (138 h after insemination).

5.3 **Removal of cumulus cells**

Detailed examination of the pre-ovulatory oocyte is obscured by the large cumulus mass which surrounds it (*Figure 3a*). Although some inference of the stage of maturation can be made from the state of expansion of the cumulus mass, the exact stage of maturation cannot be assessed without removing the cumulus cells. Recent data suggest that removal of the cumulus cells prior to insemination is not detrimental to fertilization and will allow the development of the oocyte to be assessed by an examination for extrusion of the first polar body. In this way a decision can be made about the duration of incubation required for complete maturation prior to insemination (27,28). Hyaluronidase can be used for cumulus removal (see Chapter 2, Section 2.1). It should be noted that in some mouse strains, exposure of the oocytes to hyaluronidase can cause parthenogenetic activation.

In most therapeutic IVF programmes, the cumulus mass is not removed prior to the addition of the spermatozoa. Once the oocyte has been in the presence of the spermatozoa for some twelve hours or more the cumulus cells of most oocytes have usually fallen away spontaneously under the action of the spermatozoal hyaluronidase, or they can be removed easily by gentle pipetting up and down using a glass Pasteur pipette which has been drawn to a diameter just larger than the oocyte. Alternatively the remaining cumulus cells can be removed mechanically by careful dissection with syringe needles (27 g).

5.4 **Removal of the zona pellucida**

For certain analyses of embryonic or oocyte structure (i.e. for karyotyping or for immuno-cytochemical localization), it may be desirable or necessary to remove the zona pellucida. The human zona is more resistant than the mouse zona but will dissolve in warmed acidified media (e.g. acid Tyrode's solution) or by incubation for 5 min in 0.2 mg/ml pronase (see Chapter 2, Section 3.2.1).

6. SPERMATOZOA

6.1 **Collection of the semen sample**

6.1.1 *Containers*

Collection of the semen sample should be by masturbation into a wide mouth sterile plastic specimen pot (Sterilin 60 ml). The container should be marked clearly with the patient's name, hospital number, time of production and date of most recent ejaculation or coitus.

6.1.2 *Instructions*

Each patient should be given an instruction sheet which outlines clearly the methods to be used for the collection and delivery of the sample to the laboratory. Condoms should not be used for collection of the sample as they usually contain spermicides. Nor should lubricants such as KY jelly be used as they can also be spermicidal. Patients should not try and collect the sample by coitus interruptus as often the first spermatozoal-rich aliquot may be lost before withdrawal. The presence of large numbers of squamous cells in the ejaculate often indicates that coitus interruptus has been the means used during collection. Where semen volumes are known to be large (>6 ml), a split ejaculate should be requested. Two specimen containers are supplied and the first portion of the ejaculate should be collected into one container and the rest into the other. The patient should be told that the semen-rich first portion is normally small as there is a tendency by patients to try and provide two equal sized samples.

6.1.3 *Delivery of the sample to the laboratory*

The semen sample should be delivered to the laboratory as soon as possible after production, ideally within 2 h. The sample should be protected from extremes of temperature during transport (i.e. no less than 15°C and not more than 38°C) but need not be kept at body temperature. Indeed, this may be deleterious to the spermatozoa due to high metabolic activity at this temperature (leading to an increased production of metabolites and possible decreased viability) and possible bacterial growth. Spermatozoa separated from the semen and kept in suitable media (*Tables 2* and *3*) will remain viable at room temperature for more than a week whereas they will only last for a day or so if kept in the seminal plasma.

6.1.4 *Abstinence*

The period of abstinence from coitus does affect the quality of the semen sample. Although the total count may be increased with prolonged abstinence mainly due to an increase in the volume of ejaculate, usually the percentage of motile spermatozoa and their viability is decreased. A period of 48−72 h seems optimal.

6.2 Semen analysis

A detailed account of the clinical significance of the physico-chemical properties of semen and its cellular constituents is beyond the scope of this chapter. However, a brief resumé of some of the normal characteristics will be given to assist the embryologist in the preparation of the spermatozoa for IVF. Further details can be obtained from the excellent monograph by Mortimer (29) and the WHO manual (30).

6.2.1 *Seminal fluid*

Seminal fluid makes up the vast proportion of the ejaculate and is composed of the secretions mainly from the prostate gland and the seminal vesicles. Shortly after ejaculation, the semen forms a coagulum which liquifies within 30 min of production. The normally liquified ejaculate has a watery consistency such that it can be expelled from a Pasteur pipette in drops. Increased viscosity will be noted as a mucus-like string when the pipette tip is removed after insertion into the semen. Increased viscosity does not necessarily equate with non-liquification as liquified semen may be viscous. Viscosity of the semen for sperm counting or preparation can be reduced by repeated pipetting through a Pasteur pipette or a narrow gauge injection needle (22 or 23 gauge). High viscosity of the semen can lead to inaccuracies in the counting of the spermatozoa because of the difficulty in accurately dispensing a small aliquot of the viscous sample.

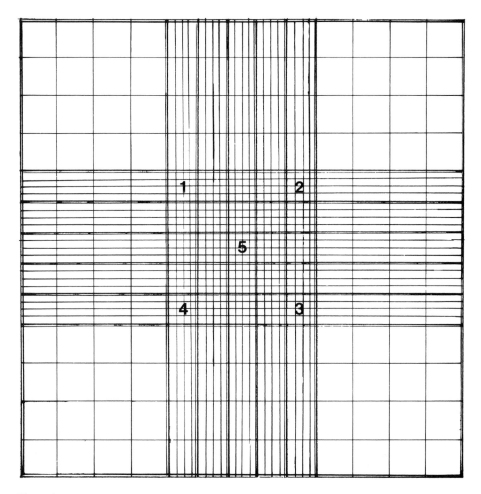

Figure 4. Chamber markings on an improved Neubauer Haemocytometer. There are two chambers on each haemocytometer. The total number of spermatozoa in the five squares (1−5) are counted for each chamber. For a dilution of 1:20, the average of the count for the two chambers represents the concentration of spermatozoa in millions per ml.

6.2.2 Counting chambers

(i) *The Neubauer haemocytometer.* The standard chamber for counting spermatozoa is the improved Neubauer haemocytometer (A.R.Horwell). This has two counting grids of 3 × 3 mm. The ruled central area (see *Figure 4*) is divided into 25 large squares (5 × 5) defined by triple rulings. Each large square is 1 mm² and is divided into 16 small squares. With the coverslip in position, the depth of the chamber is 0.1 mm. Thus the total volume over each large square is $1 \times 1 \times 0.1$ mm³ $= 10^{-4}$ ml. The method for counting using this haemocytometer is given in *Table 5*.

(ii) *Specialized counting chambers.* Two specialized counting chambers which permit direct estimations of the motile spermatozoa concentration are available. The Makler

Table 5. Protocol for semen analysis.

1.	Decant the semen sample from the collection container into a graduated conical plastic centrifuge tube (Falcon 2095). Note the viscosity of the sample on pouring. Record the volume and viscosity of the semen.
2.	Using a sterile plastic tip on a fixed volume pipette, transfer 50 μl of semen for counting to a 950 μl aliquot of 10% buffered formalin (1:20 dilution). Using a new sterile tip, place 10 μl of semen on a microscope slide and cover with a standard 22 × 22 mm coverslip for motility and morphology examination (steps 6 and 7).
3.	Fix the haemocytometer coverslip to the counting chamber (Improved Neubauer Haemocytometer). Interference fringes should be seen between the opposed glass surfaces when the coverslip is correctly positioned.
4.	Transfer about 10 μl of the formalized sample for counting to the haemocytometer using a Pasteur pipette and allowing the chambers to fill by capillary action.
5.	Leave the counting chamber in a humid environment for 10 min to allow the spermatozoa to settle onto the counting grid.
6.	While waiting, estimate the spermatozoal motility. Score one hundred spermatozoa on the slide prepared in step 2 above under phase contrast at 250× magnification. Those spermatozoa which show movement should be scored as motile.
7.	Score the morphology of the spermatozoa subjectively (see *Figure 5*) by examining them under phase contrast at 400× magnification. Score spermatozoa with one or more defects as abnormal.
8.	Now estimate the concentration of spermatozoa by counting the number of spermatozoa present in the four corner large squares and the centre square of each of the two grids in the haemocytometer chamber (one grid is shown in *Figure 4*). Spermatozoa touching the lines at the top or left of the square to be counted are included in the count, but those touching the bottom or right of the square are excluded.
9.	For a dilution of 1:20, the concentration of spermatozoa present in the original semen sample is the average of the totals for the five squares for each of the two grids in the chamber, expressed in millions/ml. If the count recorded is less than 20 million/ml, then this method of estimating the spermatozoal concentration may be inaccurate. In that case, it is advisable to count all 50 squares in the chamber (i.e. the 25 squares in each grid). The concentration of spermatozoa present in the original sample scored by this means is the total count of the 50 squares divided by ten and expressed in millions/ml.
10.	If the count is very low, perform a repeat count using a lower dilution of the spermatozoa and correct the count accordingly.
11.	Also count the round-cells (i.e. white blood cells and immature spermatogenic forms) present in 50 squares. If the count is greater than 2 million/ml then stain a slide for leucocytes to detect pyospermia and culture an aliquot of the sample to check for the presence of bacteria. The best method of staining for white blood cells is the benzidine-cyanosine peroxidase stain (30) but because benzidine is carcinogenic it is difficult to obtain. Testsimplets (Boehringer Mannheim) are prestained slides which are effective and easy to use.

chamber (Sefi Medical Instruments) is well made but expensive. The Horwell Fertility Counting Chamber (A.R.Horwell) is cheaper and based on similar principles. Having a chamber depth of 10 μm and a ruled area of 1 square millimeter divided into one hundred 100 × 100 μm squares, the number of spermatozoa contained in any 10 of these small squares represents the concentration of spermatozoa in millions/ml. A 3−5 μl aliquot of the neat semen sample is placed on the counting area and covered with the coverslip. The concentration of the motile spermatozoa can be estimated directly when viewed by phase contrast microscopy. This chamber is especially useful where there are low numbers of spermatozoa present in the sample and only an approximate count

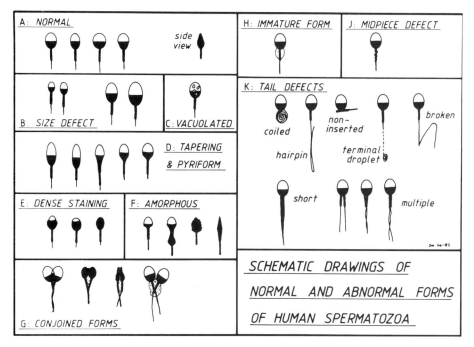

Figure 5. Schematic drawings of normal and abnormal forms of spermatozoa. From: Mortimer,D. (1985). Current Problems in Obstetrics, Gynaecology and Fertility. **7**. Year Book Medical Publishers. Reproduced with permission.

of motile spermatozoa is required, for example, in the sample for insemination after Percoll separation or 'swim-up' (see below).

6.2.3 *Normal values for semen analysis*

Although there is still uncertainty about the minimum values associated with fertility or with the ability to fertilize human oocytes *in vitro*, the following are the World Health Organization standards for 'normal' semen samples (30).

Volume	2−5 ml
Liquification time	Within 30 min
Concentration	20−200 million/ml
Motility	Greater than 40% motile
Morphology	Greater than 40% normal forms (see *Figure 5*)
White blood cells	Less than 1 million/ml

6.2.4 *Bacteriological screen*

Some laboratories screen all semen samples routinely for bacteria prior to an attempt at therapeutic IVF. Although this is a sensible practice as it may detect previously undiagnosed sexually transmitted diseases, the difficulty of treating some infections (e.g. *Chlamydia* and *Ureaplasma*) and the possibility of re-infection after screening,

Table 6. Swim-up protocol for preparation of spermatozoa.

1.	Allow the semen to liquefy at room temperature for 30 min. Add about 2 ml to a sterile plastic tube (Falcon 2001).
2.	Estimate concentration and motility.
3.	Add 10 ml of EBS + BSA or EBS + 10% heat-inactivated serum. Mix thoroughly by gentle pipetting up and down with a sterile Pasteur pipette.
4.	Centrifuge at 200 g for 5−10 min. Remove all but 0.5 ml of the supernatant using a Pasteur pipette. Suspend the pellet by gentle mixing with this remaining supernatant.
5.	Add 1−2 ml of fresh medium, running the medium VERY CAREFULLY down the inside of the tube so as to layer on top of the spermatozoal suspension and not to cause turbulence. If the semen sample is poor, the area of the semen/medium interface can be increased by the use of small volumes of semen (0.5 ml) placed in a number of tubes (Falcon 2002) and overlain with about 0.5−1 ml of medium. The supernatants are pooled to achieve the desired motile spermatozoal concentration.
6.	Stand the tube in a 5% CO_2 incubator at 37°C for 20−30 min. During this period the motile spermatozoa should migrate into the overlying medium leaving the dead spermatozoa, debris and cells at the bottom of the tube. Standing the tube at a slight angle increases the surface area of the interface between the spermatozoa suspension and the medium allowing better recovery of motile spermatozoa.
7.	Carefully aspirate 0.5−1 ml of the clear supernatant. Take 50 μl and 10 μl for counting, and motility estimation, respectively.
8.	Adjust the concentration of the suspension by dilution, or by centrifugation and resuspension, in order to achieve a motile spermatozoa concentration of between 2 and 5 million/ml. (100 000−150 000/ml after inseminating 20 μl−50 μl into the 1 ml drop, Section 7.1).

does not guarantee an uninfected sample at the time of the IVF procedure. Samples with a high round cell count (*Table 5*) always should be screened for infection.

6.3 Preparation of spermatozoa for *in vitro* fertilization

The object of these techniques is to recover a fraction of the semen sample which is enriched for motile sperm, and free of seminal plasma, cells and debris. The swim-up technique (*Table 6*) is probably the most widely used method and provides a simple means of enriching the recovered fraction for motile spermatozoa. There is little change in the percentage of morphologically normal spermatozoa recovered using this technique. Separation of spermatozoa by buoyant density centrifugation on self-generating gradients (31) (*Tables 7* and *8*) not only provides a relatively rapid method for obtaining a highly motile spermatozoa population, but also selects for spermatozoa with normal morphology. Bacterial contamination of the sample is also reduced using this technique (21).

7. FERTILIZATION AND EMBRYO CULTURE

7.1 Insemination

The oocyte that has been recovered from the follicular aspirate is transferred to a 1 ml drop of fertilization medium (*Table 3*) under oil. The maturity of the oocyte is assessed by the degree of expansion of the cumulus mass (see *Figure 3*), or, after removal of the cumulus by hyaluronidase, by the presence or absence of a germinal vesicle or the first polar body (Section 5). The oocytes are left in the 5% CO_2 incubator at 37°C for 4−5 h after retrieval. An aliquot containing 100 000 motile spermatozoa prepared

Table 7. Percoll gradients in EBS for buoyant density centrifugation.

Work in a laminar flow hood to maintain sterility of solutions.

1.   Dissolve 1.25 g of fraction V BSA in 25 ml of EBS 10× concentrate (10× EBS + BSA stock). Filter sterilize (Millipore 0.22 μm).
2.   Place 90 ml of Percoll (Pharmacia) in a sterile container and label '90% PERCOLL IN EBS + BSA'.
3.   Place 90 ml of culture grade water in a separate sterile container and label 'EBS + BSA'.
4.   To each container of 90 ml add:

      10 ml of 10× EBS + BSA stock (see step 1)
      100 μl 10× concentrate stock sodium pyruvate (11 mg/ml)
      0.2 ml stock gentamicin (*Table 1*)
      10 μl stock penicillin (*Table 1*)

5.   To 'EBS + BSA' container ONLY, add 1.3 ml of 7.5% sodium bicarbonate (FLOW Labs or make up).

To prepare 60% Percoll gradients:

6.   Working in a laminar flow hood, place into 12 ml autoclaved polycarbonate centrifuge tubes (Sorvall, Du Pont, 03115)

      6 ml of 90% Percoll in EBS + BSA
      3 ml of EBS + BSA

7.   Mix thoroughly by inverting the capped tubes.
8.   Centrifuge for 15 min at 20°C (29 000 g) in an angle head rotor (Sorvall, SS34 head at 15 000 r.p.m.).
9.   Store centrifuged Percoll columns in a fridge.
10. The density of the various layers of the gradient can be tested on a sample gradient by adding a mixture of density marker beads (Pharmacia Cat. No. 17-0459-01).

Table 8. Separation of spermatozoa on Percoll gradients (21).

1.   Allow a Percoll column to reach room temperature.
2.   Load 1−3 ml of semen onto the column.
3.   Centrifuge for 20 min in a swing-out head at 400 g in bench-top centrifuge (2000 r.p.m. with 10 cm radius).
4.   Aspirate the lowermost 1 ml from the centrifuge tube by sliding a 19G spinal needle (Becton Dickinson) down the inside of the tube and aspirating with a 1 ml syringe. Alternatively if polyallomer tubes are available these may be punctured through the base after disinfecting with alcohol.
5.   This lowermost layer should contain highly motile, morphologically normal, spermatozoa, in sufficient numbers for *in vitro* fertilization. Aliquots of the higher layers can be removed in the same manner but the motility and the morphology of the spermatozoa in these layers is decreased. Non-motile spermatogenic cells, leucocytes and debris will usually be found in the upper layers below the semen/Percoll interface.
6.   Spermatozoa prepared in this way can be inseminated directly into the drop for fertilization, or if preferred, they can be washed by the addition of 10−15 ml of fertilization medium and centrifugation.

by 'swim-up', or Percoll separation, is added to the drop containing the egg using a sterile micropipette tip. The dish is returned to the incubator and 14−20 h later the egg should be examined for fertilization.

7.2 Examination for fertilization

The cumulus cells are removed from the zona pellucida by gently drawing the egg up and down a glass Pasteur pipette drawn to a diameter just larger than the egg, or by manual dissection with syringe needles (27 g). The presence and number of pronuclei

Table 9. Laboratory procedure for embryo replacement (non-touch technique).

1.	Make approximately 20 ml of Hepes-buffered replacement medium containing 10% patient's plasma and warm to 37°C.
2.	Prepare a large drop of Hepes-buffered replacement medium under oil and warm in air (not CO_2) incubator. To do this, place about 500 μl medium into a plastic Petri dish (Falcon 3002) and cover with about 8 ml oil. Add a further 750−1000 μl of medium in order to make a 'fat round drop' which makes aspiration into the replacement catheter easier.
3.	Meanwhile, the patient is placed in the lithotomy position and tilted slightly head downward (or the knee chest position if preferred). The cervix is exposed using a Cusco speculum and gently swabbed free of mucus using a swab soaked in replacement medium. A further 3−5 ml of medium is sprayed over the cervix using a syringe and needle. The excess medium pools in the posterior fornix of the vagina and buffers the vaginal secretions.
4.	When the patient is in position for the replacement procedure, place the dish of warmed medium on the heated stage of the stereo-microscope and transfer the three 'best' embryos to the warmed 'fat round drop'.
5.	Pour the contents of one tube of replacement medium into a Falcon Petri dish.
6.	Fill a sterile 1 ml syringe with this medium and place on one side, making sure that the end is free from contact with non-sterile surfaces.
7.	Using a clean pair of scissors, expose the replacement catheter (Bourn-Wallace; HG Wallace Ltd) by cutting the ends off the sterile packet. First cut the wrapper at end containing the pink connectors (pull on the white plastic plug to expose the pink connection). Remove and discard the white plastic plug and attach the syringe filled with medium. Cut through the wrapper at the catheter end and push the cannula towards the new opening in the wrapper so that the flexible catheter end is exposed.
8.	Flush through all but 0.2 ml of the replacement medium present in the syringe.
9.	When the surgeon is ready, aspirate a small bubble of air into the tip of the catheter, then collect the embryos in about 10−20 μl of medium. Aspirate another small bubble of air to mark the position of the embryos in the catheter.
10.	Remove the syringe, catheter and introducer from the wrapping and hand it all carefully to the surgeon, helping with manipulation of syringe if necessary.
11.	Using the external white sheath as a stabilizer, the surgeon passes the catheter gently through the cervical canal until the two pink connections lock. The syringe plunger is carefully depressed to deliver the contents up to the 0.2 ml mark. After about 1 min the catheter is withdrawn completely and handed back to the embryologist.
12.	Under microscope vision, squirt the remaining contents of the syringe into a clean Falcon dish to check that the embryos have not been left in the catheter.

are scored. If the eggs have been fertilized, normally two pronuclei will be seen between 14 and 20 h after insemination (*Figure 3c*). Because of the potential risk of molar pregnancy occurring from tripronucleate eggs (32) (*Figure 3d*), these should not be transferred back to the patient.

7.3 Culture of fertilized eggs

Oocytes with two pronuclei are transferred to 50−100 μl drops of growth medium (*Table 3*) which have been equilibrated under oil overnight in the 5% CO_2 incubator. If development proceeds normally, cleavage to the two-cell stage (*Figure 3e*) can be expected to occur between 28 and 36 h after insemination and cleavage to the four-cell stage (*Figure 3f*), between 32 and 48 h. In most therapeutic programmes the embryos are transferred to the uterus at this stage (*Table 9*) and hence the further timings for cleavage in human pre-implantation embryos are less well documented.

7.4 **Development to the blastocyst stage**

The range of cell doubling times for individual embryos is large [8.1−51.0 h (33)] with a mean of between 18 h and 23 h depending on the age of the patients and the stimulation regime used. This variation probably reflects the differences in the quality of individual embryos for it is clear that the more rapidly cleaving embryos are more likely to establish pregnancies after transfer (34). Development to the fully expanded blastocyst stage (*Figure 3h*) in humans occurs in approximately one-third of the fertilized eggs where further growth has been attempted (19,35). Most fail after the four-cell stage, but it should be remembered that the majority of embryos cultured *in vitro* to these stages are 'spare' to therapeutic requirements once the 'morphologically best' embryos have been replaced. Thus their developmental characteristics may not be representative of 'the best' embryos.

Compaction in human embryos occurs at the sixteen cell stage with fully expanded blastocysts being found between 96 and 116 h after insemination. This is approximately the same time interval as in *in vivo* development, i.e. between insemination and recovery of expanded blastocysts from the uterine flushings of embryo donors (36).

8. ETHICAL AND LEGAL CONSIDERATIONS

The use of human gametes and embryos for research has been a central focus of worldwide public and political attention over the past few years. It is essential that those who intend to work with human gametes or embryos should be familiar with the regulations governing their use, both in research and in clinical practice, in their respective countries (37−42) and with the historical background that led to the implementation of those rules.

As the author is most familiar with the political and legal moves made in the UK, and since these have provided a basis for many of the deliberations and formulations in other countries, they are discussed further.

8.1 **The law and research on human embryos in the UK**

At the time of writing, the human embryo *per se* has no legal status. Under UK law it is not afforded the same legal status as a child or adult nor does the law acknowledge that the human embryo has a right to life. The Offences against the Person Act of 1861 and the Abortion Act of 1967 provide that abortion is a criminal offence except under the four circumstances provided in the latter Act. However, this applies to an implanted embryo *in vivo* and does not address the question of the embryo *in vitro*. Similarly, some protection to the embryo *in vivo* is accorded by the Congenital Disabilities (Civil Liability) Act 1976 which allows for damages to be recovered where an embryo or fetus has been injured *in utero* through negligence of a third party. There is no provision for the embryo *in vitro*.

8.2 **The Warnock committee**

Despite repeated calls from pioneers in this field (43), no official action was taken until 1982 when a government Committee of Inquiry was set up to examine the issues related

Table 10. Guidelines for both clinical and research applications of human *in vitro* fertilization[a].

Introduction

The purpose of these Guidelines is to set out the principles which the Joint Medical Research Council/Royal College of Obstetricians and Gynaecologists Voluntary Licensing Authority believes should guide those whose clinical practice or research involves the use of *in vitro* fertilization with human gametes. They have been based on the recommendations of the 'Committee of Inquiry into Human Fertilization and Embryology' under the Chairmanship of Dame Mary Warnock (the Warnock Committee), the Medical Research Council Statement of 'Research Related to Human Fertilization and Embryology' (MRC Statement) and the 'Report of the RCOG Ethics Committee on *In Vitro* Fertilization and Embryo Replacement or Transfer' (the RCOG Ethics Committee Report).

During their discussions the VLA considered it was important to define the term 'pre-embryo' used in these Guidelines. The term 'embryo' has traditionally been used to describe the stage reached in development where organogenesis has started, as shown by the appearance of the primitive streak and the certainty that thereafter a single individual is developing rather than twins or a hydatidiform mole, for example. To the collection of dividing cells up to the determination of the primitive streak we use the term 'pre-embryo'.

Guidelines

(1) Scientifically sound research involving experiments on the processes and products of *in vitro* fertilization between gametes is ethically acceptable, subject to certain provisions detailed in Sections 2−10 below.

(2) Any application made to the Authority must give reasons why information cannot be obtained from studies of species other than the human.

(3) The aim of the research must be clearly defined and relevant to clinical problems such as the diagnosis and treatment of infertility or of genetic disorders, or for the development of safe and more effective contraceptive measures.

(4) Pre-embryos resulting from or used in research should not be transferred to the uterus, except in the course of clinical research studies designed to enhance the possibility of establishing a successful pregnancy in a particular individual.

(5) Suitable signed consent to research involving human ova and sperm should be obtained in every case from the donors; sperm from sperm banks should not be used unless permission for its use in research has been obtained from the donor. Approval for each project should be obtained from the local ethical committee prior to seeking approval from the VLA.

(6) When human ova have been obtained and fertilized *in vitro* for a therapeutic purpose and are no longer required for that purpose it would be ethical to use them for soundly based research provided that the signed consent of both donors was obtained, subject to the same approval as in the preceding section.

(7) Human ova fertilized with human sperm should not be cultured *in vitro* for more than 14 days excluding any period of storage at low temperature (see Section 8) and should not be stored for use in research other than that for which local ethical committee and VLA approval has been obtained.

(8) Where a pre-embryo has been preserved at low temperature, whether donated for research purposes at the time of preservation or subsequently, it may continue to be grown to the equivalent of 14 days' normal development provided that approval has first been obtained from the local ethical committee and the VLA. Storage of individual pre-embryos at low temperature should be reviewed after **two years** and the maximum storage time should be ten years.

(9) The means of disposal of the pre-embryo should be carefully considered before the start of each project. At the end of a study steps must be taken to stop development of the pre-embryo and the appropriate disposal must be considered in discussion with the local ethical committee and details given to the VLA. The means of disposal will depend on the nature of the particular study that the pre-embryo has been used for. In view of the scarcity of the material, it would be inappropriate to discard any pre-embryo without thorough examination.

(10) Studies on the penetration of animal eggs by human sperm are valuable in providing information on the penetration ability and chromosomal complement of sperm from subfertile men and are con-

sidered ethically acceptable provided that development does not proceed beyond the early cleavage stage.

(11) The clinical use of ova at low temperatures, involving subsequent *in vitro* fertilization should not proceed to transfer to the uterus until such time as scientific evidence is available as to the safety of the procedure.

(12) Consideration must be given to ensuring that whilst a woman has the best chance of achieving a pregnancy the risks of a large multiple pregnancy occurring are minimized. For this reason:

   (a) if the IVF procedure is used no more than three pre-embryos should be transferred in any one cycle, unless there are exceptional clinical reasons when up to four pre-embryos may be replaced per cycle,

   (b) if the GIFT procedure is used no more than three or exceptionally four eggs should be introduced to the fallopian tubes.

(13) The following general considerations must be taken into account when establishing clinical facilities where *in vitro* fertilization or GIFT is carried out:

   (a) each centre should have access to an ethical committee[b], and no procedure should be undertaken without the knowledge and consent of the ethical committee,

   (b) detailed records should be kept along the lines recommended in the Warnock Committee Report, and should include details of the children born as a result of *in vitro* fertilization; the records should be readily available for examination by duly authorized staff and for collation on a national basis for a follow-up study,

   (c) where the director either does not have accredited consultant status, or the equivalent, or is a non-clinician full clinical responsibility must be assumed by a Consultant Adviser who takes an active role in overseeing the centre's treatment protocols and emergency procedures; all other medical, nursing and technical staff must have appropriate experience and training,

   (d) specialist medical, surgical and nursing facilities appropriate for the specific techniques used for the treatment should be available,

   (e) arrangements for emergency treatment should be made,

   (f) there must be adequate arrangements, where appropriate, for the transfer of gametes and pre-embryos between the clinical facilities and the laboratory,

   (g) centres should have appropriate counselling facilities with access to properly trained independent counselling staff,

   (h) all patients entering an IVF programme should be tested for hepatitis B and HIV antibodies,

   (i) donor sperm should be obtained only from a bank where all appropriate screening tests are undertaken including those recommended by the DHSS AIDS Booklet 4, *AIDS and Artificial Insemination — Guidance for Doctors and AI Clinics* (CMO(86)12),

   (j) egg donors should remain anonymous and for this reason donation for clinical purposes from any close relative should be avoided.

(14) The following general considerations must be taken into account when establishing laboratory facilities where *in vitro* fertilization is carried out:

   (a) each centre should have access to an ethical committee (see footnote to para 13(a)),

   (b) detailed records should be kept and should be readily available for examination by duly authorized staff,

   (c) laboratory staff must have appropriate experience and training in the techniques being used,

   (d) laboratory conditions must be of a high standard (e.g. good culture facilities, facilities for microscopic examination, appropriate incubators and training in 'non-touch' techniques),

   (e) where gametes and pre-embryos are cultured and stored there must be a very high standard of security and of record keeping and labelling.

---

[a]Ref. 38. Reproduced with permission from the Voluntary Licensing Authority.
[b]Refer to the Royal College of Physicians of London's 'Guidelines on the Practice of Ethics Committees in Medical Research' — revised edition November 1984 — for details of appropriate committee structure. The Chairman of an ethics committee should have no financial interest in the centre for which that committee is responsible.

to human fertilization and embryology and to make recommendations to government for legislation (37). The committee, chaired by the philosopher, Mary Warnock (now Baroness Warnock) produced a report which covered most aspects of assisted reproduction such as AID, egg donation and IVF. A key recommendation was the establishment of a new statutory licensing authority to regulate both research and those infertility services which the committee recommended should be subject to legal control such as AID, IVF, embryo donation and embryo cryopreservation. Two issues, namely surrogacy and embryo research, created special problems (44). The majority report of the Warnock Committee recommended that 'the human embryo now should be afforded some protection in law' in that 'research conducted on human *in vitro* embryos and the handling of such embryos should only be permitted under licence', and that 'no human embryo derived from *in vitro* fertilisation may be kept alive if not transferred to a woman beyond fourteen days after fertilisation'. Handling of human *in vitro* embryos without a licence or beyond fourteen days after fertilization would be a criminal offence. Furthermore, the report acknowledged the existence of 'spare' embryos and that 'as a matter of good practice no research should be carried out on a 'spare' embryo without the informed consent of the couple for whom the embryo was generated'. Two expressions of dissent about embryo research were included as part of this report. The first recommended that experimentation on the human embryo should not be allowed under any circumstances, and that the embryo of the human species be afforded special protection in law. The second expression of dissent drew a distinction between 'spare' embryos and those created for research, permitting the use of the former but prohibiting research on embryos 'brought into existence specifically for that purpose or coming into existence as a result of other research'.

8.3 The MRC/RCOG Voluntary Licensing Authority

With no immediate prospect of government legislation based on the Warnock Report, a Voluntary Licensing Authority for human *in vitro* fertilization and embryology (VLA) was set up jointly by the Medical Research Council and the Royal College of Obstetricians and Gynaecologists. This body, constituted along the lines recommended by the Warnock Committee for its statutory licensing authority, consists at present of three scientists nominated by the MRC, four obstetrician/gynaecologists and seven lay members including a lay chairman. It has the following terms of reference:

(i)   to approve a code of practice on research related to human fertilization and embryology;
(ii) to invite all centres, clinicians and scientists engaged on *in vitro* fertilization to submit their work for approval and licensing;
(iii) to visit each centre prior to its being granted a licence;
(iv) to report to the Medical Research Council and the Royal College of Obstetricians and Gynaecologists;
(v)   to make known publicly the details of both approved and unapproved work.

The VLA met first on the 26th March 1985 and shortly thereafter circulated its guidelines for both clinical and research applications of human *in vitro* fertilization. The first report of the VLA was published in April 1986 by which time it had visited

25 centres of which 24 had been approved. By the time the second report was published (April 1987, see *Table 10*) a further six centres were licensed and the VLA guidelines modified to incorporate the GIFT procedure. Of the 30 approved centres, nine mention projects which involve research on human embryos.

Although the VLA has no formal legal powers, it can exert formidable pressure for controlled research. Firstly, all but one of the centres registered currently as practising IVF have a member or fellow of the Royal College of Obstetricians and Gynaecologists as part of the team. As such, their practice must fall within the code of professional conduct of the College and is open to their disciplinary measures. Second, because of the respected scientific and medical composition of the VLA it would be extremely unlikely that any research on human embryos would be funded by granting agencies without VLA approval. Thirdly, the publication of both approved and non-approved projects will expose those IVF centres to public and especially media scrutiny. It is hoped that this voluntary arrangement will demonstrate to the public the ability of the scientific and medical fraternity to responsibly regulate research on human embryos and will provide a framework for a statutory body to tackle this difficult and emotive problem.

9. ACKNOWLEDGEMENTS

The techniques described in this chapter are based on methods used and developed over the past five years in the Department of Obstetrics and Gynaecology at the University of Cambridge Clinical School, and in Dr Martin Johnson's laboratory in the Department of Anatomy. I wish to express my thanks to all members of the Embryo and Gamete group for their support and help, and particularly to Dr Virginia Bolton and Ms Gin Flach for their help in developing and implementing the methods and for keeping *Brenda's Brother* informed and under control. I thank Sheena Glenister for typing the manuscript and Martin Johnson for his helpful criticism. The original work described in the text was supported by an MRC programme grant to PRB and Dr M.H.Johnson.

10. REFERENCES

1. Braude,P.R., Bolton,V.N. and Johnson,M.H. (1986) In *Embryo Research — Yes or No?* Bock,G. and O'Connor,M. (eds), Tavistock Publications, p. 63.
2. Edwards,R.G. and Steptoe,P. (1983) *Lancet*, **8362**, 1265.
3. Jones,H.W., Jones,G.S., Andrews,M.C. *et al.* (1984) *Fertil. Steril.*, **38**, 14.
4. Quigley,M.M., Schmidt,C.L., Beauchamp,P.J. *et al.* (1984) *Fertil. Steril.*, **42**, 25.
5. Lancaster,P.A.L. (1985) *Br. Med. J.*, **291**, 1160.
6. Asch,R.H., Balmaceda,J.P., Ellsworth,L.R. and Wong,P.C. (1985) *Int. J. Fertil.*, **30**, 41.
7. Ockenden,K., Bolton,V.N. and Braude,P.R. (1985) *Lancet*, **I**, 452.
8. Templeton,A., Van Look,P., Lumsden,M.A., Angell,R., Aitken,J., Duncan,A.W. and Baird,D.T. (1984) *Br. J. Obstet. Gynaecol.*, **91**, 148.
9. Braude,P.R., Bright,M.V., Douglas,C.P., Milton,P.J., Robinson,R.E., Williamson,J.G. and Hutchinson,J. (1984) *Fertil. Steril.*, **42**, 34.
10. Lenz,S. (1983) *Clin. Obstet. Gynaecol.*, **12**, 785.
11. Craft,I. (1985) in *In Vitro Fertilization and Donor Insemination*. Thompson,W. *et al.* (eds), RCOG, London, p. 143.
12. Steptoe,P.C. and Webster,J. (1985) *Ann. N.Y. Acad. Sci.*, **442**, 178.
13. Purdy,J.M. (1982) in *Human Conception in Vitro*. Edwards,R.G. and Purdy,J.M. (eds), Academic Press, NY, p. 135.

14. Quinn,P., Warnes,G.M., Kerin,J.F. and Kirby,C. (1984) *Fertil. Steril.*, **41**, 202.
15. Menezo,Y. (1976) *C.R. Acad. Sci. Paris*, **282**, 1967.
16. Dandeka,P.V. and Quigley,M.M. (1985) *Fertil. Steril.*, **42**, 1.
17. Hillier,S.G., Dawson,K.J., Afnan,M. *et al.* (1985) in *In Vitro Fertilization and Donor Insemination*. Thompson,W. *et al.* (eds), RCOG, London, p. 125.
18. Caro,C.M. and Trounson,A. (1984) *J. IVF and Emb. Trans.*, **1**, 183.
19. Bolton,V.N. and Braude,P.R. (1987) In *Current Topics in Developmental Biology*. Monroy,A. (ed.), in press.
20. Fleming,T.P., Pratt,H.P.M. and Braude,P.R. (1987) *Fertil. Steril.*, **47**, 858.
21. Bolton,V.N., Warren,R.E. and Braude,P.R. (1986) *Fertil. Steril.*, **46**, 1128.
22. Hewitt,J., Cohen,J., Fehilly,C.B. *et al.* (1985) *J. IVF and Embryo Trans.*, **2**, 105.
23. Wright,R.W., Anderson,G.B., Cupps,P.T. and Drost,J. (1976) *Biol. Reprod.*, **14**, 157.
24. Trounson,A., Mohr,L., Wood,C. and Leeton,J.F. (1982) *J. Reprod. Fertil.*, **64**, 285.
25. Osborne,J.C. and Moore,R.M. (1985) In *In Vitro Fertilization and Artificial Insemination*. Thompson,W. *et al.* (eds), RCOG, p. 101.
26. Bomsel-Helmreich,O. (1985) *Oxford Review of Reprod. Biol.*, **7**, 1.
27. Templeton,A.A., Van Look,P., Angell,R.E. *et al.* (1986) *J. Reprod. Fertil.*, **76**, 771.
28. Veeck,L.L. (1985) *Ann. N.Y. Acad. Sci.*, **442**, 357.
29. Mortimer,D. (1985) *Current Prob. Obstet. Gynaecol.*, **7**, 1.
30. WHO Laboratory manual for the examination of human semen and semen—cervical mucus interaction (1987) Cambridge University Press.
31. Bolton,V.N. and Braude,P.R. (1985) *Arch. Androl.*, **13**, 167.
32. Jacobs,P.A., Szulman,A.E., Funkhouser,J. *et al.* (1982) *Ann. Hum. Genet.*, **46**, 223.
33. Fishel,S.B., Cohen,J., Fehilly,C.B. *et al.* (1985) *Am. N.Y. Acad. Sci.*, **442**, 342.
34. Mohr,L.R., Tounson,A.O., Leeton,J.F. and Wood,C. (1983) In *Fertilization of the Human Egg In Vitro*. Beier,H.M. and Lindner,H.R. (eds), Springer-Verlag, Berlin, p. 211.
35. Fehilly,C.B., Cohen,J., Simons,R.F. *et al.* (1985) *Fertil. Steril.*, **46**, 638.
36. Buster,J.E., Bustillo,M., Rodi,I.A. *et al.* (1985) *Am. J. Obstet. Gynaecol.*, **153**, 211.
37. Warnock,M. (1984) *Report of the Committee of Inquiry into Human Fertilization and Embryology*. HMSO. Cmnd. 9314.
38. The Second Report of the Voluntary Licensing Authority for Human *In vitro* Fertilisation and Embryology (1987) Medical Research Council, London.
39. Ethics Committee of the American Fertility Society. (1986) *Fertil. Steril.*, **46**, Suppl.
40. Report and Conclusions of the Ethics Advisory Board of the Department of Health Education and Welfare: HEW Support of Research Involving Human *In Vitro* Fertilization and Embryo Transfer (1979) US Govt. Printing Office, Washington.
41. Senate Select Committee on The Human Embryo Experimentation Bill 1985: Human Embryo Experimentation in Australia (1986) Australian Government Publishing Service, Canberra.
42. Council of Europe, Parliamentary Assembly (1987) *Human Reprod.*, **2**, 67.
43. Edwards,R.G. (1974) *Quart. Rev. Biol.*, **49**, 368.
44. Braude,P.R. Pratt,H.P.M. and Johnson,M.H. (1984) *Bioessays*, **1**, 232.

Suppliers of specialist items

Aldrich Chemical Co., Poole, Dorset, UK; Aldrich Chemical Co. Inc., 940 West St
Pauls Avenue, PO Box 355, Milwaukee, WI 53201, USA.

Amersham International plc, White Lion Road, Amersham, Bucks HP7 9LL, UK;
Amersham Corp., 2636 S. Clearbrook Drive, Arlington Heights, IL 60005, USA.

Badger Airbrush Co., 150 Stanley Crescent Road, Poole, Dorset, UK.

Baird and Tatlock (London) Ltd, PO Box 1, Romford, Essex RM1 1HA, UK.

Bantin and Kingman, The Field Station, Grimston, Hull HU11 4QE, UK.

BDH Chemicals Ltd, Broom Road, Poole, Dorset BH12 4NN, UK; BDH Chemicals,
Gallards Schlesinger Chemicals Mfg Corp., 584, Mineola Avenue, Carle Place,
NY 11514, USA.

Beaudouin, Paris, France. (Beaudouin microforge supplied in the UK by
Microinstruments Ltd).

Becton Dickinson (UK) Ltd, Between Towns Road, Oxford OX4 3LY, UK; Becton-
Dickinson Labware, 1915 Williams Drive, Oxnard, CA 93030, USA.

Boehringer Corp. (London) Ltd, Boehringer Mannheim House, Bell Lane, Lewes,
East Sussex BN7 1LG, UK; Boehringer Mannheim Biochemicals, 7941 Castelway
Drive, PO Box 50816, Indianapolis, IN 46250, USA.

Boots Hospital Products plc., Nottingham, UK.

W.H.Brady and Co. Ltd, Daventry Road Industrial Estate, Banbury OX16 7HU, UK.

Calbiochem-Behring (CP Laboratories Ltd), PO Box 22, Bishops Stortford, Herts
CM22 7RD, UK; Calbiochem-Behring Corp., 10933 N. Torrey Pines Road, La Jolla,
CA 92037, USA.

Casmed UK, 36 Kenley Walk, Cheam, Surrey SM3 8ES, UK.

Clark Electromedical Instruments, PO Box 8, Pangbourne, Reading RG8 7HU, UK.

Collaborative Research, 12−14 St Anns Crescent, London SW18 2LS, UK;
Collaborative Research Inc., 128 Spring Street, Lexington, MA 02173, USA.

Crown Chemical Co. Ltd, Lamberhurst, Kent TN3 8DU, UK.

Day-Impex Ltd, Station Works, Earls Colne, Colchester, Essex CO6 2ER, UK.

Downs Surgical Ltd, Church Pass, Mitcham, Surrey CR4 3UE, UK.

Drummond Scientific Co., 500 Pkwy, Broomall, PA 19008, USA.

Dupont (New England Nuclear) UK Ltd, Wedgewood Way, Stevenage, Herts
SG1 4QN, UK; Dupont, NEN Research Products, 549 Albany Street, Boston,
MA 02118, USA.

Eastman Kodak Co., 343 State Street, Rochester, NY 14650, USA.

Falcon, supplied by R.L.Slaughter Ltd, 14 Bridge Close, Romford, Essex RM7 0AS,
UK.

Fisons Scientific Equipment, Bishop Meadow Road, Loughborough, Leics LE11 0RG,
UK.

Flow Laboratories Ltd, PO Box 17, Second Avenue, Industrial Estate, Irvine, Ayr-
shire KA12 8NB, UK.

Gibco Ltd, PO Box 35, Trident House, Paisley PA3 4EF, UK; GIBCO Laboratories, 3175 Staley Road, Grand Island, NY 14072, USA.

Glaxo Laboratories, Greenford Road, Greenford, Middx UB6 0HE, UK.

Th.Goldsmidt Ltd, York House, Station Road, Harrow, Middx, UK.

Go-Medical Industries, 20 Denis Street, Subiaco, WA 6008, Australia.

Gow-Mac Instrument Co., PO Box 32, Bound Brook, NJ 08805, USA.

Grant Inst (Cambridge) Ltd, Barrington, Cambridge CB2 5QZ, UK.

Harlon-Olac Ltd, Shaw's Farm, Blackthorn, Bicester OX6 0TP, UK.

Helena Laboratories, PO Box 752, 1530 Lindbergh Drive, Beaumont, TX 77704, USA.

Holborn Surgical Instruments Co. Ltd, Dolphin Works, Margate Road, Broadstairs, Kent CT10 2QQ, UK.

A.R.Horwell Ltd, 73 Maygrove Road, London NW6 2BP, UK.

Hoslab Ltd, 12 Charterhouse Square, London EC1M 6BB, UK.

IMV, L'Aigle, France.

Intervet Laboratories Ltd, PO Box 5830AA, Boxmeer, The Netherlands.

Kinematica GmbH, Lucernestrasse 147A, Lucerne, Switzerland.

Raymond A.Lamb, Sunbeam Road, London NW10, UK.

Leeds and Northrup Ltd, Wharfedale, Tyseley, Birmingham B11 2DJ, UK; Sumneytown Pike, North Wales, PA 19454, USA.

E.Leitz Inc., 24 Link Drive, Rockleigh, NJ 07647, USA.

Macarthy's Surgical Ltd, Selina's Lane, Dagenham, Essex, UK.

May and Baker Ltd, Liverpool Road, Eccles, Manchester M30 7RT, UK.

Merrell Dow Pharmaceuticals Ltd, Meadowbank, Bath Road, Hounslow, Middx TW5 9Q7, UK.

Microinstruments Ltd, 7 Little Clarendon Street, Oxford OX1 2HP, UK.

Miles Scientific Labs (ICN Biochemical Ltd), Free Press House, Castle Street, High Wycombe, Bucks HP13 6RN, UK; 30 West 475 North Aurora Road, Naperville, IL 60566, USA.

Millipore Co., 11−15 Peterborough Road, Harrow, Middx HA1 2YH, UK; Millipore Corp., 80 Ashby Road, Bedford, MA 01730, USA.

NBL Enzymes Ltd, South Nelson Industrial Estate, Cramlington, Northumberland NE23 9HL, UK.

Nunc (see Gibco).

Organon Laboratories Ltd, Cambridge Science Park, Milton Road, Cambridge CB4 4BH, UK.

Oxoid Ltd, Wade Road, Basingstoke, Hampshire RG24 0PW, UK; Oxoid (USA) Inc., 9017 Red Branch Road, Columbia, MD 21045, USA.

Paynes and Byrne, Greenford, Middx UB6 7HG, UK.

Pharmacia Fine Chemicals, PO Box 175, Uppsala 1, Sweden; Pharmacia Inc., 800 Centenial Avenue, Piscataway, NJ 08854, USA.

Polaron Equipment Ltd, Watford Business Park, Herts WD1 8XG, UK.

Promega Biotech, 2800 S. Fish Hatchery Road, Madison, WI 53711, USA.

Roche Products Ltd, Broadwater Road, Welwyn Garden City, Herts AL7 3AY, UK.

Rocket of London Ltd, Imperial Way, Watford WD2 4XX, UK.

Sefi Medical Instruments Ltd, PO Box 7295, Haifa, Israel.

Serono Chemical Co., 2 Tewin Court, Welwyn Garden City, Herts AL7 1AU, UK.

Sigma Chemical Co., Fancy Road, Poole, Dorset BH17 7NH, UK; Sigma, PO Box 14508, St Louis, MO 63178, USA.

South London Electrical Equipment Ltd, Lanier Works, Hither Green Lane, London SE13, UK.

Sterilin Ltd, Sterilin House, Clockhouse Lane, Feltham, Middx TW14 8QS, UK.

Steriseal, NI Medical, PO Box 3, 26/27 Thornhill Road, North Moons Moat, Redditch, UK.

Stortz (Rimmer Bros), 18 Aylesbury Street, Clerkenwell, London EC1 0DD, UK.

Surgikos Ltd, Kirkton, Livingston, UK.

Sutte Instrument Co., PO Box 3592, San Rafael, CA 34912-3592, USA.

Technation, 58 Edgeware Way, Edgeware, Middx, UK.

Arthur H.Thomas Co., Philadelphia, PA 19105, USA.

H.G.Wallace Ltd, Whitehall Road, Colchester CO2 8JH, UK.

Watkins and Doncaster, PO Box 5, Cranbrook, Kent TN18 5EZ, UK.

Wellcome, Temple Hill, Dartford, Kent DA1 5AH, UK.

Whatman Ltd, Springfield Mill, Maidstone, Kent ME14 2LE, UK; Whatman Chemicals Inc., 9 Bridewell Place, Clifton, NJ 07014, USA.

Wheaton Scientific, 1000 North Tenth Street, Millville, NJ 08332, USA.

Williton Box Co. Ltd, Williton, Somerset, UK.

INDEX